Approaches to Controlling Air Pollution

MIT Bicentennial Studies

The Social Impact of the Telephone, Ithiel de Sola Pool, editor

The New International Economic Order: The North-South Debate, Jagdish N. Bhagwati, editor

Approaches to Controlling Air Pollution, Ann F. Friedlaender, editor

Approaches to Controlling Air Pollution

Ann F. Friedlaender, editor

The MIT Press
Cambridge, Massachusetts,
and London, England

Second printing, 1979

This book was set in VIP Times Roman by The Composing Room of Michigan, printed and bound by Halliday Lithograph Corporation in the United States of America.

Library of Congress Cataloging in Publication Data
Main entry under title:

Approaches to controlling air pollution.

(MIT Bicentennial studies)
Includes index.
1. Air quality management—United States.
2. Air-Pollution—Law and legislation—United States.
I. Friedlaender, Ann Fetter.
TD883.2.A76 363.6 77–25484
ISBN 0–262–06064–7

Contents

Series Foreword

As part of its contribution to the celebration of the U.S. Bicentennial, MIT has carried out studies of several social and intellectual aspects of the world we inhabit at the beginning of our third century. Our objective has been to inquire how human beings might deal more intelligently and humanely with these factors, most of which are closely linked to developments in science and technology.

The papers prepared for these inquiries are being published in a Bicentennial Studies Series of which this volume is a part. Other studies in the series deal with the future of computing and information processing, linguistics and cognitive psychology, the economics of the new international economic order, the social impact of the telephone, and world change and world security.

It is our hope that these volumes will be of interest and value to those concerned now with these questions and, additionally, will provide useful historical perspective to those concerned with the same or similar questions on the occasion of the U.S. Tricentennial.

Jerome B. Wiesner

Preface

During the U.S. bicentennial year 1975–1976, MIT held a number of workshop/conferences on some of the major technical and social problems facing the world as the United States entered into its third century of nationhood. In the development of this conference series, an effort was made to find a confluence between a significant problem area—energy, food, environment, communications, economic conditions, institutional structures—and a faculty group at MIT that has recognized competence in a major aspect of the problem. In view of the massive amount of public and private resources that has been devoted to controlling air pollution in recent years MIT's economics department thought it appropriate to organize a workshop/conference dealing with problems of air pollution and administrative control as part of this bicentennial series. Thus, this volume presents the proceedings of a public conference on Air Pollution and Administrative Control held at MIT on 2 and 3 December 1976.

The Clean Air Act of 1970 marked a dramatic change in the nation's approach to controlling air pollution. From exhortations and modest implementation plans, the federal government moved toward a policy of massive intervention in the area of air pollution control with the passage of this act. Thus the Environmental Protection Agency (EPA) has far-reaching powers to set and/or enforce ambient air quality standards and emission levels from stationary and mobile sources.

In view of the broad powers delegated to the EPA and the wide-ranging impact of its actions, it is surprising that relatively little work has been done in assessing the Clean Air Act of 1970 and its implementation through the EPA, the courts, and the various state agencies. Thus, in organizing a workshop/conference on Air Pollution and Administrative Control, MIT's economics department hoped to begin to fill this gap and to initiate a dialogue between academicians and practitioners about the appropriate methods to use in controlling air pollution. The goal of the conference was not to present the conventional economic view concerning effluent fees as a means of controlling pollution but rather to assess a wide range of methods for controlling air pollution where the costs and benefits of cleanup are uncertain, where implementation is imperfect, and where administrative costs may be substantial. By explicitly considering problems posed by uncertainty and administrative feasibility, it was hoped that a more realistic view of the appropriate control mechanisms would emerge. To this end MIT commissioned a number of scholarly papers dealing with various aspects of controlling air pollution to form a basis of discussion at the public conference.

Because problems posed by administrative control are nonacademic by their very nature, considerable effort was made to ensure that people in government and industry actively involved with controlling air pollution were included in the planning and execution of the conference. An advisory board, composed of members of government and industry intimately concerned with problems of implementation, interacted with the authors of the papers in a two-day workshop held in June 1976 and commented extensively on the various drafts of the papers. To ensure a meaningful dialogue, every effort was made to include in the conference a broad spectrum of representatives from the EPA, Congress, state air pollution control agencies, public interest groups, and industry, as well as academicians from the fields of economics, law, political science, and engineering.

This volume contains the papers prepared for the conference, comments by invited discussants, and some summary comments prepared to provide some guidance for amending the 1970 act. Taken together, these papers not only provide a provocative assessment of current measures used to control air pollution but also present a number of concrete proposals for change. It is hoped that this volume will stimulate considerable discussion and contribute to more effective means of controlling air pollution.

A number of people helped organize the workshop/conference. E. Cary Brown, Norman Dahl, Peter A. Diamond, James A. Fay, Paul L. Joskow, Jerome Rothenberg, Jack Ruina, and Robert M. Solow provided counsel and guidance throughout. The Advisory Board was particularly helpful, and we are especially grateful to Donald Allen of Yankee Atomic Electric Power, Frederick Bowditch of the General Motors Environmental Activities Staff, Douglas M. Costle, then of the Congressional Budget Office, Albert E. Fry of the Environmental Protection Agency, Steven Jellineck of the Council on Environmental Quality, Steffen Plehn of the Environmental Protection Agency, and James Tozzi of the Office of Management and Budget (Nuclear and Environmental Affairs). Finally, we are grateful for the financial assistance provided by the National Science Foundation and the Ford Foundation, which made the workshop/conference possible.

Ann F. Friedlaender

Approaches to Controlling Air Pollution

1 Introduction

Economists have long argued that effluent fees or pollution taxes are the appropriate mechanisms to use to control air pollution. Because polluting firms and individuals do not bear the costs imposed by their offending emissions, in the absence of governmental intervention they have no incentive to reduce them. Hence air pollution is a classic externality, in which private costs are less than social costs. By imposing a fee equal to the costs imposed by the pollutant, the government will make the polluters take the full social costs of their actions into account and behave accordingly. Thus economists generally believe that by causing private and social costs to coincide, effluent fees will lead to an efficient allocation of resources and a socially desirable discharge of pollutants.

As an actual policy tool, however, effluent fees have been looked upon as an academic curiosity with no practical applications. Instead, specific emission standards have been set for mobile and stationary sources, the burning of coal with high sulfur content has been banned, and specific travel or incendiary restrictions have been imposed to ensure that ambient air quality standards are met. Thus the control of air pollution has generally taken the form of the heavy, visible hand of government regulation instead of the invisible hand of the pricing system to ensure that private and social costs coincide with respect to air pollution.

Whereas economists have tended to view existing practices with considerable puzzlement because of their apparent irrationality, lawyers, bureaucrats, and legislators involved in air pollution control have tended to view effluent fees with equal puzzlement because of their apparent impracticality. Of course, in a world in which the costs and benefits of cleanup are known with certainty for each kind of pollutant over its entire range of discharge, it does not matter whether fees or standards are imposed.[1] As long as cleanup occurs until the marginal costs of cleanup equal the marginal benefits, a socially optimal amount of pollution will result. This goal can be obtained either by setting standards that equal the optimal level of discharge or by setting an effluent fee equal to the difference between the private and social costs of discharge at the optimal level.

In the real world, however, the costs and benefits of cleanup are highly uncertain, implementation is imperfect, and administrative costs are typically substantial. Thus the theoretical equivalence between fees and standards does not hold, and decision makers have generally thought that standards are preferable to the imposition of fees. Standards have the virtues of being clear and unambiguous, they are borne directly by the polluters, and they are relatively easy to understand. However, standards also have a zero-one quality. If the standards are not met, offenders are typically faced with threat of shutdown or punitive fines. Thus standards tend to have a sort of brinkmanship quality that brings their desirability into question. Polluters and administrators tend to wage a perpetual game of chicken to see whether the polluters can force the administrators to relax the standards or the administrators can force the polluters to meet the standards. The most recent example of this contest is given by the behavior of the automobile companies with respect to the 1978 emission standards. Claiming that the 1978 emission standards cannot possibly be met, the automobile companies have challenged the government to shut down the automobile industry.

The challenge posed by the 1978 automobile emissions standards illustrates the primary problem with standards. If they are set too high, the costs of meeting them may far exceed the benefits. In the absence of recontracting, society is faced with two equally unattractive alternatives: either it must bear the costs of excessive cleanup or it must be willing to impose upon the polluter punitive sanctions that are generally felt to be excessive, such as "shutting down Detroit." Faced with these alternatives, society generally turns toward recontracting. This choice, however, has undesirable social consequences because it usually gives the impression that the polluters are implicitly setting the standards.

Instead of analyzing the merits and weaknesses of effluent fees and standards in detail, the authors in this volume address the issue of alternative approaches to controlling air pollution. As such, they are not concerned with questions of economic doctrine or issues of social justice but rather with entirely pragmatic questions of implementation and whether it is possible to devise some scheme that permits simplicity of standards but contains a safety valve of fines or fees that obviates the need for recontracting if the standards turn out to be excessively stringent. Thus, by examining in some detail the record of air pollution control since the passage of the Clean Air Act of 1970 and by considering some concrete proposals for change, the authors in this volume make a major contribution toward the adoption of more rational methods of controlling air pollution.

The first three papers in this volume are explicitly concerned with the

Clean Air Act of 1970 and the interactions among the courts, the states, and the EPA in its implementation, and the next four papers concern alternative approaches to controlling air pollution and consider issues related to whether a mixed fee-standard system would be preferable to the current system of standards-cum-sanctions.

The opening paper by Helen Ingram concerns the genesis and the resulting structure of the Clean Air Act of 1970. She argues that contrary to conventional wisdom, the Clean Air Act of 1970 represented a definite political rationality in that the legislation was the product of a political process that was "rational, patterned, and explainable." Consequently, although the regulations promulgated by the Clean Air Act of 1970 are widely denounced by economists as being inequitable and inefficient, by lawyers as being unworkable, by engineers as being unrealistic, and by health scientists as being too simplistic, Ingram believes that the act was a logical outgrowth of the political process and represented a major piece of innovative legislation.

It is difficult to disagree with Ingram's claim that the Clean Air Act of 1970 was innovative. The standards set by the act and the implementation procedures by which they were to be carried out were considerably more stringent than could have been foreseen on the basis of previous legislative history. Thus the bold new goals and the mechanisms prescribed for their implementation that were contained in the Clean Air Amendments of 1970 represented major departures from previous policy. As Ingram argues, with the passage of this act, "the pragmatic, functional definition of air quality, restricted to what was economically and technologically feasible, was abandoned, and clean air was legislated a fundamental national value."

In general, innovative legislation, which makes a major break with past practices, rarely occurs because the costs resulting from change are typically focused upon a small vocal group, although the benefits are usually diffused throughout society. Thus, unless there is a widely held sentiment for change, the status quo will usually be maintained; any change that does occur will be incremental or marginal.

In the case of the Clean Air Amendments of 1970, however, Ingram argues convincingly that the ingredients needed for innovative legislation were present. Clean air was an issue that had captured the public's mind at that time, and it was generally believed that all people had a right to pure air and accompanying good health. In contrast, the polluters in general and the automobile manufacturers in particular were viewed as being selfish and insensitive to the public good. Consequently, Senator Muskie and his allies were able to sell the Clean Air Act in symbolic and moralistic tones. The good guys were for clean air and the public good; the bad guys were for profits and their own selfish interests. Once the choices were put into such terms, it is not surprising that the act passed with little scrutiny or dissent.

The problem with such a simplistic and moralistic view of the world is that implementation procedures are neither simple nor based on principles of morality. Thus the expectations raised by the Clean Air Act have largely not been fulfilled. Although the air is indeed somewhat cleaner than it was in 1970, the difference is not dramatic. Automobiles and stationary sources emit considerably less pollutants than they did in 1970, but they still emit considerably more than the standards legislated in the 1970 act permit.

Part of the problem lies in the legislation, which probably set unreasonably high standards, but part of the problem also lies in the implementation of the act by the EPA, the courts, and the states. As outlined by Ingram, innovative legislation typically requires new mechanisms of implementation that are extremely difficult to deliver. Thus, in spite of the best efforts of the EPA, the courts, and the states, the difficulties inherent in implementing such drastic changes were so great that the millennium implied by the Clean Air Act of 1970 is still not at hand. Nevertheless, the interaction between the courts and the EPA has generally been fortuitous and led to more rational decision making on the part of the EPA. Moreover, whereas the experience of the state air pollution control agencies has generally been weak, there are a few instances of enlightened enforcement that may serve as a guide to future policy.

Richard Stewart argues that the pluralistic nature of environmental problems and objectives made it impossible for Congress to legislate the full range of controls that are appropriate in all situations. Consequently, considerable latitude was given to the EPA, the agency empowered to enforce the Clean Air Act. Yet because the EPA is generally constrained by bureaucratic inertia from weighing the full range of implications involved in any major environmental decisions, the courts have played a substantial role in "checking the parochial tendencies of administrative agencies . . . and, where necessary [they have strained] the relevant statute to enlarge the factors which the agency may take into account." Thus, Stewart's basic thesis is that the courts have helped the rational implementation of the Clean Air Act.

In particular, Stewart gives the courts high marks for encouraging the use of paper hearings that require the EPA to consider a wider range of factors than it has done in the absence of court intervention. Although the actual outcome of the administrative action may not have changed as a result of these hearings, they generally led to more rational decision making and ensured that the various interests obtained a full hearing. Thus Stewart believes that through these hearings the courts have helped the EPA weigh economic and environmental objectives more rationally.

In a few cases, the courts not only have helped the EPA to weigh the costs and benefits of its actions more rationally but also have taken a quasi-Constitutional role in interpreting the language of the statute. In particular, Stewart argues that court decisions regarding nondegradation are entirely consistent with the goals of the Clean Air Act, although they are not explicitly contained in the law. Thus the maintenance of ambient air quality standards in excess of those legislated in the law can be defended by the principle of diversity that underlies the First Amendment. Consequently, the intervention of the courts to preserve the principle of environmental diversity that Congress had not confronted and that otherwise would have gone largely unrecognized was entirely appropriate, according to Stewart. Moreover, because the intervention took a form that was subject to subsequent legislative review, the courts' actions represented a highly desirable interaction among the courts, Congress, and the EPA.

In similar actions, the courts have rejected the EPA's efforts to impose far-reaching transportation plans upon states to ensure that they reach the legislated ambient air quality standards. In the absence of clear legislative sanctions, the courts have generally upheld the principle of federalism and supported the states in their refusals to implement the EPA's transportation plans that include drastic measures such as gasoline rationing, restrictions on automotive use, high parking and access fees, and compulsory use of mass transit. Thus Stewart believes that the courts should assume an innovative role where there are exceptional justifications for judicial intervention in the political process, as in the nondegradation case, but that where the demand on judicial resources would be excessive, as in the transportation case, they must play a more deferential role.

Whereas Stewart is concerned with the implementation of the Clean Air Act by the courts and the EPA, Roberts and Farrell are concerned with its implementation by the states. Unlike the case of the EPA and the courts, however, in which interactions between the courts and the EPA generally led to improved implementation, the interactions between the EPA and the state air pollution agencies have not been particularly fortuitous, according to Roberts and Farrell.

Roberts and Farrell feel the problem is primarily not with the EPA or the state agencies but with the legislation that required the states to develop state implementation plans (SIP's) in an excessively short time. With limited resources at their disposal, state agencies have tended to develop simplistic pollution control plans that generally ignore problems posed by uncertainty, economic efficiency, or growth. Furthermore, in view of the substantial technical uncertainties and ambiguities inherent in these plans, political pressures have often influenced their particular specification. Consequently, Roberts and Farrell feel that it is "uncertain whether or not state plans will achieve air quality goals, even if they are effectively implemented or enforced."

Thus, according to Roberts and Farrell, implementation has often been and will continue to be highly uneven. Because state agencies have substantial discretion in determining the action that must be taken by a specific source and because they face substantial procedural difficulties in compelling sources to comply, the specific cleanup requirements for any given source often result from a bargaining process instead of an evenhanded application of the law. Similarly, because different states often have different attitudes toward pollution and economic development, behavior that may be acceptable in one state may be unacceptable in another. Consequently, the ambient air quality standards legislated in the Clean Air Act are unevenly implemented not only within states but also between states.

Because of diversities in political attitudes and technical capabilities, Roberts and Farrell feel that differences in implementation standards are inevitable. Nevertheless, because these differences cause people to bear the costs of cleanup unevenly, they tend to undermine the legitimacy of the Clean Air Act.

Roberts and Farrell are quite pessimistic about the ability of states to develop adequate implementation plans, but it is interesting to note that Connecticut has developed an innovative implementation mechanism that effectively combines standards with a system of effluent fees. Under the Connecticut plan, violators of emissions standards are subject to civil assessments exactly equal to the amount of money the firm saves by not being in compliance. This assessment incorporates savings on capital equipment, operating and maintenance costs, the rate of return in the industry, and the particular firm's tax status. This assessment is easy to calculate and is imposed without litigation. Thus the Connecticut plan is easy to implement and avoids the zero-one problem inherent in implementation plans based on standards alone. Instead of being threatened with shutdown in the face of noncompliance, firms are now faced with a fine equal to the presumed savings from noncompliance. Instead of a stick, Connecticut offers firms a definite financial incentive to comply.

Consequently, the experience of all state implementation plans is not all bad. Indeed, the Connecticut plan is an example of pragmatic decision making that was largely absent from the Clean Air Act of 1970 and the early efforts at its implementation. If the experience of the Clean Air Act and its implementation have taught us anything, it is that, symbolic legislation notwithstanding, pollution control requires a pragmatic weighing of the costs and the benefits. If the costs of enforcing legislated standards are excessive, the standards will be either implicitly or explicitly relaxed. Thus, instead of apparently rigid standards with penalties for noncompliance, what is needed is a more flexible approach that provides a mechanism for society to recontract to find the point where the marginal social benefits of pollution abatement equal the marginal social costs of cleanup.

Spence and Weitzman argue that such a mechanism will probably be found in a system that combines standards and effluent fees. According to Spence and Weitzman, in the absence of perfect information neither standards nor effluent fees in themselves provide the appropriate control mechanism. In the presence of limited information, standards may be preferable to effluent fees if the marginal costs of cleanup are relatively constant and if marginal damages increase rapidly with the level of effluents. On the other hand, standards are rigid and inefficient if marginal costs are higher than expected and marginal damages do not grow with the level of discharge.

Ideally, the control mechanism should respond to high cleanup costs for individual sources but should not create an incentive for noncompliance with the standard. Thus a tolerable approximation to an optimal control mechanism probably consists of a standard combined with a specified penalty per unit of emission in excess of the standard. In this way, the penalty will act like a high effluent charge for sources that have high cleanup costs but will not operate for sources whose cleanup costs approximate those implied by the established standards. For the majority of polluters, then, the control mechanism will operate like a system of standards. But for polluters with high cleanup costs, the control mechanism will act like a fee system. Consequently, the combined standard-fee system will act like a standard system with a safety valve. Instead of the zero-one characteristic of most of the standard systems that threaten noncompliers with shutdown, the standard-fee system will be less rigid and impose a fine related to the degree of noncompliance. This system, of course, is quite similar to the Connecticut plan.

There is a substantive difference between the standard-fee system and the Connecticut system, however. Whereas the former bases the fees or penalties on damages caused in excess of those permitted by the standard, the Connecticut system bases the fees or penalties on the costs of compliance.

Therefore, unless the marginal damages are equal to the marginal costs of cleanup, the two systems will lead to different results. In particular, if marginal costs of cleanup rise rapidly as zero discharge is reached and if marginal damages rise slowly with movements away from zero discharge levels, it is likely that the Connecticut plan will lead to more cleanup than the standard-fee plan because the penalties for noncompliance will be greater under the former than the latter. Of course, generalizations about the impact or desirability of the two plans are difficult without knowing the shape of the marginal damages or benefits curve and the shape of the marginal cost-of-cleanup curve.

Although the Connecticut plan is politically attractive precisely because it requires no direct knowledge of the marginal damage curve, an evaluation of its efficiency must depend upon some assessment of the relative costs and benefits of meeting the given standards. Thus the Connecticut plan is like all control systems in that we must have some notion of the nature of the marginal costs and benefits of cleanup before we can assess its desirability as a policy tool.

Although data on marginal costs and benefits of cleanup are relatively scarce, Rubinfeld and Dewees indicate that more statistics are available than is commonly believed, and they provide considerable documentation about the shapes and magnitudes of the marginal cost and marginal damage functions.

Rubinfeld is primarily concerned with the usefulness of market studies to evaluate individual willingness to pay for clean air. Using data on property values, Rubinfeld finds that perceived marginal damages from air pollution increase with pollution levels and increase with income. However, he finds that specific estimates of the shape of the damage function as well as the magnitude of the benefits associated with improvements in air quality are quite sensitive to the specification of the property-value equation. Thus, although one can be reasonably certain of the shape of the marginal damage function, its magnitudes are considerably more uncertain. Nevertheless, Rubinfeld feels that the variability in the estimate of benefits may be small relative to the uncertainty associated with other parameters of a policy decision. For example, he argues that if one were assessing the desirability of instituting an inspection-maintenance scheme to reduce pollution, the uncertainties associated with measuring the costs of the program would probably be considerably greater than the uncertainties associated with its benefits. Thus, Rubinfeld is guardedly optimistic about one's ability to infer marginal damage functions from property-value data and feels that the resulting estimates provide a reasonable basis for policy decisions.

It is important to stress, however, that the area of benefit measurement is an extremely controversial one. In his comments on the Rubinfeld paper, Lester Lave argues that correlations between property values and pollution levels are highly spurious. People do not pay more for houses in unpolluted areas; they pay more for houses with environmental amenities. Although one of these amenities, in fact, may be clean air, Lave argues that it is impossible to isolate the value attached to clean air. Thus, just as Rubinfeld was unable to distinguish among the values placed upon reduction in specific pollutants, Lave feels that it is not possible to distinguish between the value placed on clean air and parks.

As an alternative to assessing air pollution damages through property-value studies, Lave argues that one should make direct estimates of the costs of pollution in terms of property damage, health effects, and so forth. Countering Lave, however, Speizer argues that the data are insufficient to permit a meaningful assessment of the impact of pollution upon health, except in a few pathological cases.

Thus we are left with an uncertain assessment of our ability to measure the benefits of pollution control. On balance, Rubinfeld's position seems reasonable, But it is clear that considerably more work is needed in the area of benefit measurement before we can begin to feel at all secure about predicting the benefits associated with any pollution control program.

Considerably less work has been done to estimate the marginal costs of cleanup for any given pollutant, but Dewees presents convincing evidence that these costs rise rapidly as the degree of discharge is reduced toward zero. Thus Dewees finds that the marginal costs of controlling particulates, sulfur dioxide, and automobile emissions are all relatively low at high levels of discharge, but rise rapidly as zero discharge is approached.

Dewees makes the important point that although the general shape of the marginal cost curve seems to be relatively certain, its specific level is highly variable and changes dramatically with the pollutant under consideration and the technology used to control the pollutants. In particular, Dewees finds evidence of dramatic technical change in the case of automotive emissions. Between 1970 and 1976, the curve representing the marginal costs of emission control shifted sharply to the right. Thus, in 1976 it was possible to achieve 80 percent reduction in emissions at the same marginal cost required to obtain 40 percent reduction in emissions in 1973. Stated alternatively, with 1976 technology, 50 percent reduction in pollutants was possible at one-fifth the cost of a similar reduction using 1973 technology.

Dewees' findings with respect to technical change in automobile emissions control are interesting and raise a fundamental question concerning the impact of regulation upon technical change. It is clear from Dewees' study that the Clean Air Act of 1970 stimulated massive technological change on the part of the automobile industry in their effort to meet the standards imposed by the law. Yet it is widely believed that the rate of technical change was not as fast as it might have been. Not only have foreign automobile manufacturers been able to meet standards Detroit claimed were impossible, but the rate of technological change had also been painfully slow until the tough provisions of the 1970 act were imposed.

The experience with automotive emissions illustrates both the strengths and weaknesses of setting rigorous standards. Standards may encourage technical change (as was true in the early 1970s), but they may also encourage bluffing and delay. Because the costs of meeting the 1978 standards were apparently thought to be too high by the automobile manufacturers, they have announced that they cannot meet them, even though several foreign manufacturers have produced cars that do. Thus there appears to be a limit to how far setting standards can induce technical change.

This consideration raises the question of whether there would have been more technical change if the government had adopted a system of effluent fees instead of standards to control automobile emissions. In their paper, Mills and White argue that if a system of effluent fees had been adopted in 1970, there would be fewer automotive emissions at lower cost than we now experience. Thus Mills and White argue that a system of effluent fees would have encouraged more experimentation and lower-cost technology and would have led to engines that generated fewer emissions than those now manufactured.

The Mills–White paper in some sense brings us full circle in the fees/standards controversy and provides a useful case study to present the pros and cons of controlling pollution by setting standards or by imposing fees. Certainly the experience with automobile emissions has not been entirely happy. In particular, the rigid NO_x standards have undoubtedly forced the automobile manufacturers to adopt a higher-cost technology than one that would have permitted a more flexible trade-off between this and other emissions. Insofar as the Mills–White fee proposal permits this greater flexibility, it is relatively easy to demonstrate its potential superiority over the established standards.

It is important to note, however, that Mills and White have the advantage of hindsight, which was not available in 1970. The automobile emissions standards are not bad because they are standards per se but because their authors incorrectly perceived the relative costs of controlling different kinds of emissions and because they have not proved easy to recontract. Thus the problems are not so much with the standards as with the uncertainties concerning costs and benefits and with the difficulties of recontracting.

Thus we are brought back to our initial position that desirable control mechanisms must be sufficiently flexible to be revised in accordance with experience. Because uncertainties concerning costs and benefits are high, because implementation is imperfect, and because there are substantial administrative costs associated with control, it is important to establish control mechanisms that are sufficiently flexible that they will not lead to excessive costs or confrontation if initial assessments of the costs or benefits prove to be highly inaccurate. In this respect some system combining standards and fees appears to be very promising and should be the subject of considerably more research.

In conclusion then, perhaps the main thrust of the conference was that we should abolish reliance on symbolic standards and instead concentrate on pragmatic ways of getting the job done. These thoughts are echoed in the concluding comments of Davies, Krier, Strelow, and Kneese, who all argue for greater flexibility in administrative procedure, control mechanisms, and standards.

Clearly we are not at the millennium promised by the Clean Air Act of 1970, but since the passage of the act we have learned a great deal about the costs and benefits of various kinds of emissions controls and the difficulties associated with implementing various control systems. Thus not only is the air somewhat cleaner as a result of the 1970 act, but we are considerably wiser about the appropriate techniques to apply in controlling pollution, as well. The problem at hand is therefore to translate this wisdom into specific, workable control mechanisms that provide the flexibility that is lacking in the 1970 statutes.

Note

1. Strictly speaking, there is a difference between fees and standards in that under a fee system the polluters not only bear the costs of pollution reduction but also must pay the fees. If the fees are returned to the polluters, however, the two systems are exactly equivalent.

2 The Political Rationality of Innovation: The Clean Air Act Amendments of 1970

Helen Ingram

Introduction

The prevailing wisdom is that the current federal air pollution policy was chosen irrationally. Of course, what is rational is a function of one's goals and values, expertise and situation; yet a wide array of critics fault the existing law for not cleaning up the air. Legal scholars complain that the provisions of the Clean Air Amendments of 1970 lack precision and clarity and that consequently needless expense, confusion, and seemingly interminable litigation impede the implementation process. Scientists argue that the legislation is based on misconceptions about biological and physical systems. Atmospheric chemists suggest that the notion of threshold levels of emissions that are damaging to health is not supported by research. Physicists believe that the diffusion of pollutants and their chemical interaction are much more complex than can be appropriately handled by the simple standards in the 1970 act. Engineers claim that the deadlines set by the legislation are not technically realistic. Economists, the most outspoken of critics, condemn federal regulations as inefficient, inequitable, and ineffectual. What must Congress have been thinking?

Many blame "politics," as though politics were senseless and random and without patterns of predictable behavior. Politicians are viewed as souls lost to reason who are pushed about by political "forces" and not as individuals who are as subject to discipline and perspective as lawyers, scientists, physicians, engineers, and economists. Yet politics is rational, patterned, and explicable. As politicians see them, the choices that they make are reasonably related to individual and social welfare. In the real world of conflicting and unclear values and severely limited time and information, it is unreasonable to expect political decision makers to follow classical rules for rational decision making; as Braybrooke and Lindblom have argued, society might well suffer dysfunctional consequences if they tried.[1] The character of the issue about which a decision is made, how a policy choice relates to past policy choices, the policy alternatives actually available, the motivations and resources of participants in the policy-making process: such considerations affect what is politically rational. From a political perspective, federal air pollution policy is rational.

The examination of political rationality is justified not because political rationality is more important than other sorts of rationality or because showing that the Clean Air Act Amendments of 1970 made political sense would excuse other limitations that the legislation contains but because it helps explain behavior of participants in the legislative process. The calculus of rationality that political decision makers use must be understood and accommodated if legal, scientific, economic, and other kinds of insights are to be adopted in policy analysis. An understanding of the Clean Air Act's political underpinning is important in explaining why certain types of logically and factually correct information and inference were not entertained in 1970 and why the difficulties experienced in implementation since 1970 were not foreseen or otherwise failed to deter policy makers in 1970.

Defenders of political rationality have usually championed incremental policy making. Under conditions that usually prevail for political decision makers, it has been argued that it makes sense to suboptimize rather than to optimize and that small, successive policy changes that differ only marginally from existing policy are rational. The contention is made in the second section of this chapter that when issues have certain characteristics, innovative policy can also be rational. Under certain circumstances, postulating a symbolically appealing goal that goes substantially beyond what has been achieved in the past is preferable to endorsing more limited ambitions. Legislating stringent regulations, the costs and implementation of which are uncertain, can be justified in the belief that large change is desirable and that those affected by the legislation will act at some time in the future to prevent whatever unintended and damaging consequences have not been foreseen. It is consequently "rational" to shift the burden for correction of innovative policies to future ipso facto more experienced policy makers.

It may be argued that the 1970 act is not truly innovative. Although the results are not yet in, dramatic improvements in air quality seem not to have been accomplished. Yet most public policies fall far short of their stated goals when implemented. Furthermore, determinations about the innovativeness of a policy statement can be separated from measurements of policy impact. From an idealized perspective, the legislation is not especially novel or imaginative. Economists have criticized the Clean Air Act Amendments for lack of creativity in not instituting strategies such as emission charges, which are still fairly rare in this country. However, creativity is probably an unrealistic measurement of policy change through political institutions.

Without question, the Clean Air Act Amendments are nonincremental in the sense that the legislation is more than an incremental step from the base of past experience. The act set far more ambitious air pollution control goals, which were to be accomplished more quickly and under a more demanding regulatory regime than could possibly have been projected from previous policy evolution. Although difficult to measure, conceptual change is the most important indicator of policy innovation. The understanding of clean air changed significantly through the legislative process in 1970. The pragmatic, functional definition of air quality, restricted to what was economically and technologically feasible, was abandoned, and clean air was legislated a fundamental national value.[2]

The second section of the chapter will examine the logic of policies that involve large change (air pollution is an example of such a policy). The ideas of a number of students of public policy will be employed to show that the nature of the air pollution issue—the way in which the risks and stakes in the issue were perceived—ruled or elicited the sort of political response that occurred in the Clean Air Act Amendments.

The third section of the chapter examines more closely the political rationality of several classes of interests. To a political activist, rationality means seizing favorable opportunities. An activist must construct support for decisions and be watchful for opportunities to exercise influence. When an issue emerges that promises maximum return of credit and political prestige (which strengthens the activist's position in decisions to come) and, at the same time, requires only those resources the activist possesses and can afford to invest, then it is reasonable for the activist to become a policy entrepreneur. The Clean Air Act Amendments were fashioned by congressional entrepreneurship. The section sets out the conditions of successful entrepreneurial activity and illustrates these propositions through the legislative history of the Clean Air Act Amendments of 1970.

Experience in implementation of the act since 1970 has modified public and congressional perceptions of the air pollution issue. Legislative actors have changed and new relationships have emerged. The fourth section focuses on the new legislative setting and on patterns of perceptions that have emerged over the past seven years and concludes with some predictions about the sort of policy alterations that are presently politically feasible.

The paper draws from a broad literature. Public policy scholarship, particularly that of Lowi, Wilson, Schulman, and Braybrooke and Lindblom is especially relevant.[3] Authors from many disciplines have contributed to the very large number of books and articles written on air pollution. The analysis of congressional behavior draws upon public documents, congressional hearings and reports, newspaper and journal articles, and selected interviews with congressional staff members. Political scientists have already written a number of excellent case histories and analyses of the legislative handling of air pollution, and their work—especially that of Ripley, Davies and Davies, and Jones—undergirds much of what appears here.[4]

The Air Pollution Issue Characterized

Government, in acting on a public issue, confers benefits upon some individuals and groups, withholds them from others, and perhaps imposes costs on still others. How the stakes are perceived determines how the issue is acted upon in the political process. One of the important theoretical contributions of Theodore Lowi is his suggestion that we turn the usual models for studying policy making (which treat the character of policy last, as a result or output of a process) upside down. Policy should be considered first, as an independent variable that determines the patterns of political action.[5] Thus, following Lowi's theory, the characteristics of the air pollution issue become all important because they set the limits upon how the issue can be practically handled in the political process. With the help of Lowi and a number of other political scientists who have theorized about the character of political issues, this section shows how the air pollution issue, like a number of other issues with similar characteristics, rationally suggests innovative action.

Perceptions of Benefits and Costs

A necessary condition of legislative change is that supporting individuals and groups perceive that they will benefit from change. This support must overwhelm or at least counterbalance the opposition of individuals and groups who believe that change will worsen their situation. Stress must lie on perceptions, because actual benefits and costs matter little in politics until they are correctly apprehended. Public policies regularly impose burdens upon the unsuspecting segments of the public that fail to participate effectively in the policy-making process because they are unaware that they have anything to lose.

Some sort of analytic scheme that classifies policies according to the perceptions they evoke is essential to analysis. Lowi distinguishes among issue areas according to whether any costs are perceived (in distributive policies all participants perceive only gains) and, if so, whether they are perceived by fairly specific groups (regulatory policies) or by broad social classes (redistributive policies).[6] The limitations of these categories become obvious when they are applied to actual pieces of legislation and fail to embrace their complexity and dynamism. Modern legislation is often a composite of different kinds of costs and benefits. A piece of legislation may benefit some individuals and groups monetarily, some may gain or lose access to decision making, others may lose their jobs, and there may be still others who perceive costs and benefits in terms of nothing more tangible than affirmation or denial of certain values and beliefs. Particular provisions may promise benefits and appear costless, whereas other provisions impose obvious burdens. Dean Mann has described how the "distributive" waste treatment grant program in the 1972 water quality legislation acted as a sweetener to soften and counteract opposition to the "regulatory" features. Further, as policies change over time from mainly "distributive" to mainly "regulatory" or mainly "redistributive," new legislation ordinarily carries along some of the baggage of the past by continuing firmly ensconced programs and agencies.[7]

Based on the work of Larry Wade, Murray Edelman, Robert Salisbury, James Wilson, and others, a classification of different kinds of costs and benefits can be formulated that comprehends complex and dynamic legislation.[8] Benefits and costs can be identified as dimensions of at least four varieties: focused, diffuse, structural, and symbolic. Possibly other dimensions of costs and benefits could be distinguished, but these four are sufficient to deal with the air pollution issue. These are the "building blocks" out of which legislators put together a legislative package that is politically feasible. The task is more or less difficult depending on the prevailing perceptions. Some benefits and costs are obvious; others, more removed, can be stressed or ignored. When the building blocks of the Clean Air Act Amendments of 1970 are sorted out by examining the perceptions of the principal actors (as expressed in the public record of hearings, newspapers, reports, and so on, associated with the Clean Air Act), it becomes clear that symbolic benefits were overwhelmingly important, whereas costs of all sorts were far less significant.

Focused Benefits and Costs The most obvious kinds of benefits and costs conferred by public policies are those that are focused directly toward aiding some groups and/or burdening others. Because focused benefits and costs are targeted toward certain groups, they are likely to be highly salient to the members of those groups and are likely to motivate political action. The Clean Air Act Amendments imposed heavy direct costs upon polluting industries. Both stationary and mobile sources had to meet demanding emissions-reduction standards before deadlines that left little time to perform and few ways to delay or escape. For example, ninety percent of the emissions levels of 1970 autos had to be controlled in the manufacture of 1975 models. Electric utilities across the country had to equip all new plants with expensive new technology and cut back on emissions of existing plants to meet much more exacting primary standards. Oil companies, airlines, steel mills, and hosts of other industries faced increased pollution control costs and more regulatory restraints. Statements of industry representatives excerpted by Charles O. Jones indicate that polluters were aware of and very worried about the focused costs embodied in the Act.[9] Prior to 1970, the political activity generated by industry's determination to avoid federal regulation, which was perceived as too burdensome, forestalled the imposition of focused costs. Industry's record of legislative success can be explained by the absence of countervailing perceptions of the benefits of air pollution control.

Focused benefits usually come in the form of grants, subsidies, loans, and other direct monetary payments. There is little in the way of targeted benefits that make some groups obviously better off than other groups in air pollution legislation. The fact that air pollution control is inexpensive to the public coffers appears on the surface to be a plus, but it has another side less favorable to the building of support. The disagreeable impact of focused costs can be made more acceptable by the distribution of subsidies, grants, and loans. In some pollution areas such distribution of sweeteners can be very extensive (waste treatment grants for water pollution is an example), but it is impossible to "treat" air pollution in public waste-treatment facilities. Consequently, the air pollution program does not have an extensive public works program to distribute targeted benefits; instead, action is essentially private. Extensive focused benefits to private entities are of questionable value and are politically difficult. Although subsidies can make waste treatment technology less onerous for an industry to apply, they cannot make pollution controls attractive. Consequently, a focused cost or specific regulation is a prerequisite to prompt polluters to clean up.

The only focused advantages that air pollution policy has to give out are program and technical grants to states and localities and research funds to public and private entities. Such monetary transfers are embodied in the act, but they are hardly sufficient to create much support. Thus focused benefits were not and could not be much of a building block in the 1970 legislation.

Diffuse Benefits and Costs Diffuse benefits and costs are those that are spread out, widely shared, and difficult to relate to distinct individuals or groups. Diffuse benefits and costs include the ultimate environmental, social, and economic consequences to the community at large. The benefits of a new regulatory regime upon the common-property resource of air are bound to be perceived as diffuse. Because the air is regarded as a free good belonging to everyone, improving its quality makes everyone better off without much improving the position of any group relative to other groups. Particular individuals and groups see diffuse benefits and costs as important community matters but not necessarily as the highest priority among their own concerns. Up until 1970, clean air was often everyone's business without being many people's particular business. Long before 1970, people were aware that dirty air is dangerous to health, but impacts tended to be seen as impersonal matters of statistics and probability—such as the number of respiratory illnesses per 100,000 population. Those most vulnerable to air pollution, and thus those who should prize clean air the most, were the old, the young, and the chronically ill. Even if these groups have been concerned that they live more dangerously in dirty air, they have not traditionally been politically active segments of the public.

Considerable property damage is attributed to air pollution by experts. Yet it is hard to get home owners who are swamped by general repairs to react politically to the necessity of repainting the house a year earlier because of air pollution. A most convincing indication of the low priority placed on the shared and diffuse benefits of clean air by the public is that industrial interests were able to keep the issue of air pollution off the governmental agenda in such dirty cities as Gary and Chicago throughout the 1960s.[10] It was only when the more general environmental movement transformed clean air into a moral issue of good and bad, pure and impure that political momentum developed.

The diffuse and indirect impacts of air pollution regulations upon the economy, transportation, and life-styles were not much discussed in 1970. Major participants' abilities to predict ripple effects of air pollution control upon the economy were limited by lack of experience. Further, however great the social and economic costs might be, they were seen as less important and less immediate than certain benefits. On the infrequent occasions when concerns about the economy and jobs were mentioned in the public record, they were refuted as speculative. The only real challenge to the Senate committee version of the Clean Air Amendments on the Senate floor came from Senator Griffin:

Mr. Griffin:. . . I would not say this bill plays Russian roulette—let me say it plays economic roulette with millions of jobs in the automobile industry. Without adequate expertise, without the kind of scientific knowledge that is needed—without the hearings that are necessary and expected the bill would write into legislative concrete requirements that can be impossible—and that will force industry out of existence. . . . I want to remind the Senate that a great many jobs are involved. One job out of seven in the United States depends directly or indirectly on manufacture, sale, or service of automobiles.

Mr. Muskie: The Senator complains when he says I distort what he says. I thought I had made it eminently clear that I was not saying what he just put in my mouth.

What I said—I will repeat to make it clear—is that the judgment of this committee—includes the Senators from the Senator's side of the aisle—some of a pretty conservative political persuasion—that Congress has a duty to say, "This is what ought to be done in the interests of the health of the country." If it cannot be done, if the industry has made a good faith effort, it can come back to Congress.

We speak of Russian roulette. If it is really that choice—and I do not agree that it is—I would rather play Russian roulette with the automobile companies than with the trapped inhabitants of urban America. Their health is involved.[11]

Although the Senate Committee Report suggested that traffic of motor vehicles might have to be reduced as much as 70 percent to meet the ambient air quality standards, there was little or no discussion about the likely cost of or opposition to the implementation of transportation plans with such provisions. Because much of the expenditure for air pollution control devices was to be private and because it was uncertain how much of that cost could be passed on to consumers, there was little perception of indirect costs in higher taxes or higher prices.

Structural Benefits and Costs Structural benefits and costs arise from perceived changes in the allocation of administrative responsibility and in the procedures prescribed for carrying out that responsibility. Administrators perceive gains and losses in budget, power and prestige, job security, and so on according to where the responsibility for making decisions is located. The access of groups to administrators is also affected. For instance, giving a great deal of decision-making authority to state officials reduces the influence of parties who have less access at statehouses than at the federal level. Also, agencies and officials benefit or suffer accordingly for receiving a greater or lesser grant of authority.

The structural decisions in the Clean Air Act of 1970 that gave EPA the authority to set uniform primary and secondary air quality standards also gave it the power to disapprove state implementation plans. The granting of broad enforcement authority to the federal government was a significant shift of responsibility from the state to the federal level. Interestingly, this loss of power was not resented, perceived, or articulated in 1970 by state administrators and their clientele. Although the previous pattern had been to leave primary decision-making authority to the states, Congress felt they had failed to move rapidly under the 1967 act to set up air quality regions. It may be that state officials were relieved to be taken off an uncomfortable hook. On the one hand, the public demanded more stringent controls from their state pollution control boards, but there was also pressure to prevent loss of industry to states with weaker regulations. The national uniform standards held out the promise of preventing industry flight. In addition, perhaps state agency officials appreciated that despite their relative loss of influence, the whole program was to become more important and more generously funded. For whatever reasons, few states testified in hearings, and none argued forcefully for retaining primary decisions at the same level.

Symbolic Benefits and Costs Symbolic benefits and costs are the implicit value premises in legislation that may be either reassuring or threatening to the loyalties and attachments of mass publics. Both diffuse and symbolic benefits and costs are widely shared. However, because the affirmation of values is at stake—good over evil—the symbolic dimension of policy is perceived as much more immediate and personal than diffuse, remote consequences to the economy, the environment, or society at large. Many political battles are fought in the name of symbolic values. Larry Wade observes that "politics is more than a struggle over the distribution of material values. It is also a social process through which symbolic values, representing needs for self-esteem, dignity, and personal rectitude are distributed and validated."[12]

Symbolic rewards are at least as important as real rewards, in some ways more so—and their pursuit is not necessarily irrational. As Daniel Moynihan concluded about policies to abolish poverty and inequality, symbolic awards are immediate: material rewards in the best of circumstances are long delayed and often never do come to pass.[13] At a time when tastes and preferences are in flux in a particular issue area, the conduct of politics in symbolic terms may perform the function of educating values and defining moral commitment. Often it takes an initial appeal to symbols to create a public sufficiently self-disciplined and well organized to comprehend and balance material costs and benefits.

Certain styles of debate typify policies in which perceptions of symbolic benefits are crucial to support. Murray Edelman describes the hortatory style as astute use of ambiguous key words that have a positive connotation and conceal an emotional appeal under the guise of defining issues.[14] Throughout the legislative hearings and debates concerning the Clean Air Act, legislative leaders stressed benefits of environmental quality and public health in absolute terms, that is, as a priceless value that is without trade-offs. The identification of a scapegoat or enemy is also important in symbolic appeals. When Muskie introduced the proposed air pollution legislation to the Senate floor, he stated that it was "a tough bill" but "a necessary bill because the health of our people is at stake." He pointed out that the costs of air pollution "can be counted in death, disease and disability, . . . in the billions of dollars of property losses . . . , in the discomfort of our lives." He complained that "some industries have not exerted their best efforts to control air pollution," and that "oftentimes, funds which should have gone for pollution control have been spent on advertising and public relations designed to reduce the pressure on companies to do what is necessary."[15] Specific targets and deadlines became part of the symbolic appeal of the legislation. The mass public could easily grasp numbers such as "ninety percent of emissions by 1975."

Aaron Wildavsky explains the lack of interest of many environmentalists in least-cost approaches to pollution control—including effluent charges—by their preoccupation with symbolic benefits. He argues: "Environmentalists wish to change man's moral relationship to nature. Without law, there can be no sin, without sin no crime, without crime no punishment, and without punishment no repentance. For them the symbolic level—Thou shall not pollute—is the real one."[16]

Economists' notions of efficiency were not overlooked or rejected so much as considered irrelevant by the legislative leaders fashioning the Clean Air Act. Senate staff members maintain that the members of the Public Works Committee did not seriously consider emission charges in 1970 because they were committed to a regulatory approach.[17] Congress was preoccupied with pursuing the symbolic dimension of policy.

The building blocks with which Congress worked to construct air pollution policy can be clearly identified. On the one hand, powerful groups perceived a focused cost about to be imposed. On the other hand, the public appeared receptive to symbolic arguments that labeled polluters wicked. Considering what Congress had to work with, the Clean Air Act Amendments of 1970 are a logical and rational result. James Q. Wilson has identified air pollution policy as one of those difficult legislative areas in which specific, easily identifiable groups bear the cost of a program that confers widely distributed, diffuse benefits.[18] The groups bearing the costs are likely to feel their burdens keenly and thus are also likely to organize in order that their burdens be reduced or, at the very least, not increased. On the other hand, the general community, which is benefited, may see its stake as marginal. Under such circumstances, successful adoption of policy requires the mobilization of a broad political constituency. Mobilization of groups is made easier by a dramatic crisis that puts opponents at a hopeless disadvantage. Success can also be achieved by the appeal—often through adroit use of the mass media—of a policy entrepreneur to a mass public in ways that make the "goal being sought appear incontrovertibly good and the groups being opposed seem utterly self-serving."[19] Both elements for mobilization were present during the consideration of the Clean Air Act Amendments. During Senate Air and Water Pollution Subcommittee deliberations in late July 1970, the Washington, D.C., area was experiencing its most severe air pollution episode ever. The overarching goal of public health was balanced against the private economic goals of "polluters"—particularly the auto industry, which has lost credibility. As Jacoby and Steinbruner see the situation:

The political climate that ultimately precipitated the stringent emissions standards began to develop in 1965, when Ralph Nader published his famous indictment of the industry for safety hazards and was treated to a personal investigation at the industry's expense. Seldom has an attempt at intimidation backfired so spectacularly. The Nader affair led to a dramatic set of hearings in which the president of General Motors was forced to apologize to Nader in front of a congressional committee and a national television audience. Serious and lingering damage was done to the political credibility of the automobile manufacturers—damage soon compounded by allegations concerning their handling of the air pollution problem itself. In January 1965 the Los Angeles County Board of Supervisors requested that the Attorney General investigate collusion by the industry to withhold pollution control equipment. The supervisors charged that the committee of the AMA set up to conduct joint research was in fact a collusive arrangement to prevent the introduction of controls. As evidence that industry developments were being suppressed rather than propagated, they cited the package of control devices developed by Chrysler but kept off the market until California legislation forced its introduction. The resulting Justice Department investigation ended with a consent decree in 1969 that provided for an end to the conspiracy without officially conceding its existence. This incident unquestionably added to the public's impression of recalcitrance and bad faith on the part of the industry.[20]

The characteristics of the air pollution issue in 1970 foreordained a policy-making process that concentrated on setting high goals without giving much attention to efficient implementation. Innovative legislation—setting targets considerably more ambitious than what was known to be achievable—made sense. Just such a legislative strategy was recommended by the political scientist Thomas Vitullo-Martin for water pollution control:

A law that would bring about a radical change needs political support sufficient to overcome the support of the incumbent policies. Only a strong, effective law that envisages a revolution-by-law could deserve such support. The support needed will be manifest in political conflict during the time in which the new policy (law) is seeking its adoption. The law must be strong enough to affect, in the case of pollution, the way of life of most of the members of society. . . .

The laws at issue must have a vision or principle worth fighting for.[21]

Dynamics of Policy Change

Federal policies very frequently follow a pattern of escalation in which the extent of federal involvement in an issue becomes greater and greater in successive steps. Randall Ripley has suggested that government has a variety of techniques that it can apply to problems and that these techniques require the exercise of different degrees of coercion. Government enters an issue area with a mild technique, but, as the initial and successive techniques are tried and fail to accomplish specified goals, a more stringent method is substituted. For Ripley, techniques of subsidy are likely to be followed by regulations. Regulatory rules, in turn, are followed by manipulation when government takes over or directly interferes with private activity.[22]

There is much in the history of air pollution legislation to indicate that the ashes of each unsuccessful piece of legislation contain the seeds of a more stringent law. Table 2.1 documents the policy escalation that occurred in the succession of air pollution laws passed between 1955 and 1970. Column 3 shows that initial techniques had practically no coercive element. Research (the favored initial strategy) was intended simply to generate information that state and local governments could employ or ignore as they chose. The 1963 act contained a hint of coercion to come. The cumbersome conference procedure used to discourage water pollution was instituted to deal with air pollution. However, for the most part conferences depended upon the publicity that they generated to shame polluters into action. In 1965 and 1967 subsidy-based approaches gave way to regulation-based ones. Standards for mobile and stationary sources became mandatory. In 1970 the swing away from research and grants was complete, and henceforth the burden of research fell upon the polluter. Technology was to be "forced" by specific targets and deadlines.

Column 4 in table 2.1 documents the step-by-step reallocation of responsibility from the state and local levels to the national level. The House committee stressed in 1955 that "the bill does not propose any exercise of police power by the Federal Government, and no provision in it invades the sovereignty of states, counties, or cities."[23] In 1963 the federal government tried to entice states to take action by offering matching grants for the development of programs. The federal government stopped encouraging and demanded action by states and localities in 1967. Finally, in 1970 the federal

Table 2.1 Escalation of Air Pollution Policy, 1955–1970.

Legislation	Major Provisions	Main Techniques	Distribution of Administrative Authority
Air Pollution Control Act of 1955	Research funding for 5 years, $5 million per year.	Research grants (subsidy).	Federal research leading to exercise of state and local authority.
1959 Extension of the Air Pollution Act	Research grants continued for 4 years, $5 million per year.	Research grants (subsidy).	Federal research, state and local authority.
Motor Vehicle Exhaust Study Act of 1960	Additional federal research, concentrated on vehicle emissions.	Continuation of research. Standards setting considered and rejected.	Federal research, state and local authority.
Clean Air Act of 1963	Grants to state and local control programs. Extension of research to publication of "criteria" documents. Conference procedure for abatement.	Research grants. Institution building grants (subsidy). Remote possibility of regulation.	Gentle federal prodding toward state and local action.
Motor Vehicle Air Pollution Control Act of 1965	Accelerated federal research and demonstration. Secretary of HEW given authority to set standards.	Research grants (subsidy). Regulation of auto emissions.	First federal regulatory program—restricted to auto emissions. HEW secretary given flexibility. States retain most authority.
National Emissions Standards Act of 1967	Large increase in research funds ($185 million for 1969). HEW secretary required to set air quality control regions, issue criteria, and recommend	Research grants (subsidy). Standards required (regulation).	Federal government requires states to establish state, local, and regional air pollution program.

Table 2.1 continued

Legislation	Major Provisions	Main Techniques	Distribution of Administrative Authority
	techniques. States required to set standards.		
1969 Amendments	Extension of research provisions.	No new provisions.	No new arrangements. Frustration expressed at pace of progress.
1970 Clean Air Act Amendments	Continued federal research, but "technology-forcing" provisions. Grants to states. National emissions standards. Specific timetables and deadlines.	Research and institutional grants (subsidy). Mandatory national standards (regulation). (regulation).	Federal government makes most decisions. States and localities have a role in implementation.

Sources: John E. Bonnie, "The Evolution of 'Technology-Forcing' in the Clean Air Act," *Environment Reporter* 6 (1975 July 25):1–29. Charles O. Jones, "Speculative Augumentation in Federal Air Pollution Policy-Making," *Journal of Politics* 36 (1974):438–464.

government took over the major responsibility by establishing standards and deadlines; states and localities were relegated to the implementation process.

Frustration was the major impetus that pushed policy along the escalation course. John Bonine documents congressional impatience with foot-dragging by states and industry in requiring and installing emissions devices.[24] Had more concrete progress been made, there would have been less reason to toughen air pollution controls. A jaundiced view of industry behavior formed from past experience may have caused Congress to remove as far as possible economic and technical feasibility as legitimate excuses for failure to abate pollution. Because it was likely that politically powerful polluters who had thwarted previous policies would continue to exploit their technological expertise, Congress perhaps was attempting to rig the legislation to improve the chances of future implementation.[25] More positively, what was actually achieved under previous legislation also provided part of the dynamic of policy change. Paul Sabatier argues that the citizen participation program established under the 1967 act created grass roots constituencies that demanded more forceful action.[26]

Ordinarily policy escalation occurs in an incremental process of nibbling away at a public problem. The Clean Air Act Amendments of 1970 can be distinguished from incremental policy making by the rapid pace of change. Instead of relying upon the base of past experience in determining what was possible, legislators rejected experience as not nearly good enough.[27] The air act was innovative not only in its dramatic shift of power to the federal level but also in its deliberate attempts to get polluters to meet standards for which the technology had not even been developed. Charles O. Jones coined the phrase "speculative augmentation" to apply to situations in which the legislature, in order to satisfy public demands that extend beyond technology or institutional capacity, gives the force of law to its educated guess about how well industry can perform if the pressure is applied.[28] Paul Schulman's work with manned space exploration helps to explain the dynamic that occurs in issues that develop holistically through an innovative break with the past rather than incrementally by merely advancing the ratchet a notch in an established policy approach.[29] Schulman observes that nonincremental policies must be cast as large-scale, risk-taking endeavors if they are to begin to achieve their goals. First, he notes that nonincremental leaps in policy occur in areas in which policy content and public support are poorly synchronized. Schulman's analysis also helps explain why, once legislated, nonincremental policies lack support for wholehearted implementation. Initially, mounting public concern quickly overcomes the limits of existing policy, which is underscaled, and eventually new policies are generated in response to escalating public pressure. "Frequently the appearance of government activity itself can contribute to a dissipation of public arousal. Secure in the symbolic reassurance that 'something is being done,' the public shifts its fleeting attention to other issues of greater currency and fashion."[30] Consequently, underscaling of policy is replaced by overscaling, and more is done than the public really wants. The pre-1970 complaint that government did too little about air pollution and the post-1970 complaint that it does too much seem to fit.

Schulman's second point is that nonincremental policies must expand greatly if they are to expand at all. They must overcome the inertia, external resistance, and internal start-up problems that act as barriers. The notion that a program must reach a threshold before it takes off is obviously true in space activity, in which a critical mass of budget, personnel, and research is essential. A threshold may also exist in air pollution because a lesser effort is simply not taken seriously by either industry or the public. Air pollution policy needed adequate funding, manpower, and serious enforcement activity before it could "take off." Figure 2.1 indicates the exponential growth

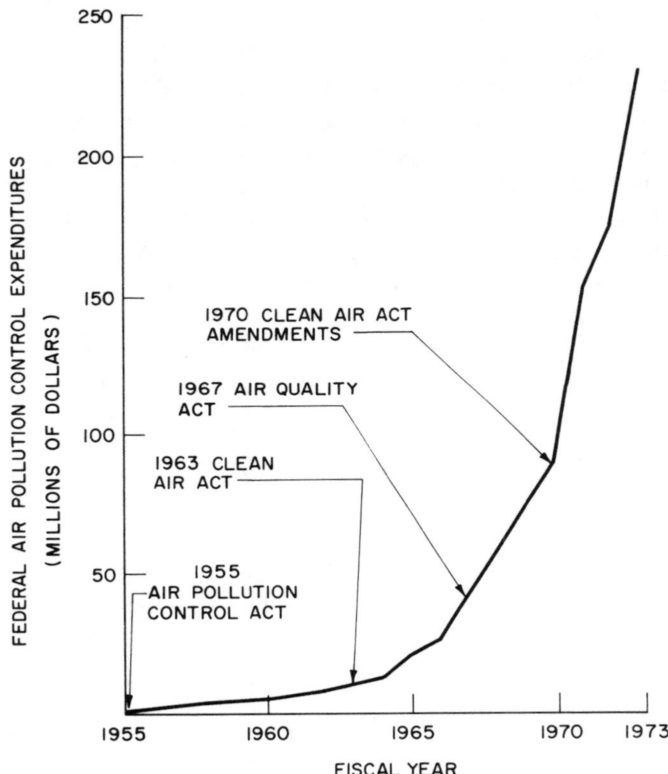

Figure 2.1. Federal air-pollution-control expenditures for the years 1955–1973. (Source: U.S., Congress, Committee on Appropriations, Subcommittee on Agriculture-Environmental and Consumer Protection, Hearings on Environmental Programs)

that occurred in federal funding of air pollution programs after the "start-up" threshold had been overcome. Figure 2.2 illustrates the slow pace of state adoption of air pollution control legislation before federal legislation made action mandatory. Fifty-state adoption of legislation, finally achieved in 1968 as a result of federal prodding, represents an essential threshold for effective enforcement thereafter.

Finally, Schulman notes that a nonincremental policy will be oversold in order to gain the support and resources thought to be essential to overcome the threshold. Once the policy has been oversold, its objectives cannot be

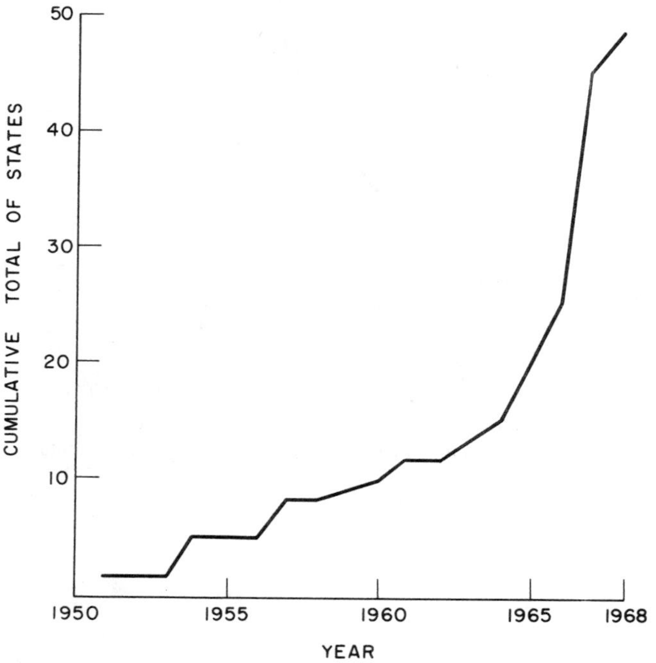

Figure 2.2. Cumulative total of states adopting air pollution control legislation, 1951–1968. (Source: National Air Pollution Control Administration, *Progress in State and Local Air Pollution Control Under the Clean Air Act,* U.S. Department of Health, Education, and Welfare, 1969)

realistically restated without threatening the political foundations upon which its support has been based. Program sponsors cannot risk credibility by admitting implicitly that the problem can be solved with fewer resources and is not as bad as originally claimed. Schulman's insight into the dynamics of nonincremental policy seems again to apply to air pollution policy. Senator Muskie repeatedly stated that, if automobile companies made a good-faith effort and still found targets and deadlines unachievable, they could appeal to Congress. The experience has been, however, that postponing deadlines has been a traumatic event that undermines confidence in the program.

Summary

This section has set out the basic rules or regulations that are the rationale of air pollution policy. The nature of the air pollution issue and public involvement in it invited—and almost predetermined—the kind of politics that resulted. The importance of symbolic benefits in the air issue led policy makers to stress the ideal of pure, life-giving air, whereas the adverse effects of pollution control were not emphasized. To some extent the lack of progress by industry and states between 1955 and 1969 invited the innovations of the 1970 Act. Had more concrete progress been made in controlling air pollution, more solid information on costs and benefits would have been available, and there would have been less reason to appeal to symbols to get the program beyond thresholds. The air act rhetoric was part of the necessary oversell of nonincremental policy, and the current erosion of public support for achieving goals at any cost is a development that rationally should be expected now that experience has exposed what, in fact, the costs are.

The Entrepreneurial Model of Decision Making

Most political science literature links innovative policy with executive leadership, but Congress created the air pollution control amendments of 1970. Much of the literature on regulatory legislation suggests that the lawmaking process involves coalition building, bargaining, and high levels of conflict, but the Clean Air Amendments passed the Senate without substantial committee or floor dissent. In this section it will be argued that the Clean Air Amendments followed an entrepreneurial pattern in which congressional activists responded positively, creatively, and competitively to opportunity. A set of propositions—drawn from the Congress and policy literature—will illustrate the conditions of successful entrepreneurial activity. The four essential ingredients for entrepreneurial activity are an issue of high political promise, appropriate goals and resources, competition for credit, and ineffectual opposition.

An Issue of High Political Promise

An issue must hold high political promise in order to prompt entrepreneurial activity. Many issues appear on the congressional agenda each session. Legislators must make choices about allocating their efforts. Issues are sorted out according to the potential they present to the individual legislator for building a record, amassing support, and extending influence. Despite the larger number of available issues, there are few that hold promise for legislators. Most issues have fairly narrow constituencies that often are confined to certain geographic areas. Many others are already staked out, and decision making is monopolized by an executive agency or a few legislators. Most important, few controversies offer legislators anything other than difficulty and frustration. The obvious solutions have already been tried and found wanting, and most available strategies already have opponents. Only rarely does a matter have untapped public support and potential for meaningful action. Such issues can command real commitment of energy, time, and resources from entrepreneuring legislators, and in 1970 air pollution was a case in point.

The environment and the value of clean air had an obvious, unexploited public appeal in 1970. The environmental concern among the general public was overwhelmingly concentrated on air and water pollution.[31] As early as 1967 public opinion surveys indicated that over 70 percent of urban residents felt there was ''some or a lot'' of air pollution. The same respondents were skeptical about whether anything effective was being done about it. When asked to rate the performance of the public and private sectors in controlling air pollution, respondents gave fairly low marks to everyone. Local industry ranked best (29 percent), followed in descending order by the federal government, local government, state government, the average citizen, and automobile manufacturers (21 percent).[32]

The growing strength of conservation and environmental groups contributed to the image of air pollution as a winning issue. The movement had accumulated an impressive series of victories, including defeat of proposed dams in Grand Canyon and the creation of Redwood National Park. The appeal of the environmental issue was dramatically demonstrated by massive participation in Earth Day, 22 April 1970.

Citizen participation programs sponsored by the National Air Pollution Control Administration (NAPCA) contributed to the creation of insistent local constituencies in many legislative districts. Federal grants had been given to a number of organizations (the National Tuberculosis and Respiratory Disease Association, the Educational Fund of the League of Women Voters, the Conservation Foundation, and the Michigan Cancer Foundation) to hold air quality workshops in cities throughout the country during 1969–1970. The program was immensely successful. Paul Sabatier concluded that viable, long-term constituencies emerged from several of the workshops, particularly those in Pittsburgh, Birmingham, and Chicago.[33]

Unlike the other major issues facing legislators in the 1970s, the problem of pollution united rather than divided people. While civil rights and the war in Vietnam sent demonstrators and rioters into the streets and divided young from old and cities from suburbs, the issue of pollution control drew support from all groups and classes. The young and affluent were particularly vocal about the issue. Only the few polluters in corporate board rooms were opposed, and this opposition could be easily discredited symbolically. Best of all, from a lawmaker's point of view, the issue was tractable. Everyone agreed that a real effort had yet to be made. The necessary strategy appeared simple: polluters would have to clean up. Presumably this solution meant the development and installation of technology.

Activist with Appropriate Goals and Resources

Entrepreneurial activity is directly related to the availability of an activist with appropriate goals and resources. Policy initiatives that come from Congress are ordinarily advanced by a legislative activist who takes up the role of legislative entrepreneur. John Johannes found that: "In the forefront of practically every congressional initiative are one or two legislators who deserve the credit for the law or for the process of gestation which led to it."[34]

An activist position is an act of will—a choice of positive action that, as Price notes, cannot be reduced to the generalities or regularities of cause-effect relationships. At the same time, entrepreneurial activities are most likely when certain goals and resources prevail. Legislators possess at least six basic goals that motivate their behavior: (1) career ladder success, (2) institutional influence, (3) constituency approval, (4) solution to public problems, (5) the advancement of ideological positions, and (6) maintenance of friendship. Because entrepreneurship means a priority commitment to an issue, the goals must be dear to a legislator. It is unlikely that most goals will prompt the kind of risk-taking activity that entrepreneurship on a nonincremental policy entails. For example, it is not probable that constituency service or friendship will prompt risk-taking activity; there are more efficient and less hazardous ways to secure these goals. Nor is a legislator likely to become an entrepreneur in order to secure institutional influence, for the folkways of Congress that secure such influence warn against high risks. Ideologists are rare in the pragmatic Congress and seldom survive long if they often act on the basis of emotional attachment to ideas. Thus entrepreneurial activity is most likely when a legislator is motivated by a desire to solve a public problem and/or by a desire to move up the career ladder.[35]

Senator Muskie's Goals Senator Edmund Muskie's commitment to solving pressing environmental problems is beyond reasonable question. His involvement with pollution issues began when, as governor of Maine, he focused on water quality. When he entered the Senate in 1959, environment was far from a popular political issue, yet Muskie steadfastly built a solid reputation in the area. Muskie was sentenced to the low prestige Public Works Committee (his next to last choice) by majority leader Lyndon Johnson as a punishment for failing to take a party line on a cloture vote. Muskie stuck with the assignment (most others transferred) and became deeply involved with the subject of water pollution. When Senator Patrick McNamara became chairman of Public Works in 1963, he created an Air and Water Subcommittee under Muskie. Between 1963 and 1970, Muskie was the prime actor in a number of air and water bills. Until 1970 Muskie's legislation in both the air and water area embodied only incremental changes. This record does not mean that he lacked an activist's proclivity for boldness. The water bill he sponsored in 1963 was a good deal more far-reaching than either the Senate committee or the House was willing to accept in 1965. Paul Sabatier explains the limitations of Muskie's 1967 air act by the absence of substantial public support for anything other than an incremental approach toward solving pollution problems.[36]

Speaking of his motivations in 1970, Muskie said:

> We had a choice: we could continue, and try to improve, past initiatives or we could change course and experiment with innovative methods which might achieve results at a more rapid pace.
>
> We had succeeded only 19 days before Earth Day in 1970 in obtaining enactment of major Federal oil pollution legislation. But that bill was lost in the fervor of environmental activism.
>
> The Clean Air Act of 1970 was a second attempt at this approach. . . .
>
> The result was dramatic and rewarding. In history there have been few laws as important or far reaching as these. The real reward was in being able to fulfill the mandate imposed on Congress by the founders—the opportunity to respond to an issue in the public interest and arrive at a result which was considerably more than an accommodation to the accumulated special interests.[37]

Without doubt Senator Muskie's career aspirations were also relevant to his entrepreneurship on the Clean Air Amendments. One of Muskie's strongest points as a contender for the Democratic presidential nomination was his record on pollution. This record was undercut by Nixon's 1970 state of the union message, which strongly emphasized the environment and proposed administration amendments to the Clean Air Act that went well beyond Muskie's 1967 legislation. Moreover, in May, a Nader task force published a report on air pollution that called into question Muskie's entire record. The senator had strong motivations to reassert his primacy in air pollution if he hoped to move on to a higher national office.

Senator Muskie's Resources Many legislators respond to the goals set out above, but whereas they may have the will to initiate action, they lack wherewithal for substantial impact. The extent of resources necessary for successful entrepreneurship depends upon the issue and the legislative setting. However, instigating, formulating, and publicizing legislation and mobilizing political support can be done only by those legislators who possess certain prerequisites. Especially important are visibility and committee support.

Visibility Legislators must have the ability to command attention and support and the ability to use this support to influence the decisions of others. For this reason, creativity in Congress often comes from the Senate, where the smaller numbers, longer terms, and larger constituencies provide an eminence that House members seldom achieve. Senator Muskie was well known in 1970 as a national party and congressional leader. He had the stature to mobilize a new constituency around innovative legislation. His long involvement in the pollution issue afforded him the reputation of an expert, and in a legislative body in which specialization is respected, reputation for knowledge means power.

Committee Support Committee agreement to an entrepreneur's position is a key to floor success.[38] The committee consensus that existed at the time of the passage of the Clean Air Amendments was partly the result of strong, bipartisan support for environmental legislation. Friendship and respect for Muskie helped consensus, and the style of the subcommittee's deliberations also promoted agreement. The group acted as a seminar with lengthy deliberations in which each Senator had a sense of impact. Senator Thomas Eagleton, for instance, contributed the notion of deadlines as a technique to force action. Fifteen days of hearings on the Clean Air Act followed fifteen days of oversight hearings. Ten markup sessions preceded the report to the floor. During that period of time, Muskie practiced the long-established logrolling style for consent building on the Public Works Committee by accommodating many of the interests of fellow committee members. The Republican insistence upon a uniform national standard emerged from the administration's proposed legislation. The compromise involved secondary as well as primary standards. Senator Robert Dole, a staunch administration supporter, filed separate views for the sole purpose of showing that the president's concerns had been satisfied.

Information Congressional researchers consistently identify access to information as a key resource to policy making. Very often it is the possession of detailed information about programs and their impact that gives administrative agencies the central advantage in formulating legislation. When a congressional entrepreneur assumes the lead in policy making, strong staff support usually exists. Price has identified committee staff size and competence as important, but more important still is staff orientation. Identifying closely with the interests and ambitions of the legislator, the staff of a congressional entrepreneur tends to be more personal than professional. An entrepreneurial staff is creative and active, continually searching out policy gaps and opportunities.[39]

The staff of Senator Muskie's Air and Water Pollution Subcommittee had considerable expertise in the environmental area. Charles O. Jones identifies superior staff resources as an important advantage of the Senate Public Works Committee over its House Commerce Committee counterpart in conference negotiations on the 1970 Amendments.[40] Among the half-dozen or so staff members who contributed to the creation of the 1970 amendments, two were particularly important: Leon G. Billings and Thomas C. Jorling. Billings had come to the Senate from the American Public Power Association in 1966. Since then he had served as principal staff architect for Senator Muskie on many environmental bills. Billings combined unchallenged superiority of knowledge about pollution legislation with complete dedication to the interests of the subcommittee chairman. In Billings's view, he and Muskie had come to think alike about the issue.[41] Jorling came close to Price's definition of a staff entrepreneur. Trained both as a lawyer and an ecologist, Jorling was firmly committed to employing legal means to achieve environmental goals. Jorling served as minority counsel to the Republicans on the committee, but his enduring interest was in fashioning innovative approaches to solving pollution problems. "He was more willing to use his position to implement his own policy preferences. . . . In the end, he valued creativity more than expertness."[42] Jorling is generally cited as the source of many concepts embodied in the 1970 Clean Air Act.

Competition for Credit

Competition for credit promotes policy innovation by congressional entrepreneurs. The fragmentation of power in Congress (that is, the wide dispersion of jurisdiction over subject matter and of influence over decision making within and between the two houses) is often blamed for deadlocks. It is interesting, therefore, that, when an issue of high political promise emerges and the jurisdiction of committees and roles of various congressional actors are fluid, fragmentation and the resulting competition between fragments facilitate greater policy change. An individual activist may be able to pursue desires for personal credit by incremental change, but numerous legislative entrepreneurs acting competitively must espouse large-scale change if they are to receive the desired rewards.

Jones and Davies and Davies have documented the one-upmanship game that resulted in amendments that were far more radical than many observers anticipated. The administration's Clean Air Amendments, introduced in February 1970, were far stronger than the incremental legislation sponsored by Muskie in December 1969. The House commerce committee modifications of the administration's bill strengthened provisions in several areas. The bill that emerged from the Muskie subcommittee summer deliberations was surprisingly tougher than the document voted on by the House. Finally, the conference version of the bill ultimately agreed to by the two houses in late December 1970 came much closer to the more extreme Senate version than to the House bill.[43] Consequently, to some extent the innovativeness of the Clean Air Act is due to the fragmented nature of the legislative process, which involved policy escalation among the administration, the House, and the Senate.

Further, several different entrepreneurs were pursuing credit for proenvironmental action and were scattered among committees. Congressmen Leonard Farbstein and Charles Vanik were supporting a ban on the internal combustion engine. Senator Gaylord Nelson was sponsoring environmental legislation in a number of policy areas. Contemporaneous with the Senate Public Works Committee's activity on the Clean Air Amendments, the Senate Interior and Insular Affairs Committee, under the leadership of Senator Henry Jackson, was fashioning the quite innovative National Environmental Policy Act and beginning consideration of land-use legislation. Only dramatic pieces of legislation were likely to get much attention in this flurry of environmental activity.

Ineffectual Opposition

Policy entrepreneurs are most likely to create innovative policy when the opposition is ineffectual in the decision-making process. According to Theodore Lowi, regulatory politics usually fit the pluralist pattern. Opposing coalitions fiercely lobby the members of Congress. Groups seeking to stop legislative proposals appeal over the heads of sponsoring legislators and committees to the floor or to the other body and are accommodated. The regulatory legislation that eventually emerges reflects a negotiated balance among adversary interests.[44] When opponents—for some set of reasons—fail to participate in policy making, the legislative process is quite different: the package of provisions formulated by the legislative activists becomes unleavened, uncompromised law. The concept of consensus politics rather than pluralist politics fits the experience of the Clean Air Act Amendments.

Automotive and other industrial interests, which would bear the burdens of most of the costs of the Clean Air Act, failed to make an effective case in 1970. Jones shows that interest groups representing stationary and mobile sources of pollution lobbied heavily against the provisions of the version that was produced by Muskie's subcommittee at the end of August.[45] However, this opposition failed to produce substantial modification through bargaining for two main reasons: poor timing and lack of effective antilegislation activists.

Importance of Timing Industrial interests really did not begin to fight until the final hour. Some sort of fairly stringent changes in air pollution law must have seemed inevitable, given the strength of the environmental movement. Industry representatives had testified on less punitive versions in the House and Senate in the spring and appeared reconciled to new controls. The extent of economic costs to industry was not made known until Muskie's pollution subcommittee reported, and by then it was difficult to build an opposition record.

Lack of Effective Antilegislation Activists Although industry lobbying was intense, it failed to secure enough effective supporters among the legislators. The following is a list of possible targets for access and the reasons why the opposition was not effective:

1. *The administration and its Republican supporters.* Because President Nixon was vying for credit with Senator Muskie on the environmental issue, administration officials and supporters in Congress were reluctant to appear less than firm against polluters. Concern about what he believed to be the adverse impact of the legislation upon industry eventually prompted HEW Secretary Elliot Richardson to send a letter to the conferees urging moderation, but this appeal was too late in the process to have much effect. Republican congressmen and senators felt the groundswell of environmental concern, especially in suburban areas and were generally anxious to prove that they, too, favored stringent controls. Electoral pressures upon legislators to take the "right" positions on issues with wide symbolic appeal is very great.[46] The House version

passed by a vote of 334 to 40, and the Senate bill had not a single opposing vote. Only Senator Robert Griffin from Michigan rose to the defense of the auto industry on the floor.

2. *The National Air Pollution Control Administration (NAPCA).* NAPCA had little leadership influence upon 1970 legislation.[47] The administration version was put together in the office of the secretary of health, education, and welfare. The Muskie version was prepared by the Senator's committee and its staff. NAPCA provided technical information for both, and its concern with implementation had some moderating influence. At the same time, the agency was on the defensive in the public eye and did not want to appear to speak for industry. Progress under the 1967 act had admittedly been poor, and NAPCA had been accused of foot-dragging. Thus, the agency was both too cautious and too understaffed to play a creative role.
3. *The states.* States may have been alarmed at what the legislation would do to their industrial tax base. They may also have been wary about future conflicts with industry, because the law would give states a major role in implementation. However, state governors and officials did not effectively articulate these concerns. Perhaps they failed to do so because, like NAPCA, they were on the defensive about their accomplishments under the 1967 act. In 1967 state air pollution agencies were still relatively underdeveloped. Further, the political strength of the environmental movement was being felt at state capitals as well as in Congress. As a result, states did not effectively exert a modifying influence in 1970.
4. *The House Committee on Commerce.* The Founding Fathers expected the Senate to be more conservative and less subject to transient public passions than the House; exactly the opposite pattern has emerged on many issues. Liberal leadership comes from the Senate because each Senator reflects the heterogeneous interests of an entire state. House committees act as boards of appeal for geographically based interest groups that have strong influence in smaller, more homogeneous congressional districts. Thus, House committees have acted as a brake on the Senate in a number of environmental issues, including land-use planning and strip-mining regulation.

   In the case of the 1970 Clean Air Amendments, however, the House Commerce Committee was outflanked and outmaneuvered. For the most part, the committee treated the act as usual legislation to be handled fairly routinely. It acted to strengthen the administration version, but did not basically modify it. The only legislator to take an activist interest was Paul Rogers, and he lacked the leadership position, the expertise, and the staff to challenge Senator Muskie successfully. It was not until after the election recess in the fall that committee members realized something important was afoot in the version reported by its Senate counterpart. It was then that Harley Staggers (D-W. Va.), the chairman of the Subcommittee on Public Health and Welfare, became really involved. House conferees were very skeptical about setting specific deadlines. However, they succeeded in modifying the Senate timetables only modestly (permitting automobile companies to request an extension of the 1975 deadline in 1972 rather than in 1973). The role of the House committee was generally to accede to the Senate, not to provide an alternative arena to disgruntled interests in industry or elsewhere.

Research has persistently indicated that Congress often can take responsibility for legislation.[48] Very frequently such legislation is incremental. At the same time, Congress can and does innovate, and, as this section of the chapter has documented, the Clean Air Amendments of 1970 are an example. Congressional proclivity for innovation often follows the model of entrepreneurship characterized in the propositions stated above. Successful entrepreneurship requires the presence of certain issues and of particular actors with specific motivations and resources. Entrepreneurship, which results in innovation, is also facilitated by a legislative setting that provides competition but not outright opposition. Such were the particular legislative conditions surrounding enactment of the Clean Air Amendments of 1970.

Revised Perceptions and Changed Patterns in the Politics of Air Pollution

Few national policies achieve their stated goals and objectives. Pressman and Wildavsky, in their book *Implementation*—subtitled in part *How Great Expectations in Washington are Dashed in Oakland*—have said that the process of implementation is so complex and difficult that policy analysts should be pleasantly surprised that anything ever works at all.[49] The gap between policy promise and actual performance is not at all unique to the Clean Air Act. The contrast has been dramatic because the Clean Air Act Amendments promised so much so explicitly and at such little cost to all but the few. Clear and healthful air is still a dream of the future despite six years of effort under the act. The simplicity and concreteness of the standards and deadlines that were so useful in the 1970 appeal for mass public support have served to highlight policy failure as they have been extended. The burdens of cleaning up the air have been heavier and borne by different people than expected in 1970. Instead of making just the polluter pay, everyone now must pay—especially the motorist and the home owner. The Clean Air Amendments were intended to overcome administrative foot-dragging by reallocating power to the federal level. The result has been more intergovernmental confusion and conflict than were ever anticipated.

Changes in Perceived Costs and Benefits

The public record of the difficulties that congressional activists anticipated in the implementation process is sparse. Both House and Senate hearings were over before the more stringent Senate draft had been formulated. The lack of real opposition on the Senate floor meant that Muskie and others were not required to say how the bill would work. In fact, the act has encountered great problems in operation, and these difficulties have changed people's notions of what is at stake. The current perceptions of the risks and rewards of the air pollution issue are very different grist for the political mill from what they were in 1970. The reasons perceptions have changed—the energy crisis, environmental backlash, interference by the courts, failure of technology, and so on—are less important here than understanding the precise nature of present perceptions. How the benefits and costs of air pollution policy are presently perceived affects the support for and opposition to policy change. In light of different dimensions of policy introduced earlier, the analysis that follows examines these changed perceptions.

The Unsuspecting Victims of Direct Regulation In 1970 people perceived that some interests were going to have to suffer from the deadlines and controls in the Clean Air Amendments. Focused costs were to be imposed upon industry, which would have to invest in research to develop pollution control devices and spend money to clean up. This stipulation seemed fair because industry behavior had been irresponsible in the past. Polluters would have to install devices or shut down.

However, experience under the Clean Air Amendments has shown that it is the motorist—especially the urban commuter—who has been threatened with the brunt of regulations. It could be argued that the Senate committee gave some warning, for its report stated in part: "If the Nation is to continue to depend upon individual use motor vehicles, such vehicles must meet high standards. The bill recognizes that a generation—or ten years' production—of motor vehicles will be required to meet the proposed standards. During that time, as much as seventy-five percent of the traffic may have to be restricted in certain large metropolitan areas if health standards are to be achieved within the time required by this bill."[50] Beyond a few such paragraphs in the report there is little indication in the record that anyone understood or was upset about the implications for ordinary people. It was not really until EPA required twenty-six cities to submit transportation plans that the impact became clear.

Today it is all too obvious who is affected by transportation controls. The urban motorist may be forced to leave the car at home, do without a parking place, pay a prohibitive fine, crowd over for bus lanes, install a tail-pipe device on an old car, submit to vehicle inspections, and so on. Transportation plans have been too unpopular to implement anywhere and have had to be so watered down that there has not been a prayer of meeting the primary air standards in major cities. Even Boston, the center of environmentalism where all the major politicians were committed to the Clean Air Amendments, has not been able to live with a transportation plan aimed at reaching the clean air goals within specified timetables.

The Indirect Adverse Effects of Air Pollution Control upon the Economy and Energy Costs The lofty and emotionally appealing benefit of the Clean Air Amendments in 1970 promised to be clean and healthful air. Everyone would breathe easier and because the government directly applied pressure to polluters. As pointed out earlier, the value of clean air so completely overshadowed considerations of jobs and prices that these aspects of air pollution control were scarcely mentioned by anyone other than the obvious vested interests.

The legislators' faith in technology swept aside considerations of the economic burden of cleaning up. In asserting that legislation could force technology, the act overestimated how much effect a law could have upon the orientation and behavior of private enterprise. The assumption was that industries would be sufficiently inventive to find low-cost, workable solutions to pollution if the government closed loopholes (such as requiring only those controls that were economically and technically feasible) and forced industry to move faster by imposing tight deadlines.

The impact of the Clean Air Act Amendments on the economy was considerably exacerbated by the 1973–1974 energy crisis. Cleaner-burning oil became both more expensive and less available to electric utilities. The impact of pollution control devices on fuel economy in automobiles became a matter of concern. The energy crisis–an event not foreseen in 1970—has now heightened perceptions of costs.

The Burden upon Administration and Intergovernmental Relations The 1970 act radically altered the structure of administering air pollution programs. The implications of these structural changes seem to have been generally overlooked. The only cost of structural change that was explicitly recognized was the need for more federal budget and staff. The Senate Committee on Public Works report stated:

The committee emphasizes that the act, the deadlines proposed, and the new programs authorized will be without meaning unless supplemental manpower (doubling present staffing of the National Air Pollution Control Administration) and supplemental funding (an increase of at least $44 million over the present budget request) are provided in this fiscal year.

Failure by the Executive Branch to request and Congress to approve these needed increases will substantially impair implementation of this legislation and make both the executive and legislative branches subject to charges of lack of commitment. Within a short period, the pressures for additional staff and funds will also be felt by the States. Should the States fail to respond to that pressure, the deadlines established by the Act would require broader Federal involvement in regional and state programs.[51]

In actuality the added weight that the act has placed upon federal, state, and local agencies has been much greater than mere budget increases could alleviate. Charles O. Jones has documented the impossibility of implementing air pollution legislation that is so demanding that it is beyond the capacity of federal, state, and local air pollution agencies. The act has reshuffled responsibilities to the extent that none of the partners knows what to expect from the others. State and local administrators do not know what federal officials want, what they will accept, or what they will disapprove. Federal officials do not know how much resistance to expect from states and localities or what to do about it when it develops.[52] Uncertainty has been exacerbated by the action of the courts in such matters as formulation and approval of state implementation plans. Yet another unsettling influence has been governmental reorganization. Administration of the federal air pollution program was severely set back by the reorganization of NAPCA into the Environmental Protection Agency in 1970. Whatever the long-term benefits of pollution program consolidation, the successor of NAPCA has experienced great difficulties in administering a number of tough programs. For instance, the EPA for a time faced continual criticism from the House Appropriations Committee, chaired by Jamie Whitten. In addition, the EPA has not had much support from the White House, the states, or the public.

State officials are now acutely aware of the risks involved in a structural arrangement under which federal law and the EPA make unpopular demands that states must implement without much power to modify. Governor Dan Walker of Illinois expressed the frustration of many state officials during the 1975 congressional hearings on the Clean Air Act Amendments:

States must have the flexibility needed to administer a program which balances our needs for safe environment and an adequate supply of energy.

Since the various State implementation plans have been instituted, there has been an endless chain of litigation on both the Federal and State level. This litigation has resulted in conflicting opinions which have only served to

> confuse the issues involved and have in themselves caused delay in compliance. . . .
>
> States must be given the clear authority to issue variances or enforcement orders with compliance dates extending beyond the May 1975 deadline. . . .
>
> During the period immediately ahead, the States should be given the option to determine which type of interim controls can best be used.[53]

In fact, there seem to be no insurmountable technical obstacles to the development of workable pollution control devices. Rather, the difficulties are organizational and political. Neither the automotive or utility industries—the major sources of air pollution—have historically made much of an investment in long-term research. The emphasis of the auto industry is upon sales. The time focus is short-term; the longest time frame is the two-to-three-year period over which a model change is planned. Technical modifications are introduced gradually and often begin with a single model line and spread to other models over a period of years.[54] Utilities have traditionally depended upon manufacturers for technological changes. The state regulatory frameworks in which they are enmeshed, along with the conservatism of their leadership, have biased utilities against taking chances. The utilities' main concern is with reliability. Such long-term orientations were bound to be difficult to change, especially when both the automobile and utility industries felt they had options to prevent implementation of the act.

Although the utility and automotive industries could be symbolically discredited as narrowly selfish during the consideration of the Clean Air Act Amendments, their very great political and economic power became obvious during the process of implementation. The federal government had the ultimate weapon—shutting down industry—but its use, like the use of the A-bomb, was unthinkable. For instance, at one point during the debate over whether or not to suspend the 1975 standards, Ford Motor Company claimed that a suspension denial would cause them to close down their factories; this drastic move would have resulted in a reduction of the GNP by $17 billion, an increased unemployment rate of 800,000 persons, and a decrease in tax receipts of $5 billion at all levels of government.[55] The tough talk of the Clean Air Act—imposing heavy monetary penalties for failure to meet the standard—was whistling in the wind when the economic consequences were considered.

The potential power of electric utilities is hardly less awesome. Utilities have a special political advantage in that they are scattered in every state and congressional district. The larger and larger monthly bill that home owners get is a constant reminder of the indirect, pass-through charges of pollution control. When the Arab oil boycott occurred, utilities claimed that sulfur standards could not possibly be met and that the effort to clean up would mean increasing rates. In turn, the increase in energy prices was blamed for the economic recession and slow recovery.

It is interesting that people's raised consciousness of diffuse costs did not cause a gutting of the Clean Air Act in 1973–1974 rather than simple suspensions and postponements. There is much evidence to suggest people still value air quality, but today there is widespread awareness that everyone shares indirectly in the expense of air pollution control, by higher electric energy rates, generally higher prices, and more expensive new automobiles.

Changes in the Legislative Setting

Along with the very great changes in perceptions of the air pollution issue, the legislative setting is very different from what it was in 1970. Individual legislators see altered and diminished possibilities for legislative record and support building upon the issue. Different actors have become involved and they relate to one another in new patterns. Interests and groups that lacked effective access to the decision-making process in 1970 have gained entry. Broadly, an entrepreneurial pattern of decision making has given way to interest-group coalition building and bargaining typical of regulatory policies. The high levels of conflict ordinarily experienced in congressional treatment of regulatory policy now exist for air pollution, as well. Policy made by Congress on the air pollution issue is now likely to be a careful balance among competing interests rather than the creation of policy entrepreneurs. Incremental changes rather than the large policy change of 1970 will probably characterize new legislation. Some of the most significant modifications of the legislative setting are addressed below.

Changes in Costs of Decision Making The air pollution issue currently promises to be costly to legislators. The public no longer simply wants Congress to act to clean up the air. Rather, the structure of public opinion suggests that the legislator must delicately balance the aesthetic and health values of clean air against the value of a healthy economy and adequate energy supply. A great deal of concern about the quality of the environment persists, but people are now quite aware of the costs of cleaning up. An Opinion Research Corporation survey conducted in August 1975 suggested that the public was sharply divided about raising the price of cars to cover the costs of emission devices, at least when people were asked if they would pay $250 for 10 percent cleaner air. Thirty-eight percent of a national sample believed that it was more important to keep the price of cars lower by eliminating pollution devices. Forty-eight percent expressed a desire to see controls installed on new cars now, even though it meant an increase in price. Similarly, the same survey showed that forty-four percent of the respondents disagreed with the statement, "Cleaning the environment is more important, even if it means closing down some old plants and causing some unemployment." Practically the same proportion—forty-three percent—agree with the statement.[56]

When values are in obvious conflict, legislators who become involved in the issue must anticipate that whatever action is taken will generate opposition. The importance of the air pollution issue makes it difficult to ignore, but legislators cannot reasonably expect to attract undivided support for any stand they may take. When the costs of decision making are likely to be high, congressmen are apt to be far more willing to share responsibility for decision making with the executive and with the states.[57] Thus, whereas in 1970 Congress was determined that the Clean Air Act Amendments be congressional legislation, Congress currently has cause to defer somewhat to other decision-making arenas.

Changes in Legislators' Goals and Resources Because of important changes in the goals and resources of Senator Muskie and of other legislators, the conditions of entrepreneurial activity no longer exist. Senators, by virtue of the number of committees in which they are involved, are less likely to be irretrievably caught up in specialization than are members of the House. Consequently, senators have some flexibility in emphasizing certain parts of their legislative careers at different times. As presidential possibilities have become more remote for Senator Muskie, his attention has turned more to building and protecting the institutional prerogatives of the Senate. In recent years a large proportion of the senator's time has been spent chairing the new and important Senate Budget Committee. With changes in the Congressional party leadership in the offing, there is speculation about Muskie and the majority leadership. Senator Muskie's motivation to solve pollution problems may well remain as strong today as it was in 1963 when the Subcommittee on Air and Water Pollution was formed. In fact, when he had to give up either Public Works or Foreign Affairs upon taking up the Budget Committee post, he opted to keep the latter, presumably because of his environmental interest. Nevertheless, the senator's allocation of time and effort has changed.

In addition, Senator Muskie no longer has the resource of a united committee. Turnover on the Public Works Committee has removed a number of members who were very supportive of Senator Muskie's leadership. Because of the reduced visibility of the Clean Air Act in the press, the new members of the committee see a lessened opportunity for policy action; it has been difficult, therefore, for the committee to achieve a quorum. Furthermore, there are currently Democratic members of the subcommittee who are willing to stake out positions quite separate from the chairman. Senator Gary Hart (D-Colo.), who opposed the committee version of the 1976 amendments, maintained that they did not adequately protect the environment. The seminar-type consensus that existed on the subcommittee in previous years was based partially upon hours and days of meeting together studying and discussing issues. Because Senator Muskie spends more time on other issues and because of the difficulty of obtaining a quorum, there is less opportunity for these sessions to take place. The new rule requiring open markups has also affected the style of the subcommittee. The physical and psychological setting is not conducive to the intellectual exchanges that once marked the seminar atmosphere. The result of the changes in committee personnel and style is that committee recommendations emerge from hard bargaining, and disgruntled members are willing to appeal to the floor of the Senate. With marked divisions in the committee, more floor fights and amendments from the floor are likely.

Changes in Competition for Credit Intercommittee bargaining has replaced congressional competition for credit for air policy. The fierce individual competition for a reputation in environmental policy has subsided. Legislators attracted to policy entrepreneurship have moved on to other fields of emerging public interest, such as energy and food scarcity. The Clean Air Act Amendments have more or less settled the question of credit for the design of federal air policy by giving the responsibility to Muskie and the Senate subcommittee. The current debate is not about how to make federal legislation more stringent but how to modify the act and to soften its negative impacts. The Senate subcommittee lacks the working consensus to entertain initiatives in different legislative directions. Senator Muskie has been placed in a defensive posture: he sees himself holding the line against gutting changes. Speaking of a subcommittee decision to uphold the concept of nondegradation, Muskie has said:

> This decision was a victory for environmental quality. It comes at a time when Congress is under heavy pressure to sacrifice all environmental initiative for economic recovery even when there is little demonstrable relationship between the two.
>
> But while we did not abandon nondegradation and while we have maintained the health standard we have had some setbacks. We will not get a clean car as soon as necessary and I fear that the industry will use the additional time they have to lobby for further extensions rather than develop new systems.
>
> We had to delay our deadlines and we had to make major compromises on urban-auto pollutant health standards because of the impossibility of making radical structural changes in our cities. Only by a narrow margin of 7 to 5

did we keep the emission limits—through permanent controls and precise timetables—as the enforcement mechanism.

To those of you here who have followed the course of environmental law and who have witnessed the anti-environmental rhetoric of the past year, these setbacks may seem acceptable, when balanced against the threat of much greater setbacks in the current economic and political climate. But I am concerned by the drift in national policy away from the basic objectives which we set forth at the beginning of this decade.[58]

The initiative to modify the act has switched to the Subcommittee on Health and the Environment of the House Committee on Commerce. Because the House Commerce Committee leadership was outmaneuvered in conference in 1970, they took less credit and are in a less defensive position about the contents of Clean Air Act Amendments. In addition (as will be discussed in more detail later), the House committee members—particularly subcommittee Chairman Paul Rogers—have accumulated new resources. Therefore, the major setting for modifying the Clean Air Act is now likely to be the House.

Changes in the Pattern of Politics Today's pattern of politics involves many actors with diverse points of view operating in a number of settings. Many of the interests that were not influential in the 1970 deliberations have emerged as effective participants. The scope of the current debate concerning modifications of the 1970 act has meant that activists and groups that have expertise in the details of implementation of the act are accorded special attention and legitimacy in formulation of policy. Because of the prevailing type of politics (that is, negotiation among opposing groups and interests), decision-making arenas that are good centers for bargaining are likely to be focal points for activity. The changed positions of industry, the environmental groups, the states, and the House Commerce Committee are worthy of special attention.

Industry has proven the necessity of its participation in fashioning pollution policy by demonstrating the essential nature of its role in the implementation process. If industries—automotive and utilities, in particular—feel they cannot meet standards and deadlines or if they can present other alternatives, implementation of targets can be thwarted. One advantage of a legislative process in which industry is an active bargainer is that polluters may become committed to a certain level of effort. A basic reason for the enlarged role of industry in current air pollution policy formulation is that the issue is no longer largely symbolic. Polluters are no longer a small band of industrialists who can be discredited; pollution is now seen as a characteristic of a system in which producers as well as consumers participate. As an essential component, industry cannot be left out of decision making.

Environmental groups are no longer in the first flush of victory, as they were in 1970. Since the peak of the environmental movement in 1970, they have experienced a number of defeats as well as conquests. In 1976, nevertheless, environmental groups continued to be very important political forces in spite of the energy crisis and economic problems, and they had every reason to be cautiously confident. Because success is no longer a surprise, these groups need no longer feel that if they do not keep industry beaten down, if they give an inch, the environmental movement will be simply swamped by the superior resources of industry. As a result, key environmental spokespersons have come to rely more on scientific data and sophisticated analysis than on rhetoric. Recent congressional testimony, filled with numbers and tables, is witness to this change. In general—and with some notable exceptions—environmental groups have become more flexible and more willing to entertain modifications to legislation that was once an inviolate symbol. A number of influential environmentalists have come to support the use of economic incentives to achieve implementation.[59]

States have at long last shed their role as poor little brothers in the air pollution regulation process. Traditionally, pollution control was a local matter. State regulatory laws and agencies were established mainly because of federal prodding. One of the consequences of the demands made upon states by the Clean Air Act Amendments and by the federal grant funds that they provided has been considerable augmentation of capability at the state level. As figures 2.3 and 2.4 indicate, both state personnel and state funding have increased dramatically since 1970. Along with these larger resources, there are new incentives for active state participation. On the firing line, implementation of unpopular decisions made elsewhere has been a frustrating role for states, and states are currently demanding a much greater say in managing air quality control in their individual geographical areas.

The Subcommittee on Health and the Environment of the House Committee on Commerce has become the center for brokering the interests concerned with the modification of the Clean Air Amendments. One cause for this subcommittee's change in status is Congressman Paul Rogers. Subcommittee Chairman Rogers has a constituency interest in the health issue because of the large proportion of retired persons in his Florida district. In 1970 he was something of a neophyte to environmental policy and lacked the prerogative for formal subcommittee chairmanship. Today, his ten terms in the House have accorded him a leadership post. The volumes of his subcommittee's hearings are filled with testimony about the experience under the Clean Air Act and establish Rogers's command of the intricacies of the issue. A great deal of disagreement exists within the Commerce Committee

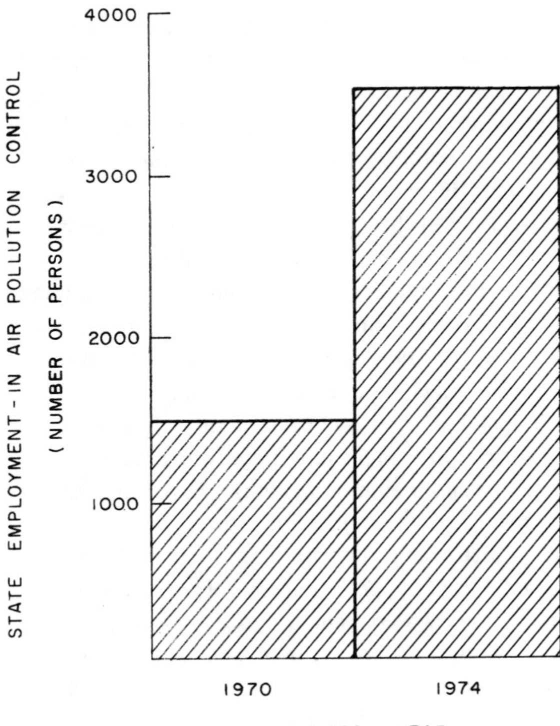

Figure 2.3. Five-year increase of employment in state air-pollution-control programs. (Source: U.S. Bureau of the Census, Environmental Quality Control, Government Finances and Employment, various years)

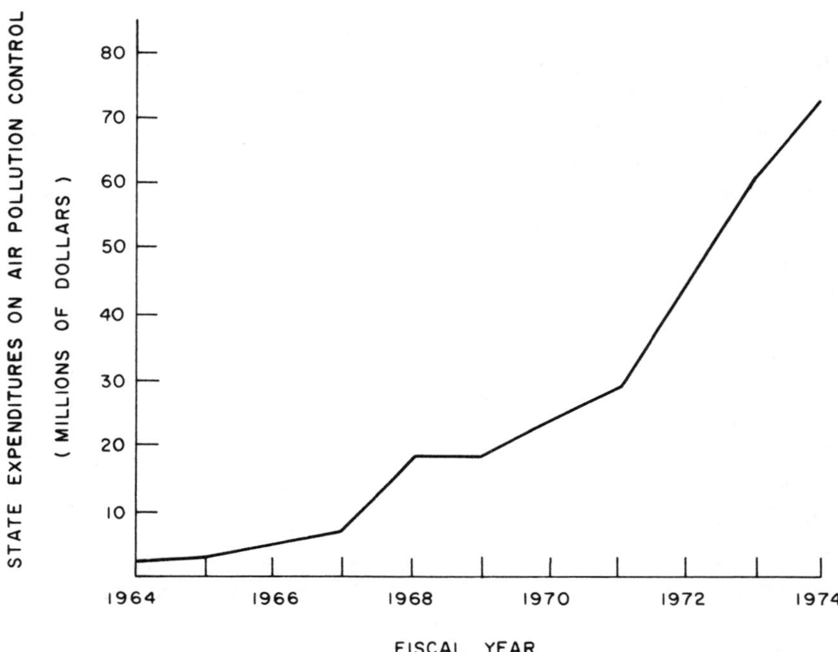

Figure 2.4. State expenditures on air pollution control for the years 1964–1974. (Source: U.S. Bureau of the Census, Environmental Quality Control, Government Finances and Employment, various years; and U.S. Congress, Committee on Appropriations, Subcommittee on Agriculture-Environmental and Consumer Protection, Hearings on Environmental Protection, various years)

about the desirable degree of federal regulation. Rogers has a reputation as a policy moderate, and his facility as a broker among different points of view is fundamental to his present strength. In addition, the staffing of the subcommittee has increased substantially. Thus, the House Subcommittee on Health and the Environment has become the center for the balancing of interests that the current pattern of politics in air pollution policy entails.[60]

Innovative or Incremental Changes Current perceptions of the air pollution issue and contemporary patterns of politics suggest that modifications of the Clean Air Act Amendments of 1970 are in order. The question is whether the modifications will be marginal or fundamental. Many economists and physical scientists believe that the difficulties encountered in the implementation of the Clean Air Act justify a completely different approach to the handling of air pollution control. One solution that has often been suggested is to devise a system of emission charges that will result in reduced pollution but leave the basic decisions of how and when to abate to the polluter.[61] A thoroughgoing change from the legal, regulatory approach to a charges approach would represent an innovation. Innovation requires a combination of issue perceptions and the presence of entrepreneurs—a combination that is rare, although it did occur in 1970. Charges were not seriously considered in 1970 for a variety of reasons. Most important, focused and diffuse costs were not very clearly perceived and, therefore, a strategy to reduce costs and to increase the efficiency of implementation had no particular appeal.

Although there is now heightened concern over costs and efficiency, the public support and legislative conditions for fundamental change that were present in 1970 do not exist today. The current pattern of broker politics in air pollution (in which the House Commerce Committee serves as a center for bargaining among coalitions of interests) is inhospitable to nonincremental changes in the current regulatory system. The regulatory approach of the Clean Air Act Amendments is generally taken as a given, and activity is concentrated on tinkering with the machinery. Legislators are open to suggestions that may make the machine run better. For instance, there has been a good deal of interest in a system of excess emission fees. This is a modified-charge approach that retains standards and deadlines and comes into operation only when targets have been missed. The fee levied against polluters taxes away any gains the polluter may have gleaned from failure to install control devices. The tax provides regulators with an attractive alternative to either shutting down the offending plant (and experiencing economic dislocation) or suspending enforcement (which undermines credibility). It is likely that legislators and other policy makers will be willing to entertain similar kinds of alterations in the regulatory system, alterations that leave its basic premises intact but promise to make it work better.

Conclusion

The central purpose of this chapter has been to argue that federal air pollution policy has been made rationally; that is, it has been predictable and explainable in that it has followed certain regularities or laws that hold for it and for other issues in the same class. It is true that enactment of the Clean Air Act Amendments of 1970 was unusual because it failed to follow the most familiar legislative pattern of incremental change. The air pollution issue belonged to a category of issues that can be labeled nonincremental and innovative. Such issues can be easily freed from a number of difficult barriers to policy formulation. In this category, severe focused costs may be imposed upon vested interests for the benefit of a diffuse public. Incremental policy usually is more attentive to those with a direct interest than the inchoate mass. Under a system in which policy failures are often obscured, innovative policies set bold goals that reach well beyond what can usually be attained. Such policies are vulnerable to criticism as they enter the implementation phase because, even if they actually accomplish more than incremental policies might, they have about them an aura of disappointed expectations. Innovative policies require quantum leaps in institutional capability, whereas incremental ones accept the inertia that is often most comfortable to bureaucratic organizations.

Innovative policies are a recessive strain in policy making because of inherent weaknesses that permit the impetus for large policy change to fizzle. Innovative policies tend to be oversold in order to mount a substantial effort and promise more than can ever be accomplished. Costs that will be incurred in the implementation of policy are minimized. Hirschman has noted that large development projects are so very difficult that, if the sponsors completely understood the morass they were getting into, they would never begin construction.[62] The same holds true for many nonincremental and innovative policies. If policy makers knew or made explicit to the constituency the potential risks, costs, and troubles, support would evaporate. Innovative policies also tend to be highly dependent upon the transitory attentions of the mass media. In order to curry media favor, policy entrepreneurs tend to create symbols that appeal to the emotions rather than to hardheaded analysis. Media coverage is important in creating the necessary public atmosphere for formulation of innovative policies but seldom persists to support the process of implementation.

Innovative policies instigated by Congress are highly dependent upon the availability of policy entrepreneurs who have the appropriate resources and who function in a favorable legislative setting. Entrepreneurship is most likely to occur in the Senate, where the size of the constituency, the breadth of interest, and the length of the term provide a substantial platform to command the necessary attention. Entrepreneurs often need committee or subcommittee chairmanships in order to manipulate the flow of business to their advantage. Access to information is very important, and most successful entrepreneurs are supported by a creative staff. Entrepreneurs can be encouraged to espouse more radical policies by competitors vying for credit.

Because nonincremental policies often fall short in implementation, an innovative tack in policy is seldom followed very long. Issues have a tendency to go through cycles of advance and retrenchment. Legislation that sets the major premises of policy is followed by efforts devoted to maintenance. Policy makers are willing to tinker with the machinery but not to make major alterations. When and if periods of major revisions occur, it is because perceptions of the issue are somehow altered and new possibilities for entrepreneurship are created.

With the hindsight that profits from knowledge of events, experience of implementation difficulty and inefficiency, new information, and broader understanding of earlier knowledge, critics have labeled the Clean Air Act irrational. Yet political decision makers in 1970 were limited by the perceptions of the issue that *then* prevailed and were frustrated by the failure of previous, less stringent laws. A political rationale has been offered for the actions of policy makers in this chapter. It has been argued that, under the circumstances that existed, it was rational to set high goals without clear knowledge of how or at what price they could be achieved. The public commitment to clean up the air through legislation was important. A change in public preferences and tastes so fundamental as to require public health and aesthetics to balance off heavily against industrial activity and economic progress probably cannot be accomplished without emotional symbols, exaggeration, and inefficiency. The burden of present policy makers is to modify the act to improve air quality more easily and efficiently. They have the opportunity to do so with the benefit of information gained through the experience with innovative policy.

Critics maintain that in the future, political rationality must deliver better policy than it did in 1970. Kneese and Schultze argue that "the legislative genius that finds ways to mold a coalition behind a piece of legislation, though still necessary, is no longer sufficient to devise the instruments of social intervention."[63] Congress must sort out policy instruments that are finely tuned to elicit the appropriate responses from highly complex industrial and organizational processes. The choice of policy techniques must be based upon a realistic assessment of their economic costs and benefits and of the likely long-term consequences. Such analysis is a difficult challenge for a political body whose job has always been to be sensitive to political support first and foremost—registering it and building upon it.

Herein lies a challenge to the professionals and experts who would like to see Congress approve a more rational policy in the scientific or technical sense. Can policy alternatives be formulated in ways that are sensitive to the political needs of decision makers? Can policy entrepreneurs be provided with sound new initiatives that appeal in simple and even symbolic terms to a mass public? When basic alterations are probably not politically feasible, can professionals suggest more sophisticated tools and more sharply focused incentives and disincentives that will be likely to achieve more within the current regulatory scheme?

Notes

While the limitations of this paper are the responsibility of the author, the following persons deserve credit for thoughtful and extensive criticisms of various drafts: Frederick R. Anderson, Aaron Wildavsky, Charles O. Jones, Paul Sabatier, Karl Braithwaite, Charles Meyers, J. Clarence Davies III, and Richard Ayers.

1. David Braybrooke and Charles Lindblom, *A Strategy of Decision* (New York: Free Press, 1963).

2. John E. Bonine, "The Evolution of 'Technology-Forcing' in the Clean Air Act," *Environment Reporter* 6 (25 July 1975): 1–29.

3. Theodore Lowi, "American Business, Public Policy, Case Studies, and Political Theory," *World Politics* 16 (July 1964): 677–715; James Q. Wilson, *Political Organizations* (New York: Basic Books, 1973); Paul R. Schulman, "Nonincremental Policy Making: Notes Toward an Alternative Paradigm," *The American Political Science Review* 69 (December 1975): 1354–1370; and Braybrooke and Lindblom, *A Strategy of Decision*.

4. Randall B. Ripley, "Congress and Clean Air: The Issue of Enforcement, 1963," in *Congress and Urban Problems,* ed. Frederic N. Cleaveland (Washington, D.C.: The Brookings Institution, 1969), pp. 224–278; J. Clarence Davies III and Barbara S. Davies, *The Politics of Pollution,* 2nd ed. rev. (Indianapolis: Bobbs-Merrill, 1975); and Charles O. Jones, *Clean Air: The Policies and Politics of Pollution Control* (Pittsburgh: University of Pittsburgh Press, 1975).

5. Theodore Lowi, "American Business;" idem, *The End of Liberalism* (New York: W. W. Norton, 1969); idem, "Decision Making vs. Policy Making: Toward an Antidote for Technocracy," *Public Administration Review* 30 (May/June 1970): 314–325; idem, "Four Systems of Policy, Politics, and Choice," *Public Administration Review* 32 (July/August 1972): 298–310.

6. Lowi, "American Business," pp. 690–691.

7. Dean E. Mann, "Political Incentives in U.S. Water Policy: Relationships Between Distributive and Regulatory Politics," in *What Government Does,* eds. Matthew Holden, Jr., and Dennis L. Dresang (Beverly Hills: Sage Publications, 1975), pp. 106–116.

8. Larry L. Wade, *The Elements of Public Policy* (Columbus, Ohio: Charles Merrill, 1972); Murray Edelman, *The Symbolic Uses of Politics* (Urbana, Ill.: University of Illinois Press, 1970); Robert Salisbury, "The Analysis of Public Policy: A Search for Theories and Roles," in *Political Science and Public Policy,* ed. Austin Ranney (Chicago: Markham, 1968), pp. 151–175; Wilson, *Political Organizations*.

9. Jones, *Clean Air,* pp. 196–197.

10. Matthew A. Crenson, *The Unpolitics of Air Pollution: A Study of Non-Decisionmaking in the Cities* (Baltimore: Johns Hopkins Press, 1971).

11. U.S., Congress, Senate, *Congressional Record,* 91st Cong., 2d sess., 1970, 116, p. 16,097.

12. Wade, *The Elements of Public Policy,* p. 14.

13. Daniel P. Moynihan, *The Politics of a Guaranteed Income* (New York: Random House, Vintage Books, 1973), pp. 156–159.

14. Edelman, *The Symbolic Uses of Politics,* p. 137.

15. U.S., Congress, Senate, *Congressional Record,* 91st Cong., 2d sess., 1970, 116, pp. 32,900–32,9001.

16. Aaron Wildavsky, "Economy and Environment/Rationality and Ritual: A Review of the Uncertain Search for Environmental Quality," *The Yale Law Journal* 86, in press.

17. Karl Braithwaite, Professional Staff Member, Subcommittee on Environmental Pollution, Committee on Public Works, U.S. Senate to the author, 9 September 1976.

18. Wilson, *Political Organizations,* p. 334.

19. Wilson, *Political Organizations,* p. 335.

20. Henry D. Jacoby and John D. Steinbruner, "The Context of Current Policy Discussions," in *Clearing the Air,* ed. Henry D. Jacoby et al. (Cambridge, Mass.: Ballinger, 1973). pp. 10–11.

21. Thomas Vitullo-Martin, "Pollution Control Laws: The Politics of Radical Change," in *The Politics of Eco-Suicide,* ed. Leslie L. Roos, Jr. (New York: Holt, Rinehart, and Winston, 1971), p. 363.

22. Randall B. Ripley, ed., "Introduction," in *Public Policies and Their Politics: Techniques of Government Control* (New York: W. W. Norton, 1966), pp. vii–xviii.

23. Cited in Bonine, "Evolution of 'Technology-Forcing,'" p. 3.

24. Bonine, "Evolution of 'Technology-Forcing.'"

25. Richard Ayers, National Resources Defense Council, to the author, 13 December 1976.

26. Paul Armand Sabatier, "Social Movements and Regulatory Agencies: The NAPCA-EPA Citizen Participation Program" (Ph.D. diss., University of Chicago, 1974).

27. Bonine, "Evolution of 'Technology-Forcing,'" p. 21.

28. Charles O. Jones, " Speculative Augmentation in Federal Air Pollution Policy-Making," *Journal of Politics* 36 (May 1974): 438–464.

29. Schulman, "Nonincremental Policy Making."

30. Schulman, "Nonincremental Policy Making," p. 1357.

31. James McEvoy III, "The American Concern with Environment," in *Social Behavior, Natural Resources, and the Environment,* eds. William R. Burch, Jr., Neil H. Check, Jr., and Lee Taylor (New York: Harper and Row, 1972), p. 226.

32. U.S., Congress, *Congressional Record,* 90th Cong., 1st sess., 1967, 113, A3764.

33. Paul Armand Sabatier, "Social Movements and Regulatory Agencies: Toward a More Adequate—and Less Pessimistic—Theory of 'Clientele Capture,'" *Policy Sciences* 6 (June 1975), p. 313.

34. John Johannes, *Policy Innovation in Congress* (Morristown, N.J.: General Learning Press, 1972), p. 17.

35. David E. Price, *Who Makes the Laws? Creativity and Power in Senate Committees* (Cambridge, Mass.: Schenkman, 1972), pp. 306–322.

36. Sabatier, "Social Movements and Regulatory Agencies: The NAPCA-EPA Citizen Participation Program."

37. Edmund S. Muskie, "Remarks by Sen. Edmund S. Muskie" (Delivered to the University of Detroit Student Bar Association Symposium on Environmental Law, Detroit, Mich., 20 February 1976), p. 2.

38. Richard F. Fenno, Jr., "The House Appropriations Committee as a Political System: The Problem of Integration," *The American Political Science Review* 56 (June 1962): 310–324.

39. Price, *Who Makes the Laws?,* pp. 329–331.

40. Jones, *Clean Air,* pp. 55–57.

41. Leon G. Billings, interview, Washington, D.C., 18 March 1976.

42. Price, *Who Makes the Laws?,* p. 330.

43. Jones, *Clean Air,* pp. 201–205; and Davies and Davies, *The Politics of Pollution,* pp. 52–56.

44. Lowi, "American Business," pp. 695–701.

45. Jones, *Clean Air,* pp. 195–198.

46. David R. Mayhew, *The Electoral Connection* (New Haven, Conn.: Yale University Press, 1964), pp. 66–67.

47. Jones, *Clean Air,* p. 181.

48. Lawrence Henry Chamberlain, *The President, Congress, and Legislation* (New York: Columbia University Press, 1946); Ronald C. Moe and Stephen C. Teel, "Congress as Policy Maker: A Necessary Reappraisal," in *Congress and the President: Allies and Adversaries,* ed. Ronald C. Moe (Pacific Palisades, Calif.: Goodyear, 1971); and Price, *Who Makes the Laws?*

49. Jeffrey L. Pressman and Aaron Wildavsky, *Implementation* (Berkeley, Calif.: University of California Press, 1973).

50. U.S., Congress, Senate, Committee on Public Works, *National Air Quality*

Standards Act of 1970: Report with Individual Views to Accompany S. 4358, 91st Cong, 2d sess., 1970, S1196, p. 2.

51. U.S., Congress, Senate, Committee on Public Works, *National Air Quality Standards Act of 1970: Report,* p. 3.

52. Jones, *Clean Air,* pp. 87–136, 211–235.

53. U.S., Congress, House, Committee on Interstate and Foreign Commerce, Subcommittee on Health and the Environment, *Hearings on Clean Air Act Amendments—1975,* 94th Cong., 1st sess., 26 March 1975, pt. 2, p. 1308.

54. Henry D. Jacoby and John D. Steinbruner, "Advanced Technology and the Problem of Implementation," in *Clearing the Air,* p. 52.

55. Clarence M. Ditlow, "Federal Regulation of Motor Vehicle Emissions Under the Clean Air Act Amendments of 1970," *Ecology Law Quarterly* 4, no. 3 (1975), p. 506.

56. Opinion Research Corporation, "Public Attitudes Toward Environmental Tradeoffs," *Public Opinion Index* 33 (End August 1975): 1–8.

57. Robert H. Salisbury and John Heinz, "A Theory of Policy Analysis and Some Preliminary Applications," in *Policy Analysis in Political Science,* ed. Ira Sharkansky (Chicago: Markham, 1970).

58. Muskie, "Remarks," p. 4.

59. Richard E. Ayers, "Enforcement of Air Pollution Controls on Stationary Sources Under the Clean Air Act Amendments," *Ecology Law Quarterly* 4, no. 3 (1975), 441–478.

60. Arthur Magida, "Clean Air Act Deliberations—The Changing of the Guard," *National Journal* 8 (13 March 1976): 341.

61. Allen V. Kneese and Charles L. Schultze, *Pollution, Prices, and Public Policy* (Washington, D.C.: The Brookings Institution, 1975), pp. 87–104, 108.

62. Albert Hirschman, "The Principle of the Hiding Hand," *The Public Interest* 6 (Winter 1967; reprinted ed., Washington, D.C.: The Brookings Institution, April 1967): 10–23.

63. Kneese and Schultze, *Pollution, Prices, and Public Policy,* p. 120.

Comment

Charles O. Jones

Professor Ingram has taken charge of some very difficult concepts in seeking to explain the Clean Air Amendments of 1970. Two of these appear in the title of her paper, "rationality" and "innovation." Others are relied on in the text: "entrepreneurship" as a form of political rationality, "mass public opinion" as a driving force in policy development, and "symbolism" as a form of policy action. Her use of this terminology has, in truth, contributed greatly to our understanding of environmental decision making. My purpose, as I see it, is to stimulate discussion of this effort by raising questions designed to encourage conceptual clarification and thus to advance our knowledge about this most complex set of issues. My method for doing so is to draw attention to three recent events of importance for air pollution control politics: the interment, if not death, of the Clean Air Amendments of 1976; the pollution control agreement between the Clairton Coke Works of U.S. Steel and county and state air pollution control agencies; and the sixth anniversary of Earth Day (by invitation only!).

"Clean Air Amendments Die at Session's Close" (*Congressional Quarterly Weekly Report,* 9 October 1976). As has been amply documented, the 1970 Clean Air Amendments had broad support in Congress—passing in the Senate by 73–0 and in the House of Representatives by 374–1. In 1976, the clean air bill initially passed the Senate 78–13—still a large margin. But voting on amendments revealed two contending groups, one supporting more stringent automobile emission provisions (led by Senator Gary Hart, D-Colo.) and one supporting less stringent nondegradation provisions led by Senator Frank Moss, D-Utah). A third group supported the bill as reported and opposed both amendments. These votes on amendments are notable because they reveal that air pollution legislation had indeed become controversial in the Senate. Likewise, in the House, the 1976 clean air legislation faced formidable opposition. By a 224–169 vote, the automobile emission deadline was postponed to 1982. The bill itself passed, 324–68.

In fact, much of the political history of the Clean Air Amendments of 1970 since enactment has consisted of seeking modifications, or fall-back positions, from the stringent provisions included in the law. The actions taken in Congress in 1976 were symbolic of this development. The many interests affected by the legislation lobbied intensely, with the result that the compromise legislation (as reported by the conference committee) did not reach the Senate floor until *the day before adjournment* and thus conditions were perfect for a successful filibuster.

The conference report for the Clean Air Amendments of 1970 also reached the floor at the end of the session in that year. Both chambers approved the report by a voice vote on 18 December. What was the difference? In 1970, tough air pollution control legislation had broad support. In 1976, air pollution control had become a highly conflictual issue. Put another way, successful filibusters don't just happen. They tend to reflect strong sentiment in the legislative chamber.

In the closing debate in the Senate, Senator Edmund S. Muskie (D-Maine) sought to assure the Senate by reviewing the comprehensive effort of the Senate and House committees to produce an acceptable revision of the Clean Air Amendments:

> We [the Senate Committee on Public Works] have held a total of 56 days of hearings on the Clean Air Act over the past 4 years—56. . . . The subcommittee held 24 markup sessions on amendments to the Clean Air Act from June of 1975 to November of 1975. The full committee held 24 markup sessions on clean air amendments, from November of 1975 to February of 1976. . . .
>
> Six conferences were held on these amendments totaling approximately 30 hours. . . . In addition, the Senate conferees met separately in caucus nine times for approximately 18 hours. . . . In addition to that, there were around-the-clock sessions at the staff level. . . .
>
> No, Mr. President, this is not a careless piece of work; it is not a hasty piece of work; it is not an irresponsible piece of work; it is not a radical piece of work. It is, I think, one of the best examples of committee action at its best that I have seen in the Senate in my lifetime. . . .
>
> Mr. President, finally, on the question of the attention this bill has received, may I point out that on the House side the consideration of the House bill was just as thorough as here. The House actually had 66 markup sessions in full committee and 22 in subcommittee, for a total of 88 markup sessions in the development of their legislation. (*Congressional Record,* 1 October 1976, daily edition, S17,533–S17,534)

Although not unimpressed by this record, Senator Jake Garn (R-Utah), leader of the fatal filibuster, was not reassured: "So at 5:45 P.M., last night I finally got a copy of the report, and I find it a little bit difficult to believe that the greatest deliberative body on earth . . . should be given the opportunity to deliberate for only approximately 24 hours on a huge bill that looks like chickens have scratched all over it. It not only is not printed but it has notations in pencil. It has arrows and darts and things scratched out. And I am not convinced I am capable of understanding from that report what we are passing'' (*Congressional Record,* 1 October 1976, daily edition, S17,545). And so it went. Time was on the side of the opponents. A motion to postpone adjournment by one day was defeated and the Senate adjourned without voting on the Clean Air Amendments. The House did not act at all.

What conclusions can we draw from this event for Ingram's effort? At the very least it suggests the need for a clear conception of ''political rationality.'' In what sense does ''policy beyond capability'' or ''law beyond technology'' represent rational politics? In response to this puzzle, Ingram concludes, ''Legislating stringent regulations, the costs and implementation of which are uncertain, can be justified in the belief that large change is desirable and that those affected by the legislation will act at some time in the future [perhaps by returning to Congress] to prevent whatever unintended and damaging consequences have not been forseen. It is consequently 'rational' to shift the burden for correction of innovative policies to future ipso facto more experienced policy makers.''

Two versions of rationality are suggested here. The first is short run and suggests that politicians may, ''under certain circumstances'' (presumably expression of views by a mass public) be moved to act in extraordinary fashion (in a manner not normally judged to be rational?). As evidenced by what happened in 1976, however, short-run rationality or entrepreneurship may, again under certain circumstances (possibly when associated with a highly technical issue-area), result in somewhat longer-run *ir*rationality. One reasonable interpretation of what has happened since the enactment of the Clean Air Amendments of 1970 is that steady, but undramatic, progress toward pollution control in many local areas was interrupted by rather grandiose federal intervention. Of course, we cannot be absolutely certain one way or the other. What we can question is whether such legislation is ''*politically* rational'' inasmuch as the politicians must now seek to disentangle much of what has been done since 1970 without seeming to do so. Put another way, one wonders whether, in this case, the ''future . . . policy makers'' have come to be ''ipso facto more experienced.'' We cannot resolve any of these matters without either providing a somewhat more definite referent for the concept of political rationality or employing other concepts less likely to create misunderstanding. Ingram is to be commended for, in essence, demanding that we face this important issue.

''Innovation'' too deserves careful consideration for it is a term frequently used to justify policy moves of various types. Ingram judges that ''The Clean Air Act Amendments [of 1970] are innovative in the sense that the legislation is more than an incremental step from the base of past experience. The act set far more ambitious air pollution control goals, which were to be accomplished more quickly and under a more demanding regulatory regime than could possibly have been projected from previous policy evolution.'' Once again we encounter difficulties in conceptual understanding. Innovation appears to be associated only with nonincremental policies. Yet we know that innovation as change can occur incrementally over time. Clearly a referent is required. Is a program innovative if its goals are ''more ambitious'' than those of previous programs or if it introduces more stringent regulations than in the past or if it introduces different methods of enforcement and implementation or it it actually achieves definite results? Clearly ''Innovation compared to what?'' is a relevant question to raise—one that then demands specific measures for intelligent analysis.

In regard to the Clean Air Amendments of 1970, I propose consideration of a case that may raise doubts about how innovative the federal program turned out to be (at least by the last two interpretations mentioned above). I have been following the Clairton Coke Works case for some years. My clipping file on that issue begins with an item dated 6 August 1970. The headline reads: ''U.S. Steel to Seek 2 Variances.'' Elsewhere[1] I have recorded and analyzed what happened from August 1970 to November 1973. To summarize briefly: after the Allegheny County Board of Air Pollution Appeals and Variance Review refused to grant a variance to U.S. Steel, the whole matter went to court and has been there on and off ever since. What I said in my book, completed in late 1973, serves as well to describe the actions and attitudes since that time: ''one anticipates a continuing process of charge, countercharge, litigation, negotiation, adjustment, and, eventually perhaps, disinterest and barely visible continuance. One day's news is cause for optimism that real progress is, or soon will be, realized. The next day brings discouragement as this or that complication is encountered in what must be perceived by most readers as a shapeless process of enforcing incomprehensible standards set by 'them.' ''[2]

At present, we are encouraged. The Court of Common Pleas, led by Judge Henry Ellenbogen, was recently successful in getting a settlement worked out between the contending forces, U.S. Steel and county, state, and federal air pollution control agency representatives. Interestingly enough, however, the agreement was worked out in a fashion reminiscent of the agreements of the 1950s. Although the decision-making arena was different—the court rather than the offices of the mayor or big industry—the manner of bringing government and industry together was similar and familiar. Whether, in fact, the settlement will work depends on continued cooperation among all parties. Already there have been threats by environmental groups to oppose the agreement, particularly because, as in the 1950s, they were not involved in the final negotiations. But in contrast to recent public hearings on such matters, there is a vocal citizen group on the other side. A group called "Clairton Organizes to Keep Employment (COKE)" appeared at the first set of hearings to support the agreement. The acting chairperson of COKE, Priscilla McFadden, observed that "We are not against clean air, but we feel the air standards should be set by reasonable people." She also stressed that COKE wants to "make it clear we are behind our good friend and neighbor, U.S. Steel."[3]

Whatever the eventual outcome of the Clairton Coke Works case, it has already illustrated that innovation is a multidimensional concept. Ingram has stressed the policy development dimensions of innovation—those responsible for program initiatives make a break with the past. We see with Clairton, however, that such bold strokes do not always result in the kind of dramatic change intended. Indeed, a full review and analysis of the Clairton experience may well raise the question of whether greater progress in realizing clean air might not have been achieved by means other than those mandated by the Clean Air Amendments of 1970.

It might be helpful in this connection to introduce a modified version of Lindblom's famous four-fold decision-making scheme.[4] Although it is a most useful set of criteria, the scheme is admittedly imcomplete. The version offered in figure 1 attempts to fill in the blank spaces and make more evident the fact that the two dimensions are continua, and thus introduces a great number of decision-making points rather than just the four suggested by Lindblom (shown here by the light dotted lines).

For present purposes, one might observe that, whereas the *development* and *approval* of the Clean Air Amendments of 1970 might well be a case of fourth-quadrant decision making, its *implementation* and subsequent *adjustment* are more appropriately third quadrant in nature. A case might also be made that implementing and adjusting some aspects of the 1970 act might well fit into the second quadrant. Not all of that has to be resolved, however, to clarify why I have introduced the scheme here. I simply want to illustrate the complexities of innovation and suggest how important it is that we specify the referents. I take it Ingram would agree, particularly given her statement that "Because nonincremental policies often fall short in inplementation, an innovative tack in policy is seldom followed very long."

The third event, or more appropriately "nonevent," to which I wish to refer is the sixth anniversary of Earth Day. A truly extraordinary public opinion happening occurred in 1970. For various reasons, we witnessed a

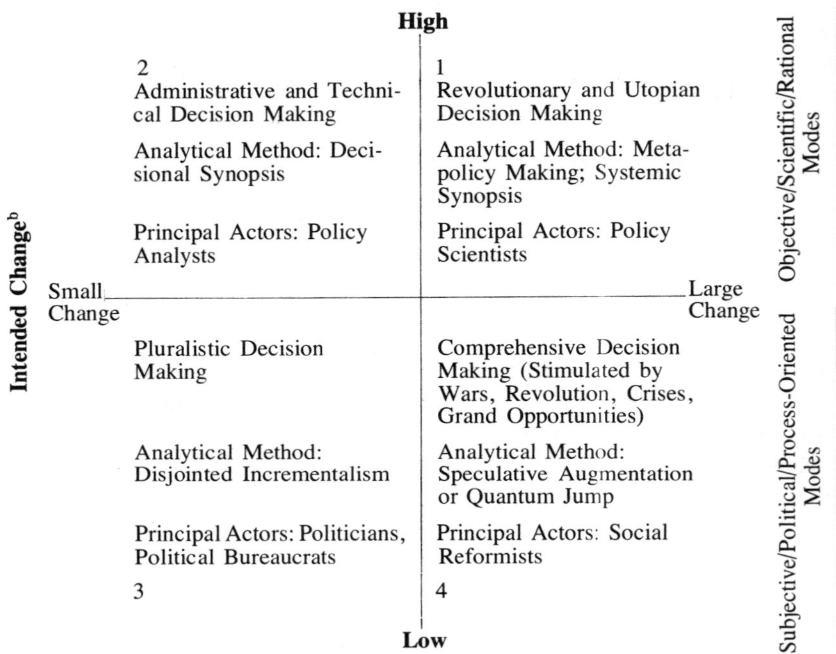

Figure 1. Four modes of decision making (variations on a theme of Charles E. Lindblom). (Source: Charles O. Jones, *An Introduction to the Study of Public Policy,* rev. ed. (Duxbury, Mass.: Duxbury Press, 1977), p. 220.

buildup of public concern for the environment during 1969–1970 with the result that in a very short period of time the federal government experienced a quantum increase in an area of responsibility previously designated a state-local concern. Whereas 53 percent of the respondents mentioned air and water pollution as one of the top three problems requiring action in 1970,[5] very few persons list it as a priority issue today. In large part, this shift is not surprising. Thus, for example, in response to the question, "Whatever happened to Earth Day?" a staff correspondent for *The Christian Science Monitor* emphasized that "The focus of the ecology movement in the U.S. has shifted from the popular rallies and demonstrations of recent years to lobbying in the corridors of Congress."[6] The fact is that a number of national, state, and local environmental lobby groups have emerged in the past six years to monitor enforcement of existing laws, press for new initiatives, resist compromises of or retreats from earlier legislative victories, and publicize the environmental records of candidates for public office.

These developments are what students of public opinion have told us to expect. Massive, fervent expression of opinion on issues does not occur all that often. Consequently elected decision makers are normally responsive when it happens. But its exceptional nature also leads us to expect either that concern will dissipate completely or that it will take the form of more traditional political action in a pluralist society, such as interest group organization and pressure.

What has all of this to do with Ingram's paper? As with the other important concepts she has utilized, mass public opinion too requires most careful explication. What are its forms? What is the context within which it occurs? What are its intentions? How does it change over time? How should decision makers respond? Is "policy beyond capability" the answer? These questions direct our attention to the complex and dynamic policy relationships over time between public attitudes and demands, and government decision makers and their programmatic responses. Any effort to respond soon drives us to consider many of the issues with which Ingram has wrestled in her paper. As I have suggested in the limited space available here, it is my own conviction that greater attention must be devoted to dissecting the many diverse elements of "mass public opinion" and its influence before drawing conclusions as to its effects. That was, in part, my intention in posing the three alternative futures for public pressure in the implementation of air pollution control programs.[7] Although quite primitive, the effort is at least directed to opinion change over time and to its possible consequences.

I am grateful to the author for producing a most stimulating paper. However, the principal thrust of my remarks (which, hopefully, will be interpreted as constructively critical) suggests two conclusions. First and foremost, we must try for greater conceptual clarity in analyzing environmental and other policies. This requirement must be met whether we intend to contribute to our own disciplinary development or wish to recommend policy change. Second, without this conceptual explicitness, one can as easily justify another, quite different title for Ingram's paper: "The Political Irrationality of Thwarting Innovation: The Clean Air Amendments of 1970."

Notes

1. David Braybrooke and Charles E. Lindblom, *A Strategy of Decision* (New York: Free Press, 1963).

2. Gallup International, *The Gallup Poll Index,* Report No. 60, June 1970.

3. *Pittsburgh Press,* 14 November 1976.

4. Braybrooke and Lindblom, *A Strategy of Decision,* 1963, p. 78.

5. Gallup International, *The Gallup Poll Index,* 1970, p. 8.

6. *The Christian Science Monitor,* 2 February 1976.

7. Charles O. Jones, *Clean Air: The Policies and Politics of Pollution Control* (Pittsburgh: University of Pittsburgh Press, 1975), p. 214.

Comment

Charles J. Meyers

"Good Laws" and Bad Societies

I mean to compliment Professor Ingram when I say that her paper makes dismal reading in this Bicentennial Year and, perhaps more to the point, in the 187th year of our ongoing, though shaky, experiment with a republican form of government.

Taking the Clean Air Act of 1970 as her exemplar, Ingram argues that "innovative legislation" is the product of a political process that "is rational, patterned, and explainable." By her definition, "innovative legislation" is legislation that makes a quantum leap beyond incremental policy making—legislation that sets bold goals and requires implementation beyond the capacity of the existing administrative machinery. Innovative change, thus defined, usually occurs, according to Ingram, when the debate is cast in symbolic terms—when actual benefits and costs are ignored and decisions are made on the basis of what is right and wrong or good and bad.

In promoting the Clean Air Act, Senator Muskie and his allies performed as rational political entrepreneurs, according to Ingram: they characterized the issue as the health and welfare of the people, who are obviously entitled to clean air and good health (at no cost to themselves) versus the self-interest of wicked polluters, who for tainted profits manufacture smelly cars and maintain grimy smokestacks. Proponents of clean air wore white hats; polluters naturally wore black hats. The issue was a moral one, as alcohol and marijuana once were; any attention to costs and benefits is out of place in a fantasy land where good always triumphs over evil.

What makes Ingram's study useful, though dismaying, is the thesis, cogently supported, that political reward comes from structuring debate on policy issues in these simplistic, moralistic terms. The rewards are specific and accrue to particular, identifiable political entrepreneurs. When innovative legislation proves to be unworkable, because the goals are unattainable and the means too expensive, the price that must be paid is social despair over the capacity of government to govern. But it is not the political entrepreneur who pays—we do.

The trouble with fantasy is that the real world won't go away. Innovative legislation, encumbered by overly ambitious goals, unrealistic timetables, and wasteful enforcement mechanisms, fails because the public eventually wakes up and discovers that costs exceed perceived benefits. When a choice must be made between some amount of air pollution and shutting down Los Angeles, the affected public prefers to keep moving. The paradox of innovative legislation is that the very conditions that allow for its enactment—the disregard of benefits and costs—also lead to its failure. The same kind of failure has occurred with the 1972 Water Pollution Control Act Amendments, another innovative act and an even worse one in that it set zero discharge in 1985 as its absurd goal. As the National Water Commission wrote in its Summary Report in late 1972 (p. 38): "This 'zero-discharge' policy has strong emotional appeal, but in the Commission's judgment it is an impractical and unattainable goal. Striving to achieve it will involve exorbitant costs, confusion in planning, misallocation of resources, and will risk public disillusionment with the entire national effort to protect the environment."

A second and related consequence of debating and deciding policy issues in moral terms is the tendency to restrict governmental action to the single option of regulation by imposing standards, to the exclusion of such other strategies as subsidies, effluent charges, and tradable pollution rights. Least-cost solutions are ignored. Kneese and Schultze[1] blame regulation on the lawyers, but I believe the trouble is more deep-seated than that. The proponents of the Clean Air Act obtained its passage by equating pollution with evil and I am afraid that most people, not just lawyers, still believe that the way to eliminate evil is to outlaw it, regardless of costs.

But there is another incentive as well. Observe that regulation not only suppresses evil but also has the not-altogether fortuitous advantage of requiring the creation of new government agencies staffed by multitudes of lawyers, economists, political scientists, and others looking for work—in the public interest. Elected officials are not the only public servants whose conduct is, to repeat Ingram's phrase, "rational, patterned, and explainable."

If Ingram's thesis is correct—if innovative legislation is the product of political rationality—the country appears to be doomed to wave after wave of laws setting forth unattainable or at least unrealistic goals to be achieved in the immediate future by the exclusive and wasteful means of regulation—with each crest of the wave being followed by ever-deepening troughs of despair over the inability of government to accomplish anything. Some argue that innovative legislation improves the social order by promulgating ideals, which, though not immediately achievable, set the tone for future effort. One giant step forward more than equals the next two steps back. I don't believe it. The cost is not just the waste of resources but also is the loss in confidence in government—and in ourselves.

Lawyers have an aphorism: hard cases make bad law. I would go further: "good" laws can make bad societies.

Note

1. Allen V. Kneese and Charles L. Schultze, *Pollution, Prices, and Public Policy* (Washington, D.C.: The Brookings Institution, 1975), p. 116.

3 Judging the Imponderables of Environmental Policy: Judicial Review under the Clean Air Act

Richard B. Stewart

The chief thing which the common sense individual wants is not satisfactions for the wants which he has, but more, and *better* wants . . . life is not fundamentally a striving for ends, for satisfactions, but rather for bases for further striving.

Frank H. Knight, *The Ethics of Competition* (1935), 22–23

This essay will examine court decisions reviewing the federal Environmental Protection Agency's implementation of the Clean Air Act as an occasion for exploring some more general issues of environmental policy and the role of the courts in their resolution.

The first section of this essay reviews the relevant objectives of environmental policy. Next, Congress' disposition of these objectives in the Clean Air Act is examined, with particular emphasis on its efforts narrowly to constrain administrative discretion in implementing environmental policy The third section of the essay traces the course of judicial review under the Clean Air Act. The allocation of responsibility among Congress, court, and agency and the courts' "quasi-constitutional" role in environmental policy are discussed. The concluding section evaluates, in the context of environmental policy, the role of reviewing courts in a mixed economy in which administrative agencies play a major role.

Objectives of Environmental Policy

Governmental measures to protect the environment may be viewed as a form of economic regulation justified by "market failure."[1] This useful perspective advances our understanding of environmental problems, points to the desirability of market-type incentives as a basic tool for their solution, and underscores the relevance of economic efficiency as a policy objective. At the same time, the richness of environmental problems exposes the limitations of economic analysis as a solvent of questions of collective choice, and dramatizes the importance of additional and apparently conflicting policy objectives. Similar complexities characterize other areas of governmental economic policy, but environmental questions present the attendant contradictions in high relief and raise insistently the question whether we possess adequate institutional or analytic means of resolving them.

Economic Efficiency

The following are truisms of our time:

1. We live in a world of scarce resources that should be used efficiently to advance human satisfactions.
2. As a basic rule, waste in the utilization of such resources to produce goods, services, and amenities can be avoided if production of a given commodity proceeds to the point where the marginal cost (in terms of foregone opportunities for utilization of the resource inputs) of producing a marginal unit of the product just equals the marginal value of that production in terms of consumer satisfactions as measured by the willingness of pay for an additional unit.[2]
3. The level of production or preservation of environmental amenities should be determined by the foregoing criterion.
4. The same criterion should determine the nature and extent of governmental intervention to correct the failure, because of the collective or "public good" nature of the amenities in question, of private market exchanges to satisfy such criterion.

Although truisms, these principles deserve forceful reiteration in the context of environmental policy. As other papers presented in this book demonstrate, governmental environmental policies all too often have disregarded the need to avoid resource waste.[3] Nonetheless, our environmental quality objectives cannot and should not be determined by a process of calculating marginal costs and marginal benefits. Quite apart from the forbidding methodological obstacles to identifying and quantifying all relevant costs and benefits and the equally formidable difficulties in instituting such a process of calculation in a legislative or bureaucratic setting, the economic calculus fails to deal with important competing objectives and also lacks any analytical basis for reconciling efficient resource allocation with such competing objectives.[4]

The Dynamic Nature of Preferences

The ultimate determinant of value in economic analysis is individual preference among various goods, services, and amenities. Economic analysis characteristically assumes that these preferences are exogenously given and fixed for purposes of determining what mix of products should be produced with given resources and production possibilities at a given time.[5] In fact, however, preferences are unstable and endogenous; they are powerfully shaped by our past experiences (consumption of products) and our expectations regarding potential future experiences (potential consumption of products). Nonetheless, the assumption of exogenously fixed preferences is a tolerable and workable one so long as we are dealing with marginal production decisions that have a comparatively short time frame, that concern only a relatively small proportion of the total products consumed or potentially consumed by given individuals, and that concern products for which relatively close substitutes are available and as to which individuals have tolerably complete information.

All of these assumptions come under great strain in the context of governmental decisions concerning environmental quality. Air pollution and other forms of environmental degradation pose health threats of which individuals may be totally unaware. Even when the effects of degradation are detectable by the lay individual—such as decreased visibility or discomfort—they may have simply been accepted as a given feature of the environment that is not realistically open to change. The very process of debate over appropriate governmental policies with respect to environmental quality will tend to change individual preferences for environmental quality by providing them with information and causing them to regard essential features of their environment as matters of social choice.[6]

The interplay between imperfect information and environmental preferences pivots on individual attitudes toward risk and uncertainty. These attitudes have great importance in the context of environmental policy because our knowledge concerning the potentially adverse environmental effects of various technologies is highly imperfect. The degree of risk aversion not only may vary widely among different individuals but also will be shaped by the structure of the risk in question.[7] Individual attitudes toward risk and uncertainty can also be affected by the provision of more information. In a quite practical context, should the environmental impact statement for a nuclear power reactor be required to disclose the possible adverse effects of a disastrous Class IX "meltdown" accident, even though the statistical probabilities of such an accident are, in the judgment of scientists, almost infinitesimally remote?[8] Such issues, in turn, raise the ultimate question whether society should honor lay attitudes toward risk and uncertainty even if they be judged "irrational" by informed scientists or economists.

The problems of ignorance and attitudes toward risk and uncertainty are aspects of the more general problem of the dynamic nature of preferences over time. Because environmental policies determine a substantial range of individual experiences for which there may be no close substitutes and because such policies, particularly as they affect land-use decisions, are likely to be relatively long range in character, the assumption that such policies should be based on existing preferences is less satisfactory than it would be in other contexts. What weight, for example, should we give to low preferences for environmental quality held by persons who have never experienced a high-quality environment? To take such preferences at face value could foreclose the possibility of new experiences that might change preferences in favor of higher-quality environments. On the other hand, simply to disregard existing preferences and dictate higher levels of environmental quality may not be justified by the fact that individuals come later to prefer a higher-quality environment; it is not obvious that the change in preferences over time should be accorded any normative significance, and to justify collective choices on the basis of such a change could be attacked as simply bootstrapping.[9]

How should choices about collective provision of environmental quality be made in view of the dynamic character of preferences? One might believe that some goods (such as environmental quality) or tastes for such goods were superior to others, regardless of what preferences individuals now hold and select governmental policies that would promote the consumption of such goods and the encouragement of associated tastes for them. But the difficulties in such an approach are obvious. By what criteria are superior products or tastes to be determined, and how are such decisions to be made and implemented consistently with the principles of our polity? Alternatively, one might conclude that the problem is hopeless and that the best we can do is to consult our existing preferences (including, perhaps, metapreferences as to future preference structures) in decisions concerning the provision of collective goods. A third possible approach is to advance those products and associated tastes that would be chosen as preferable by an individual who had experienced in an intensive fashion all of the respective candidates.[10] Assuming that this is even a sound principle for individuals, which is fairly debatable,[11] how is such a principle to be followed when the choices in question, as here, are not individual but collective in nature?

Distributional Considerations

In applied economic analysis, the value of production is measured by the willingness of individuals to pay for such production, and such willingness to pay is a function of individual wealth as well as preference. Thus economic efficiency—maximization of the value of production—is always relative to a given distribution of income. At the same time, reallocating resources to increase the value of production normally results in a redistribution of wealth that may offend relevant notions of equity.

Although evidence is sparse, plausible analysis suggests that economically efficient measures to enhance environmental quality will, on balance, benefit the wealthy more than the poor. If environmental quality is indeed a luxury good, then basing environmental policy choices on economic efficiency may entail a shift in the distribution of real income in favor of the wealthy, a result that many would regard as inequitable.

First, available evidence indicates that most of the costs of enhanced environmental quality are financed through increased prices for manufactured goods (reflecting costs imposed on manufacturers by government regulations) or increased user charges imposed by government. The net distributional incidence resembles an excise tax and is regressive.[12]

With respect to benefits, one must distinguish the distribution of physical benefits and their valuation. Some environmental programs may yield greater physical benefits to the poor than to the wealthy. For example, air pollution control provides greater physical benefits in urban areas for poorer, central-city residents than for wealthier suburbanites; however, this conclusion conceals the complications of workplace benefits and the possible offsetting effects of land-value changes,[13] as well as the fact that such programs provide very little in the way of benefits for the rural poor and impose heavy costs on them.[14] But other environmental programs, such as wilderness preservation and improved water-based recreation activities, appear to favor the wealthy, in large part because substantial resource expenditures are required in order to enjoy the benefits of such programs.[15]

Moreover, the economic value of environmental benefits is measured by the willingness of individuals to pay for them. Wealthier individuals will have a greater willingness to spend money for environmental amenities than poor people, both because they have more total wealth or income and because environmental amenities, unlike food, shelter, and clothing, are generally not regarded as necessities that command priority on the more limited resources of poor people.[16] Moreover, wealthier people are likely to value environmental quality more highly because they are normally better educated concerning the benefits of environmental quality and are more likely to have experienced high-quality environments.[17] By contrast, central city residents may not have a strong preference for better air quality because they are acclimated to poor air quality and may be unaware of the health benefits of cleaner air.[18]

The analysis of relevant costs and benefits strongly suggests that programs to enhance or protect environmental quality are likely to be substantially regressive unless strenuous compensating fiscal measures are undertaken.

A Right to Environmental Quality?

Thus far we have considered the distributional implications of environmental policies solely from the perspective of the aggregate distribution of income. Such an approach may be inappropriate if environmental amenities are regarded as not fully commensurable with other products. More specifically, one might regard provision of a certain minimum of environmental quality as an indispensable element of individual capacity for self-development and self-respect and might posit a right on the part of each individual to enjoy provision of a minimum level of environmental quality, even though such a policy, in the aggregate, would be economically inefficient or distributionally regressive.[19] Clearly there are extreme, limiting cases in which such a premise seems plausible—for example, we would probably not countenance the deliberate exposure of individuals to deadly pollution regardless of the potential gains in economic efficiency. But whether a policy of furnishing each person some minimum level of smog-free air or vernal recreation opportunities is sound raises difficulties that will be addressed below in the specific context of the Clean Air Act's provision for uniform ambient air quality standards.[20]

Moral Duties to Future Generations and to Nature

Many forms of environmental degradation pose health or ecologic risks of a long-run nature, and preservation of wilderness areas and other high-quality environments could promise long-run benefits. These circumstances pose vexing problems if we are to weigh the welfare of future generations against our own. Conventional economic analysis discounts future costs and benefits. If comparatively high private rates of return on capital were utilized in discounting future environmental costs and benefits, the interests of future generations would count for little, even in the context of irreversible harms. There is an unresolved debate among economists whether public investments in collective goods (such as environmental quality) should be assessed by a lower "social" discount rate,[21] but it is unclear whether even the use of such a lower rate would adequately discharge our obligations to future generations.[22]

Thoughtful commentators have suggested that we might also owe obligations of preservation and care to natural species and objects.[23] Any such obligation would be foreign to economic analysis (except insofar as a given individual, in fact, might entertain such an obligation and be willing to expend resources in order to discharge it). Yet the arguments that we *should* entertain such duties can not be dismissed out of hand.[24] Were such arguments accepted, serious problems would arise in determining what percentage of resources should be diverted from human preference satisfaction to the fulfillment of such moral duties.[25]

Technology and Information

Preferences are not the only dynamic factor in the economic equation. Our environmental policies should provide incentives for the development of environmentally superior technologies and improved information concerning the environmental consequences of proposed courses of action. In many cases, such incentives may be entirely compatible with economic efficiency. It is, for example, a fair hypothesis that a system of emission fees not only would promote economic efficiency but also would prove superior in many instances to regulatory approaches in stimulating the development of environmentally superior technologies and generating information concerning the desired level of environmental controls.[26] But research and development of technology and information may often require substantial governmental outlays, and economic efficiency provides an uncertain guide to where those expenditures should be made. To a large degree, the environmental problems about which we choose to develop information will be those that we end up taking collective measures to deal with. And the technologies that we choose to advance can also substantially affect the preferences that those technologies will shape and hence the future valuation of the fruits of such technologies.[27] Consider, for example, the respective implications of severing the dependence of the automobile on crude oil by promoting coal liquifaction technology, electric cars, or a radical program of mass transit.

Expectation Interests

Another dynamic aspect of environmental policy is the appropriate response to expectation interests generated by past policies that are now undergoing change. This is, to be sure, a transitional problem, but transitional problems are always with us. Resources have been invested or employment obtained under the aegis of prior policies that permitted substantial amounts of pollution or unrestricted use of the automobile. Changes in those policies may generate unanticipated windfall losses and dislocation. From both an analytic and a political view, thorny problems arise, first, in determining whether the existence of such transitional losses is a sufficient argument against adoption of policies that, from the viewpoint of economic efficiency or otherwise, would be superior to the status quo were there no transitional problems, and, second (and perhaps regardless of the answer to the first question), whether measures should be taken to compensate those losses, how compensable losses should be defined, and the method of compensation.

Decentralized Decision Making

A presumption in favor of decentralized decision making has long characterized our polity. This presumption may be explained by considerations of economic efficiency and the claim that, other things being equal, decentralization implies fewer transaction and information costs. This seems a plausible claim with respect to some questions of environmental choice—such as the allocation of abatement burdens among emission sources in order to meet given environmental quality objectives. For example, a fee or marketable permit system for allocation of abatement burdens would permit individual firms to adjust the extent of pollution control to their respective control costs and thus would seem to have important efficiency advantages over centralized regulation. But it is doubtful whether other choices—such as the choice of environmental quality goals or standards, which requires collection of considerable amounts of information about the hazards of pollution and the costs of reducing such hazards—would be accomplished more cheaply if repeated many times by decentralized decision makers than if taken once on a centralized basis.

Decentralized decision making might be more economically efficient because it better reflects the preferences of the governed for various types of collective goods (particularly when those preferences vary geographically) than does centralized decision making. Yet the experience with environmental policy in the United States has been in the direction of increasingly centralized, uniform regulation by the federal government. This development may be justified by efficiency goals on the claim that local authorities are unduly influenced by organized economic interests with a considerable financial stake in environmental degradation[28] and therefore do not mirror the preferences of the governed as effectively as the federal government. State and local bureaucracies are (in general) less well equipped, by size, staffing, and experience, to resist the influence of organized private economic interests. In addition, state and local authorities may be prey to a variant of the "Tragedy of the Commons"[29] in which any one authority is reluctant unilaterally to take steps to improve environmental quality for fear that industrial development will shift elsewhere. Finally, the total transaction costs that must be surmounted in order effectively to organize environmentally concerned individuals may be considerably higher if decision making is decentralized rather than centralized. Accordingly, centralized decision making could tend to reduce the disparity in the political and administrative influence of corporations, unions and other organized interests, and comparatively unorganized environmental interests.

But even if economic efficiency were better served by centralization, there are additional arguments for decentralization that might still carry the day. Decentralization may facilitate more extensive and informed participation by individuals and groups in the processes of decision than does centralized decision, and such participation may be viewed as a good in itself, quite apart from substantive outcomes. From this perspective, whatever duplication of information gathering and decisional transactions that decentralized choice may involve should be accounted a benefit rather than a cost if it involves participation by a larger total of individuals in collective choices.[30]

Further, decentralized decision has sometimes been associated with diversity in the decisions reached. For reasons developed later in this essay, such diversity may be an important strategy for reconciling competing environmental objectives and dealing with the problems posed by the tendency of collective decisions to shape preferences and become in that way self-fulfilling prophecies. However, it is quite unclear whether decentralized decision will indeed produce diverse levels of environmental quality. The "Commons" problem faced by local decisional authorities under pressure from organized economic interests to maintain relatively lax controls on pollution and development may lead decentralized decision to produce a relatively uniform low level of environmental quality everywhere. If so, we are faced with the paradox that centralization may be the only possible avenue to diversity.[31]

Thus the considerations involved in choosing between centralized and decentralized decisional processes are complex, they vary depending on the precise issue involved, and they involve contingent evaluation of particular circumstances. In some measure, the choice among them should and must turn on the substantive policies that one wants to advance.[32]

The Legitimacy of Decisional Processes

Any society at a given time will have a limited repertory of paradigm processes for deciding matters in which there is not well-defined agreement among the members of the society. Examples in our own society include market exchange, majority vote by legislative representatives elected on a one-person–one-vote basis, reasoned judicial application of general principles, and the processes of scientific verification. However "correct" a decision may be by other criteria, it may not be accepted if reached by processes of decision other than those that share widespread allegiance.

This analysis does not imply the absence of any logical link between sound decisions and those processes viewed as legitimate. The legitimacy of such processes is not a purely aesthetic, much less an arbitrary, matter; obviously it must bear some long-term relation to the capacity of those processes to yield decisions that accord with the interests and values of a society. Yet, at any one time the accepted repertory of processes may lag behind the need to make sound decisions on new types of questions.

The realm of environmental policy exemplifies such a lag. Because the environmental "problem," in at least some respects, is a result of market failure, the market itself cannot serve as the process for its resolution.[33] The processes of scientific verification cannot determine environmental policies because the latter involve differences in human preferences and values that are not susceptible to resolution by scientific verification. All questions of environmental choice cannot be resolved by the legislature; not only are its decisional capacities limited in a quantitative sense so that necessarily a measure of choice must be left to other agencies of government, but there are also qualitative limitations to its decisional processes that could make it unwise to reserve all such choices to the legislature even if it had the quantitative capacity to make them.[34] In the past, courts had an important role in environmental policy through their development of common-law liability rules to influence the conduct of private actors, either directly through the award of injunctive relief or indirectly through damage judgments. But the magnitude of environmental problems and the need for comprehensive prophylactic measures have outstripped the capacity of the courts. Moreover, the fact that environmental policies today involve very large resource commitments raises questions whether such policies should be decided by institutions politically more responsible than the courts.

The administrative agency is the institution of choice for filling the gap between the respective capacities and qualifications of the legislature and the judiciary in devising and implementing collective decisions about environmental quality. Had the administrative agency not already been created in order to carry out other forms of economic regulation, it would certainly have had to be invented to grapple with environmental policy. Yet the legitimacy of the administrative agency's decisional processes has remained a troubling question. Administrative agencies cannot be viewed merely as executory instruments of the legislature, nor can they feasibly replicate the legislature's legitimating processes of representative election as a method of decision. On the other hand, agencies are invested with missions, their heads are not effectively independent, and they are too involved in the political arena of ad hoc policy choice and implementation to share credibly the full measure of legitimacy that principles of reasoned decision afford to courts.

And we see plainly enough now that agency choices cannot be resolved on objective scientific principles.[35]

The courts have responded to the problem of agency discretion by expanding the right to participate in formal processes of agency decision and the right to obtain judicial review to a broader range of affected interests.[36] Serious questions as to the viability and wisdom of this response remain; its costs, in terms of both time and resources, may be disproportionate to its advantages.[37] But it does seem clear that the courts will continue, in the main, to play a "second-line" backstopping role in environmental policy questions by reviewing agency decisions, rather than the "front-line" role they once occupied in setting common-law liability rules.

The discussion thus far has demonstrated that issues of environmental policy are fraught with imponderables that must be clarified and resolved at two related levels. The first level is analytic: is it possible to develop substantive rules of policy choice that adequately reflect the richness of environmental policy issues and provide a framework for deciding among competing objectives? The second level is institutional: how do we devise procedures and institutional arrangements that will ensure informed, reflective, sound environmental choices? This essay seeks to cast some light on these difficult questions by examining the courts' role in the implementation of the Clean Air Act.

Commentators who have reflected on the analytic imponderables of environmental policy have, in the main, come down in favor of presumptive reliance on economic efficiency as the basic yardstick of choice for determining appropriate levels of environmental quality, subject to ad hoc adjustments in favor of competing objectives.[38] This conclusion is fully consonant with the practice of reviewing courts under the Clean Air Act, which has been primarily directed at reshaping the exercise of administrative discretion so as to increase the range of relevant factors that the agency considers and to improve the quality of that consideration; the net result has been to soften excessive rigidities in legislative policy prescriptions and to promote the adoption of measures that are more likely to be economically efficient. However, a handful of decisions—concerning the principle of nondegradation and the issue of federal authority to compel state enforcement of federal controls on polluting activities—invite a more venturesome judicial role. As this essay will argue, serious structural defects in legislative and administrative processes of decision justify strong judicial intervention, on grounds other than constitutional, to mandate measures that are economically inefficient but protect fundamental environmental opportunities that might otherwise be foreclosed.

The Federal Legislative Structure for Air Pollution Control Policy

Before examining the role of the federal courts in reviewing the implementation of air pollution control policy over the past six years, the statutory structure and political setting in which the courts have discharged their responsibilities must be outlined. The Clean Air Act Amendments of 1970 represent a remarkable effort on the part of the Congress to constrain the administrative discretion of a major regulatory agency. Compared, for example, to the Federal Communications Act, which directs the FCC to regulate broadcasters "in the public convenience, interest or necessity"[39] the provisions of the Clean Air Amendments of 1970 contain many detailed requirements and commands that channel and restrict EPA policies.[40]

Congress's determination in the Clean Air Amendments to mandate administrative action along legislatively predetermined lines reflected profound dissatisfaction with the lack of tangible achievement in air pollution control programs over the previous decade. Despite an increasing degree of public concern over air pollution, most air pollution indices showed a continuing increase in emissions and a decline in ambient air quality. This trend was attributed in part to the weakness of state governmental units that had been primarily responsible for pollution control policy and their vulnerability to "blackmail" by industrial and labor interests. In addition, many congressmen viewed the National Air Pollution Control Administration (NAPCA) of HEW (the agency responsible for federal air pollution control programs) as lacking zeal, particularly in the discharge of its discretionary authority to regulate emissions from new automobiles. The diagnosis of undue laxity on the part of NAPCA was, in the case of Democratic members of Congress, reinforced by partisan political considerations. A Republican administration could be blamed for such ineffectiveness which would in turn justify legislative imposition of stringent controls. Such a response was also wholly consonant with the presidential ambitions of Senator Edmund Muskie, the chief architect of the Clean Air Amendments.[41]

Moreover, the legislators' perception of administrative laxity and the need for corrective measures matched a more general sense of growing disillusionment with administrative agencies that was evident not only among "reform" figures (such as Ralph Nader) but also among legal commentators and judges. The notion that agencies were habitually "captured" by or otherwise unduly deferential to the interests that they were charged with regulating was achieving widespread currency,[42] and such distinguished commentators as Judge Henry Friendly urged Congress to deal with the problem of agency discretion by greater statutory specification of agency policies.[43]

While Congress was determined, for the reasons just outlined, to press for straightforward action to improve air quality, the complex nature of air pollution problems and associated political considerations resulted in a relatively complicated statute. The bedrock of the Clean Air Amendments was provision for uniform federal ambient standards of air quality in lieu of the near-exclusive reliance in prior legislation on state-set air quality standards.[44] The statute required that the EPA (the successor agency to NAPCA created by executive reorganization in 1970) promulgate within 120 days primary standards for six major pollutants specifying permissible concentrations "requisite to protect the public health" but "allowing an adequate margin of safety" and secondary standards to protect against adverse "welfare" effects of pollution.[45]

Following adoption by the administrator of ambient standards, the states were required to devise, within nine months, implementation plans to limit emissions within the state in order to ensure attainment of the federal standards. The amendments directed the administrator to approve such plans if they would assure attainment of the primary standards within three years and of the secondary standards within a "reasonable time." If state plans were inadequate, the federal administrator was required to promulgate a plan for the state that would achieve the federal standards within the applicable time deadlines. The amendments authorized the administrator to grant limited extensions of the deadlines for achieving the ambient standards if a state could show that it was infeasible to meet the statutory deadlines.[46]

Speedy federal adoption of uniform ambient standards could be expected to eliminate the "Commons" problem posed by a system of decentralized state-set standards. The "Commons" problem does not necessarily require *uniform* federal standards for its resolution and considerations of economic efficiency might dictate geographically varying ambient standards depending on the costs and benefits of pollution control in various regions, but the political and administrative problems in permitting some regions of the country to have dirtier air (and more industrial development) than other regions argue powerfully for uniformity. At the same time, the distribution of the abatement burden required to meet the federal standards in any one region was left to the states, which, as decentralized decision makers are presumptively better equipped to assemble the information and to make the often delicate (or not so delicate) political trade-offs required in determining such a distribution.

In addition, the amendments provided for three sets of nationally uniform *emission* limitations to be established by the administrator.

First, the act provided for nationally uniform federal emission limitations for new stationary sources of air pollution that would require "the application of the best system of emissions reduction which (taking into account the cost of achieving such reductions) the Administrator determines has been adequately demonstrated."[47] By requiring new sources to meet the same level of control regardless of location such limitations deprive "clean" regions already in compliance with the federal standards of a locational advantage that otherwise could entail significant shifts (in comparative terms) of industrial development and population away from the heavily polluted and heavily unionized northeastern states—shifts clearly inimical to the interests of organized labor and of House members from such areas, who would be exposed to the threat of redistricting.[48] Moreover, it was expected that the federal emission limitations for new sources would be quite stringent and would effectively prevent deterioration of air quality in "clean" regions whose ambient concentrations were well below the federal standards.[49]

Second, the Clean Air Amendments authorized nationally uniform emission limitations for "hazardous air pollutants" that "may cause or contribute to an increase in mortality or an increase in serious, irreversible, or incapacitating reversible illness."[50] Congress contemplated controls on highly dangerous pollutants that should be prohibited entirely or restricted to the greatest extent technologically feasible. There is a strong case for centralized control of such pollutants. Central authorities are more likely to have the detailed knowledge of health hazards that is required to determine which pollutants should be subject to stringent controls. Moreover, the stringent nature of the controls to which extremely hazardous pollutants should be subjected tends to nullify the value of regional variations in controls and the associated justifications for decentralized decision.

The third set of federal emission limitations in the Clean Air Amendments applied to new motor vehicles.[51] Federal controls on emissions from new motor vehicles had been authorized by Congress in 1965 in terms that afforded broad discretion to the Administrator as to the level and timing of controls. In 1967, Congress provided that such controls would preempt more stringent state regulations (with the exception, subject to administrative approval, of California) because of fear that multiple and possibly contradictory state controls would create an undue burden on the automobile industry.

The Clean Air Amendments of 1970 drastically curtailed the Administrator's discretion by requiring that a 90 percent reduction in existing automotive pollutant levels be achieved by 1975–1976, with limited provision for a one-year administrative extension of the time deadlines. These drastic constraints, according to Senator Muskie, were required in order to achieve the federal primary ambient standards for automotive pollutants within the statutory timetable, and, he added, the deadlines would probably force the development of the necessary control technology. In tightening the screws to force action by the administration and by Detroit, the Congress in 1970 unfortunately did not explicitly reconsider whether federal emission limitations for new automobiles should be geographically uniform, or whether the states (other than California) should be denied the right to impose emission limitations more stringent than those required by the federal government. Some measure of diversity in controls on new auto emissions would be far preferable to the pattern of almost total uniformity mandated by Congress. For example, there is strong evidence[52] that a "two-car" strategy imposing stringent controls on new cars in polluted areas and looser controls elsewhere would be superior to the present statute in terms of both economic efficiency and distributional equity.[53]

Finally, the 1970 amendments included novel measures to ensure that Congress's mandates were carried out. The past record of state and federal administration of pollution controls revealed an enforcement effort that was patchy and protracted. Although this history might have suggested to Congress that it consider alternatives to regulatory controls—such as emission fees—that would substantially undercut the incentive and ability of polluters to interpose debilitating procedural obstacles to prompt implementation and enforcement, Congress instead sought to make the regulatory approach more stringent and effective by enlisting the energies of environmental litigants and the resources of the courts.

In addition to providing generally for judicial review of the administrator's actions in implementing the statute,[54] in section 304 of the act Congress authorized suits by "any citizen" in the federal district courts to require the administrator to carry out the mandatory duties imposed upon him by the statute. To encourage such suits, the statute authorized the discretionary award by courts of attorneys' fees. These measures were designed to enlist the courts' help in forcing the administrator to meet the deadlines set forth in the statute and also to ensure that the administrator would carry out his obligation to require state implementation plans adequate to guarantee attainment of the federal ambient standards.[55] The "citizen suit" provisions also authorized private litigation to enforce state or federal emission limitations against polluters when state and federal authorities failed to take enforcement action.[56]

The stage is now set for an examination of the judicial role in the implementation of the Clean Air Act Amendments. How have reviewing courts dealt with EPA's implementation of the statute? What does the judicial record tell us about Congress's ultimate success in controlling administrative discretion and the wisdom of its efforts to do so through judicially enforced mandates? These are the questions that this essay next addresses.

The Record of Judicial Review

This section of the essay divides the discussion of judicial decisions under the Clean Air Act into three parts. First, there are the "action-forcing" decisions, in which litigants have invoked the citizen suit provisions of the act to force compliance by EPA or polluters with congressional mandates and deadlines set forth in the act. The record here is one of straightforward judicial enforcement of the legislative directives. Second, there is a large class of cases arising under the act's general judicial review provisions in which action by the Administrator is challenged as contrary to law. The decisions in this category have emphasized the substantial discretion enjoyed by EPA in implementing the act and have attempted to promote reasoned and careful exercise by EPA of that discretion. This class of decisions is accordingly referred to as "discretion checking." Third, there is a handful of exceptional "quasi-constitutional" cases in which the courts are invited to disregard normal principles of statutory interpretation and judicial deference to the reasoned exercise of administrative discretion and to intervene to protect basic environmental objectives that might otherwise be disregarded.

"Action-Forcing" Decisions

There have been several actions brought by environmental groups under the statutory citizen suit provisions charging EPA with failure to adhere to statutory deadlines or to initiate an entire category of control measures having nationwide applicability.[57] The most notable of these actions was *NRDC* v. *EPA*,[58] which sustained the contentions of environmental groups that EPA had violated the Clean Air Amendments by permitting all states to postpone for two years the statutory deadlines for submission of transportation control plans and by granting seventeen states an extension until 1977 of the deadline for attainment of the ambient air quality standards for automotive-type pollutants. EPA believed that additional time was required because transportation control plans would involve controls on automobiles that would be not only difficult to devise and implement but also politically controversial. In the *NRDC* v. *EPA* decision, the Court of Appeals for the District of Columbia Circuit insisted upon strict compliance with the literal terms of the statute. It required prompt submission by the states of transportation control plans to EPA, required EPA to follow the statutory extension procedures,[59] and also held that state plans must ensure continued maintenance of ambient standards once they have been achieved.[60]

The enforcement of these statutory requirements has been criticized as obliging EPA and state governments to resolve technically complex and politically controversial issues in a hurried manner that discredited the effort to improve air quality and resulted, in the end, in less progress toward clean air than had these requirements not been enforced immediately. In at least some jurisdictions, including Massachusetts and California, the requirements of earlier submission of transportation control plans apparently surprised state officials who had been working on a more generous timetable. In part as a result of the changed timetable, and perhaps in part as a result of the court's ruling on extension procedures,[61] officials in many states declined to assume responsibility for devising transportation control plans. EPA was accordingly obliged to prepare such plans—often draconian in character and of questionable administrative feasibility—that, in turn, provoked strong local opposition. Moreover, EPA plainly lacked the necessary resources itself to enforce measures such as exclusive bus lanes, parking regulations, and inspection and maintenance programs. As a result, EPA was forced the relax the stringency of many of its measures in order to placate local opposition or to institute litigation against state governments to force them to carry out the transportation control plans drafted by EPA.[62]

The outcome of EPA's forced-draft implementation of measures to control new "indirect sources" of pollution, such as shopping centers, in order to maintain ambient standards was even more unfortunate. The EPA launched a series of regulatory measures to require review of new construction that involved it in essentially local land-use decisions, in which it had little competence, and pitted it against politically powerful developer interests. In 1974 Congress inserted a provision in EPA's appropriation bill precluding agency expenditures for imposition of measures to limit parking supply as a means of dealing with the indirect source problem.[63] By 1975 the entire indirect source program had become so complicated and controversial that EPA indefinitely suspended its efforts to carry out such a program.

If one believes that greater flexibility in timing and content could have produced more viable transportation control and indirect source programs, the blame for failure to provide flexibility must be directed at the congressional statute rather than at the court's decision in *NRDC* v. *EPA*. The requirements of the statute were plain.[64] The judges displayed no enthusiasm in implementing the terms of the statute,[65] but it seems clear that they had no other choice. Given the statutory language and the action-forcing logic that underlay that language, the court would have been remiss in its constitutional obligations had it refused to heed the legislature's directives, however firmly persuaded it might be of their unwisdom.

It is arguable that the Congress's imposition of deadlines and inflexible mandates were justified, despite the unhappy fate of the transportation control and indirect source programs. The draconian commands of the Clean Air Act clearly produced a new attitude of seriousness and urgency on the part of governmental agencies and polluters alike and have apparently resulted in some substantial improvements in air quality. Even EPA officials have (informally) welcomed citizen suit enforcement of deadlines and the statutory mandates as a necessary prod to a sluggish bureaucracy. Moreover, the transportation control and indirect source programs might very well have turned out badly even had more time and flexibility been afforded; local opposition to such programs would have arisen regardless of the timing of their implementation because they squarely contradicted the entrenched habits of motorists and the interests of powerful economic groups. Yet the history does suggest that a more gradual introduction of these programs might have achieved greater success. Surely Congress could have imposed deadlines and mandates to force the general pace of air pollution control efforts while providing the EPA a limited opportunity, subject to judicial review, to decline to enforce deadlines and mandates if, on full consideration of the facts in a particular circumstance, the interest in effective pollution control would be better served thereby.

Two other important ''action-forcing'' decisions deserve mention. In *Sierra Club* v. *Ruckelshaus*,[66] environmental groups invoked the citizen suit provisions to require EPA to assure preservation of existing levels of air quality in regions where the air quality was better than that required by the federal ambient standards. In resolving this litigation, the courts went beyond the letter of the statute into a quasi-constitutional domain that is discussed later in this essay. The other relevant decision, *NRDC* v. *Train*,[67] was a successful citizen suit to compel the EPA to promulgate ambient air quality standards for lead. The court held that the EPA's previous action in removing lead from gasoline under section 211 of the act on the ground that ambient lead was a health hazard[68] and the administrator's promulgation of new source performance standards for lead, pursuant to section 111, demonstrated that EPA had already determined that lead had adverse health effects attributable to emissions from numerous sources and therefore that sections 108 and 109 required that ambient standards for lead be established.[69]

EPA had argued that lead was a hazard better handled by emissions limitations on mobile and new stationary sources than through the more cumbersome approach of ambient air quality standards, which would trigger the necessity for submission by the states of new implementation plans. Giving due regard to these administrative considerations, one might well conclude that action under other sections of the statute should not be decisive on the issue whether ambient standards should be promulgated for the same pollutant. But the more significant point is the limited reach of the court's decision, which concedes a largely unreviewable discretion in the Administrator to decide whether to promulgate ambient standards for new pollutants where he has not taken equivalent action against them under other sections of the statute. In this aspect, the ruling rather confirms the broad scope of administrative discretion at key points in implementing the statutory scheme.[70]

Decisions Checking Administrative Discretion

This part of the essay examines the great bulk of judicial decisions reviewing EPA implementation of the Clean Air Act, in which statutory directives impose some limits on EPA but nonetheless afford considerable agency discretion in shaping policy.

The ''discretion-checking'' decisions discussed here are of three basic types: decisions reviewing EPA's adoption of ambient standards and of federal emission limitations for stationary sources; cases involving EPA's approval or disapproval of particular state implementation plans or EPA's promulgation of such plans; and decisions reviewing implementation of the federal emission limitations on automobiles. In each type of case, EPA's discharge of its responsibilities requires it to decide detailed and complex scientific, engineering, and other technical questions and, at the same time, to assess the broader costs and benefits of alternative measures; despite Congress's effort to limit agency discretion, the relevant statutory language often fails to provide detailed guidance for such decisions.[71]

The lack of detailed guidance in applicable statutes places the reviewing courts in a delicate role. The court must first demark such outer bounds on agency choice as can be discerned in the relevant statute in order to determine whether the agency has conformed to applicable legislative directives. Within these outer bounds, the judges most ultimately respect the discretion that the statute leaves to the agency and attempt simultaneously to exercise a measure of oversight designed to promote the rational and evenhanded exercise of that discretion.[72] These judicial tasks are made more difficult by permeation of complex technical issues in the agency's decisions.

The exercise of substantial agency discretion and the special problems that it presents reviewing courts have special implications for the procedures that the agency should follow in making its decisions. In the action-forcing cases discussed previously and the quasi-constitutional cases discussed later, the court normally has little need for a detailed factual record by which to assess the agency's action, because the controlling law (statutory and judge-made respectively) is likely to resolve the validity of the agency's conduct in a relatively straightforward manner. But evaluating whether the agency has kept within the often nebulous statutory bounds on its discretion and has exercised that discretion in a way that reflects reasoned consideration of a range of competing variables enmeshed with technical details in a given case is likely to call for an extensive evidentiary base.

Judicial review thus has procedural as well as substantive implications that merge one into the other. This part will first discuss procedural matters, in which the courts have made a pioneering effort to promote better EPA decision making without hobbling the agency. It will then consider the methods by which judges measure the agency's ultimate policy choices. Here the judges have generally sought to loosen the legislative constraints on EPA policy choice wherever the statute would plausibly permit such a reading. At the same time, they have probed the EPA's own justifications for its policy choices with considerable rigor.

Procedures The federal Administrative Procedure Act provides two paradigm procedures to be followed by regulatory agencies in taking action. First, there is the trial-type hearing in which the agency and the affected private parties have the right to introduce direct and rebuttal oral evidence through witnesses and to exercise the right of cross-examination; the agency's ultimate decision is required to be based on the evidence developed at the hearing and must ordinarily take the form of an opinion that discusses the evidence and makes relevant factual findings. Such a hearing is often required under regulatory statutes in cases in which the agency takes individual enforcement action or otherwise seeks to impose sanctions on particular parties.[73]

The second basic type of procedure is ''notice-and-comment'' rule making, under which the agency is required to give public notice of proposed regulations of general applicability and interested persons are given the right to submit written comments on the proposed regulations prior to final agency adoption.[74] Under this procedure, there is no trial-type hearing. Moreover, the agency decision is not "on the record," because the agency need not base its decision solely on the written comments submitted but may take into account any information that it deems fit. The agency, however, is required to give a "short and concise statement" of the reasons for its action.

The Clean Air Act often fails to make clear which of these traditional procedures or variants of them should be followed by EPA in taking action. Moreover, given EPA's responsibilities, neither of the traditional procedures is wholly appropriate. A general requirement that EPA observe trial-type procedures before taking action would cripple effective implementation of air pollution controls. The range of technical and economic issues involved in any given decision is often intricate and far-reaching. Pollution sources determined to put off the day of reckoning could utilize the machinery of trial testimony and cross-examination to spin out the agency decisional processes for months or years.[75] Accordingly, with the exception of a few decisions discussed below,[76] the courts have generally refused to order EPA to follow trial-type procedures unless the act plainly required it. As a result, EPA has ordinarily been able to go forward with decisions on ambient standards, state implementation plans, federal emission limitations for new stationary sources, and automobile emission control extensions without the burden of trial-type hearings.

However, merely requiring EPA to observe traditional notice-and-comment procedures in these decisions might not afford sufficient procedural protection, given the importance of the decisions in terms of resource commitments and the public health and welfare. Because notice-and-comment rule-making procedures traditionally do not require the agency to base a decision on any specified set of materials and call for only a cursory explanation by the agency of its action, such procedures are unlikely to afford affected parties a significant opportunity to affect agency policy. In addition, they provide little basis for searching judicial review.

Recognizing the inappropriateness of traditional procedures, the federal courts have creatively devised a new set of procedures to govern EPA decisions in those numerous instances in which the language of the act is not dispositive. This improvisation has been an outstanding success in improving the quality of decision making at the EPA and providing a realistic basis for judicial review without unduly hobbling the agency's ability to get its job done.

These new procedures have been developed through judge-made extensions of traditional notice-and-comment procedures. The first step was a requirement that EPA articulate the grounds for its action in far greater detail than had ordinarily been required in previous notice-and-comment rule making. In *Kennecott Copper Co.* v. *EPA*,[77] the EPA's secondary ambient standard for sulfur dioxide was challenged as unsupported by evidence of adverse effects at the prescribed concentrations. Obviously troubled by the weakness of the agency's support for the standard, the court remanded for a fuller explanation by EPA of the basis for its decision. On remand, EPA decided to withdraw the standard.[78]

The requirement of reasoned elaboration by the agency of the grounds for its action was developed further in decisions reviewing EPA postponement of automobile emission deadlines;[79] approval of state implementation plans, or disapproval followed by EPA promulgation of such plans;[80] and promulgation of emission limitations for new stationary sources.[81] These decisions not only have required EPA to articulate in detail the reasons for its choices but also have required it to respond in its decision to the criticisms and contrary evidence adduced in comments by those opposing EPA's proposed action. The requirement that EPA respond to criticism has meant that the documentary evidence and written argument submitted to EPA by opponents will be included in the materials to which the EPA must respond in its decision and to which a court will look on review.[82] In such circumstances, government lawyers eager to sustain EPA's action have not hesitated to include documents that they believed would support EPA's actions. Moreover, internal agency documents arguably contrary to EPA's decision can generally be obtained by opponents upon resort to the federal Freedom of Information Act.[83] As a result, a reviewing court will normally have before it a record consisting of all of the documentary materials relevant to the agency's decision.

In essence, what has emerged, in procedural terms, is a "paper hearing" that combines many of the advantages of a trial-type adversary process (except oral testimony and cross-examination) but avoids undue delay and cost. The development of a paper-hearing procedure and the related requirement that the agency explain in detail the bases for its decision have been outstanding successes. EPA officials, as well as environmental advocates and industrial representatives, have expressed satisfaction with these procedural innovations and their beneficial impact on the quality of agency decision making.[84] Joseph Sax correctly notes in his comments on this essay that adoption of a paper-hearing procedure does by itself resolve the thorny problem of how the ultimate responsibility for decision should be apportioned between court and agency.[85] But as a procedural model, it far better accommodates agency interest in flexibility and judicial interest in a reviewable record than either of the traditional paradigms.

Judicial Review of the Merits We turn to the substantive aspects of judicial review. To the extent that the statutory language fairly permits it, the courts have generally construed the Clean Air Act to enlarge the effective discretion of the agency by permitting or requiring it to take into account a larger range of factors than the statute might seem to encompass. In so doing, the courts have eased the draconian tenor of the act in order to deal with transitional problems and to promote economic efficiency by permitting or requiring a broad array of relevant costs and benefits to be considered by the agency. In the end, however, striking the precise balance between these respective costs and benefits has generally been left to the EPA, so long as it has considered all of the relevant factors and given a reasoned explanation for its choice.

The decision in *International Harvester* v. *Ruckelshaus,*[86] in which the court set aside the EPA's denial of an extension of the 1975 automobile emission deadlines, exemplifies this approach. The 1970 amendments require that the administrator make four specific findings prior to granting an extension—that the suspension is essential to the public interest, health, or welfare, that manufacturers have made all good faith efforts, that effective control technology is unavailable, and that National Academy of Sciences investigations and other information confirm the absence of requisite technology.[87] Although the statutory language suggests that extensions be granted only in dire circumstances and thereby seems to afford very little leeway for administrative judgment, the court held that EPA should engage in a broader assessment by utilizing a relatively sophisticated risk-benefit analysis to ascertain the consequences of possible error in denying or allowing the postponement. Applying this analytic framework, the court found that the adverse effects on public health of erroneously granting an extension were comparatively slight, whereas the consequence of erroneously denying an extension could be massive economic dislocation. In light of this assessment of risk and benefit together with evidence from the manufacturers casting serious doubt on the availability of technology to meet the deadlines, the court held that the burden was on the agency to show that a suspension was not warranted and that EPA's explanation for its decision was inadequate. On remand, the administrator granted the suspension, subject to tightened interim standards.[88]

Although the Congress never contemplated that the burden of responsibility for persuasion or of producing evidence on a suspension issue would be the administrator's, there is much to be said for the court's providing a measure of flexibility when the draconian logic of the statute fails in some degree to achieve its purpose and when application of a drastic sanction could be quite costly to the society.[89]

A judicial requirement that the EPA consider a broad range of relevant costs and benefits also emerges from decisions reviewing the EPA's adoption of federal emission limitations for new stationary sources. For example, in *Portland Cement Association* v. *Ruckelshaus,*[90] the court set aside EPA's adoption of new source performance standards for Portland cement plants because the agency had failed adequately to consider industry criticisms that the controls were too stringent and involved excessive environmental and economic costs. Although the 1970 Clean Air Amendments do not obviously authorize EPA consideration of the adverse environmental consequences of emission controls,[91] the court, relying in part upon the National Environmental Policy Act, required EPA in setting such standards to take into account "counterproductive environmental effects" and to consider the Portland cement industry's claim that the EPA standards would require additional electricity generation, which would produce more air pollution than the controls would eliminate. In addition, the court required the agency to consider industry cost-benefit analyses of the standards and also to respond to contentions that the EPA standards would render some categories of Portland cement plants uncompetitive.[92]

Although the act requires that state implementation plans be judged by the overriding criterion of ensuring compliance with the federal ambient air quality standards, here, too, the courts have sought to alleviate excessive agency zeal and bureaucratic tunnel vision by insisting on special care in EPA decision making in contexts in which its decisions have large social and economic impacts. Judicial concern that environmental quality not be pursued without some appropriate regard to the benefits rendered and costs incurred is reflected in decisions setting aside EPA-promulgated emission limitations in state implementation plans that would involve large costs and rested on shaky technical foundations. For example, in the *South Terminal* case,[93] the First Circuit Court of Appeals set aside an EPA-promulgated transportation control plan for the Boston area that imposed serious disincentives to automobile use at Logan airport and in the central Boston area. The basis for the court's action was a finding that the agency had not adequately rebutted critics' claims that the plan was based on isolated and insufficient monitoring data and was otherwise technically flawed. The court's concern with the harsh economic and social impact of the plan was an evident factor in its willingness to probe technical details. Similarly, in *Texas* v. *EPA,*[94] the Court of Appeals for the Fifth Circuit immersed itself in the abstruse details of diffusion models and emission inventories to set aside as unsubstantiated an EPA plan that imposed stringent limitations on industrial sources.

To the extent that the statute fairly permits, courts have also been prepared to sustain EPA's refusal to impose stringent control measures in situations in which the adverse social and economic consequences would be excessive.[95] In addition, the Supreme Court in *NRDC* v. *Train*[96] sustained EPA's use of streamlined procedures to grant interim variances to deal with transitional problems in implementing state plans rather than the cumbersome procedures specified in the act for extending deadlines.

Normally, courts will foreclose a broader range of considerations and more flexible policies only when they are required to do so by unambiguous statutory language and logic.[97] For example, the Supreme Court's decision in *Union Electric Corp.* v. *EPA,*[98] held that a power company could not defend against the imposition of emission limitations required to meet the primary ambient standards within the statutory deadlines by alleging that they were technologically and financially infeasible. The decision was compelled by the statute's language and its ''technology forcing'' deadline strategy. Two concurring justices plainly expressed their belief that the result was irrational but was required by the Clear Air Act.[99]

Once the reviewing court has determined that the agency has complied with statutory directives and has not ignored relevant costs and benefits, it may still consider whether the agency has adequately explained its exercise of discretion and struck a reasonable balance among competing considerations and interests. The degree to which reviewing courts will probe agency decisions in order to resolve these issues will obviously vary depending on judicial temperament and outlook, and such variation is not easily encapsulated. Controversy whether the proper standard of judicial review is the supposedly deferential ''arbitrary and capricious'' standard, the assertedly more rigorous ''substantial evidence'' test, or the ambiguous ''clear error of judgment'' principle[100] is not very helpful, particularly in the novel context of EPA rule making on the basis of a ''paper record.'' The prevailing practice is perhaps best expressed in Judge Leventhal's suggestion that the court must determine whether the agency, in exercising its discretion, has taken a ''hard look'' at the relevant evidence and policy considerations and has made a reasoned decision in the particular circumstances.[101] Such a standard of review is designed to not only impose ultimate responsibility on the agency rather than the reviewing court but also to ensure that the agency has made a careful and considered discharge of that responsibility. It requires the court to review in detail the bases for agency judgment but to acknowledge in the end that the judgment is the agency's. Such a standard imposes a difficult burden on reviewing courts—perhaps more difficult than those the court would bear in making the basic decision itself—and some courts have carried it with more enthusiasm and vigor than others.[102] But even those judges who profess diffidence and urge a more restricted judicial role are often impelled to examine and review particular agency decisions closely, if only to rebut brethren who would pursue a more searching role as a matter of course.[103] Judge Bazelon, for one, has asserted that such detailed review can and should be avoided by insisting upon procedural requirements that will ensure agency consideration of all relevant factors.[104] But such an alternative offers no escape from careful judicial scrutiny of the ingredients of decision. The mere fact that the agency is forced to go through certain procedural rituals of receiving evidence and writing opinions does not guarantee that such formalities will have any impact on the ultimate decision.[105] That impact can only be gauged by examining the justifications given by the agency for its action in the light of the evidence adduced and the contention of opponents. An inquiry into such matters is inescapable if the courts are to assume a substantial role in ensuring the fairness and rationality of agency decisions.

There are hazards in such a course. On the one side, the technical nature of many of the issues presented may strain judicial competencies and inhibit informed review.[106] On the other side, there is a danger that searching judicial inquiry into the reasons and evidence adduced to support agency policy choices could produce an undue number of judicial remands without substantially improving the quality of agency decision making or could invite the judges to decide policy issues that should be left to the agency.[107] But these dangers have not materialized to any substantial degree in judicial review under the Clean Air Act. Although Sax's comment on this essay properly expresses concern whether detailed judicial review can be reconciled with the need to accommodate the specialized agency's "intensity of experience," recent decisions demonstrate that careful review is compatible with adequate scope for agency judgment and vision. A notable example is *Ethyl Corp.* v. *EPA*, in which the Court of Appeals for the District of Columbia Circuit sustained (after a detailed review of an extensive documentary record) EPA's phase-out of lead additives in gasoline because it reflected agency "choices of policy" and "assessment of risks" that had a "rational basis" in the record.[108]

Given the important interests at stake on all sides in the environmental area, a substantial judicial role in scrutinizing bureaucratic agency decisions is warranted, and the record suggests that court exercise of such a role has materially enhanced the quality of ultimate decisions without undue costs.[109] Some of the important justifications for the courts' assuming such a role were ably summarized by William F. Pedersen, a former EPA lawyer who is well qualified to assess the impact of detailed judicial review:[110]

> The effect of . . . detailed factual review by the courts on the portion of the agency subject to it is entirely beneficial. It is a great tonic to a program to discover that even if a regulation can be slipped or wrestled through various layers of internal or external review without significant change, the final and most prestigious reviewing forum of all—a circuit court of appeals—will inquire into the minute details of methodology, data sufficiency and test procedure and will send the regulations back if these are lacking. The effect of such judicial opinions within the agency reaches beyond those who were concerned with the specific regulations reviewed. They serve as a precedent for future rule-writers and give those who care about well-documented and well-reasoned decisionmaking a lever with which to move those who do not.
>
> I see no way to relieve the courts, ultimately, of a substantial burden of factual inquiry. Courts alone have the time, the influence, and the freedom from ceremonial and "political" considerations that are necessary to a thorough, dispassionate and effective review of extremely complex and controversial matters.

Quasi-Constitutional Decisions

The essay now considers a limited number of cases in which reviewing courts are invited to override normal principles of statutory interpretation and judicial deference to agency discretion in order to protect environmental interests that do not enjoy acknowledged constitutional status. Such interposition must be justified, if at all, by fundamental defects in legislative or administrative decisional processes that jeopardize interests analogous to those enjoying constitutional protection; hence the "quasi-constitutional" title for this category of decisions.

The particular decisions in this category that we examine here deal with two sets of issues. First is the issue of nondegradation presented in *Sierra Club* v. *Ruckelshaus,*[111] in which the District Court ruled that the Clean Air Act requires EPA to prevent significant deterioration of air quality in those regions of the nation where existing air quality is superior to that required under the geographically uniform federal ambient air quality standards. This decision was affirmed, in turn, in a memorandum opinion by the Court of Appeals and by an equally divided Supreme Court.[112] The Court of Appeals recently reaffirmed its conclusion that a policy of nondegradation is mandated by the Clean Air Act.[113]

The second issue concerns EPA's authority to compel, on pain of criminal and civil sanctions, state and local governmental officials to enforce EPA-drafted controls on mobile sources of air polution. The federal courts have divided on the question; three courts of appeals have denied, in whole or in part, the existence of such power[114] and one court of appeals has sustained it.[115] The Supreme Court has granted certiorari in order to resolve this conflict.[116]

Nondegradation A requirement that states whose air quality is better than that required by the federal ambient standards preserve their existing high-quality environments would make it difficult for such states to develop industrially to the same extent as other states with air pollution levels equal to or exceeding the federal ambient standards. Under traditional principles of federalism, one should require a clear statement from Congress to override, through use of the commerce power,[117] a state's own power to choose between environmental amenities and economic growth, particularly in a context in which some states would end up seriously disadvantaged vis-à-vis others.[118] Such a requirement of legislative clear statement is necessary to trigger "political safeguards of federalism" that have particular importance in view of the courts' reluctance to impose constitutional limits on federal powers whose exercise may threaten state autonomy.[119] Yet a detailed analysis (set forth in an accompanying footnote)[120] of the statutory and other materials relied upon in the *Sierra Club* decisions fails to establish any such clear statement or to reveal any articulated congressional policy sufficient to justify an override of state autonomy. The Clean Air Amendments do not explicitly mandate nondegradation in "clean" states, and such a requirement would be inconsistent with the amendments' basic structure and logic. The *Sierra Club* decisions are thus not sustainable on normal principles of statutory construction and can be justified, if at all, by deeper considerations, which we now address. The following discussion does not purport to replicate the processes of judicial reasoning that may have underlain these decisions; rather it seeks to rationalize their result in terms of broader principles that might support a novel judicial role in protecting environmental interests.

Potential Efficiency Justifications for Requiring Nondegradation One might seek to justify nondegradation policy on the ground that the uniform ambient standards adopted by EPA are economically inefficient because they are too lax. If the ambient federal standards were uniformally too low but political opposition from the more heavily polluted states precluded any such uniform readjustment, a "second best" solution might be a freeze on pollution levels in cleaner areas. Alternatively, federal standards might be appropriate for areas that are presently heavily polluted but too lax for "clean" areas because of the special value individuals might attach to relatively pristine environments or because the costs of controlling to a given level would be less in such areas. Third, the present federal ambient standards might be appropriate in light of present control technology, but suboptimal under environmentally superior technologies to be developed in the future. If political opposition from organized interests were to make it difficult to adopt more stringent and more efficient standards in the future, dynamic considerations might justify a freeze on current levels of air quality in "clean" areas.

However, it is quite unlikely that these efficiency arguments justify the *Sierra Club* decisions. It is far from obvious that economic efficiency would dictate substantially higher air quality in less developed areas inhabited by few people than in more industrialized and heavily populated areas; a very substantial element in cost-benefit calculus with respect to air quality consists of the benefits to be achieved from reducing adverse health effects associated with pollution—benefits that are roughly proportional to the exposed population.[121] A willingness-to-pay analysis might indicate that preservation of pristine environments is more highly valued (because of the strong preferences of educated, wealthy individuals for such environments) than improvement of urban air, but any such calculation would depend on an intricate examination of empirical data that would seriously strain judicial competency. It would also be quite incredible if a calculus of relevant costs and benefits dictated a set of regionally varying air quality standards that even roughly coincided with the pattern produced by a nondegradation policy.

Moreover, if preservation of extremely high air quality in "clean" areas were economically efficient, why couldn't the states in such areas be expected to maintain such air quality voluntarily? The Clean Air Act permits states to enforce levels of air quality higher than those required under the naturally uniform federal standards;[122] a federal requirement of nondegradation can be justified only on the supposition that the states will erroneously fail to opt for a higher level.[123] There are, however, powerful arguments that might explain why a state would fail to choose a policy (nondegradation) that was economically efficient, and such arguments might justify the *Sierra Club* decision.

In the absence of a nondegradation requirement, "clean" states might compete with one another for new development, leading to a "Commons" dilemma in which each state permits more degradation than it would prefer or than it would allow if transaction costs did not preclude agreement with competing states. In such circumstances, a federal nondegradation requirement could be justified as simply "forcing" the relevant actors "to be free." But this contention also rests on highly contingent empirical judgments. Moreover, a court must confront the brute political fact that Congress dealt with the "Commons" problem by providing *uniform* federal ambient standards rather than a scheme explicitly involving nondegradation.

Alternatively, we might seek to justify nondegradation by positing a "market imperfection" that could lead a "clean" state not to choose the economically correct policy. The costs of foregone development imposed by nondegradation will be borne primarily by the state and its residents. But many of the benefits of maintaining high levels of air quality will be enjoyed by nonresidents in ways that cannot easily be taxed or recouped by the state. The commerce clause in the federal Constitution prevents a state from levying a border toll on out-of-state residents who visit the state in order to enjoy its high levels of environmental quality.[124] Moreover, a considerable portion of the benefits from pristine enjoyment consists of "long-distance" ideological statisfactions to nonresidents concerned to prevent despoilation of the Grand Canyon or the Alaska wilderness. The state is substantially precluded from "taxing" these benefits in any way. To the extent that preservation of high levels of environmental quality is justified by a national concern for the opportunities of future generations, the state is likewise precluded from reaping the benefits of such a policy. Because a comparatively undeveloped state accordingly bears most of the costs of providing a high level of air quality and reaps only a small portion of the benefits, it could be argued, the state will elect to supply lower levels of air quality than could be justified from the viewpoint of the society as a whole. The presence of such a "market failure" in decentralized state decision making could justify the conclusion that a policy of nondegradation would be more efficient than a policy that permitted each state to pollute up to the levels permitted by the uniform federal standards. However, this rationale rests again on contingent judgments, and equity considerations might preclude a court from requiring a state to sacrifice its interests for those of the nation as a whole by adopting a nondegradation policy if the court is unable to compensate the state for its sacrifice.[125]

The Justifications for Judicial Intervention Even if a nondegradation policy might achieve a more efficient use of resources than uniform ambient standards and even if we believe that there are structural defects that would prevent decentralized state decision making from achieving an optimal policy, what would justify the courts' mandating a nondegradation requirement when the national legislature has conspicuously failed to do so? Under the prevailing view, the legislature enjoys essentially plenary authority over questions of resource allocation, even as to matters touching on interests more "individual" or "fundamental" in character than those implicated here.[126]

However, the primary jurisdiction of the Congress in resource allocation matters might be rejected in the nondegradation context on the ground that the national political process suffers from structural defects that prevent the adoption of geographically more efficient, nonuniform policies. There is a powerful tendency in Congress to adopt geographically uniform policies because explicit nonuniform policies often involve a differential distribution of benefits and burdens. Congressmen would be reluctant to mandate such a distribution for fear that their constituencies might be frequent losers if such a practice supplanted prevailing legislative norms of reciprocity.[127] These political factors are reinforced by bureaucratic constraints that lead administrators to favor uniform policies.[128]

The temptation to assert a general justification for judicial interposition in these circumstances is a strong one (particularly if we assume that the judges would limit themselves to a statutory rather than to a constitutional rationale for decision) because the justification is premised on the existence of basic defects in the political process that only the court can overcome. But the temptation to yield to this argument must be resisted. It would engage the courts in a problematical and contingent weighing of the costs and benefits of alternative resource allocation policies, a task that judges are ill-equipped to undertake. Judicial imposition of differential benefits and burdens on various geographical units might also impede the legislature's incentive and ability to make compensatory provision for those unduly burdened.[129] But the more general ground for rejecting judicial interposition in resource questions must rest on the view that such decisions are peculiarly the province of the legislature. If economic efficiency is the ultimate rationale for the criterion of collective decisions about resource use, such a view has great persuasive force. Efficiency is ultimately determined by citizen preferences for environmental quality as opposed to other goods, and an elected representative legislature is, in the main, likely to be a far better mirror of such preferences than a court. As a mirror of current preferences, the courts exhibit structural defects (in terms of institutional traditions, personnel recruitment, and professional incentives) far more serious than any that might be found in the state or national political processes.[130]

A Constitutional Rationale for Judicial Intervention The *Sierra Club* decisions cannot ultimately be justified by judicial intervention to promote economic efficiency. Instead, their justification lies in principles of diversity that underlie the First Amendment and the federal structure of our political system.[131] This justification is grounded on the shaping impact of experience on individual preferences and values. Environmental degradation in general and air pollution in particular are associated with industrialization and urbanization, processes that together shape outlooks and expectations in characteristic ways. At least since the advent of the romantic movement, relatively pristine environments and the solitude or social life associated with them have been recognized as providing conditions that not only provide mental and spiritual refreshment for those seeking a temporary alternative to urbanized life but also provide fertile soil for the development of aesthetic, social, and political views that challenge the values fostered by urban industrialism.[132]

Our heritage of environmental diversity and associated diversities in economic, social, aesthetic, and political modes of living and thinking is however, challenged by pressures for industrial development and environmental degradation. For reasons detailed previously, neither decentralized nor centralized political decision making appears capable of stemming these pressures except through measures, such as the Clean Air Act, that tend to produce greater environmental uniformity. We accordingly face the prospect of a substantial impoverishment of our collective capacity for experiential diversity.[133] This prospect stirs constitutional resonances, most obviously in the First Amendment, which celebrates diversity in three aspects relevant here.

First, there is the individualist aspect of the First Amendment, which supposes each individual to be the ultimate arbiter of value for himself, a supposition that entails not only liberty of conscience and thought but also a system of experiential heterogeneity in which individuals can develop a critical view of existing preferences and beliefs and can develop, explore, and mutually test a variety of ideas and values. To a large degree, this system would remain simply formal if the social and material conditions upon which the growth of new forms of perceiving, judging, and living also depend were unduly uniform and therefore incapable of stimulating the diversity necessary for individuals to explore and test alternative modes of life and thought.[134]

A second aspect of the First Amendment relevant here is the utilitarian, which follows logically from the first aspect. If each individual is the ultimate arbiter of value, then it follows that there can be no a priori or settled judgment of the true and the good. The ultimately rewarding ends of human existence can be determined only through a process by which individuals explore and evaluate alternative pleasures, modes of being, and ideals. And in order for this continual redefinition of the good life to progress, there must be not only a formal freedom in each to follow his own way but also social and material conditions that promote heterogeneity of experience. A society that impoverishes the alternatives capable of exploration diminishes its collective chances of ascertaining, through the exertions of its individual members, the ingredients of the good life.[135]

Many have urged the preservation of unspoiled environments as a precondition to establishing a quasi-religious communion with nature that will dissolve the alienation generated by urban industrial society.[136] It has been argued also that undeveloped environments must be preserved as an embodiment of a nation's historical heritage and associated values.[137] Some would preserve nature in order to question the headlong pursuit of economic growth, because: ''In the long run utility like everything else is simply a figment of our imagination and may well be the fatal stupidity by which we shall one day perish.' ''[138] One may not agree with any of these positions and indeed may be highly skeptical of all of them and yet affirm the value of preserving diverse environments as a stimulus to testing the diverse ideals of the good life. In Mill's words, uniformity is ''the negation of the main determining principle of improvement, and even of the permanence of civilization, which depends on diversity, not uniformity.''[139]

There is a third relevant aspect of the First Amendment, which regards diversity as an end in itself rather than simply a means to the development of a critical individualism or to a utilitarian search for the good life. The judgment that diversity is an intrinsic good must rest on more than aesthetic or moral preferences for variety or disharmony as contrasted with unity, however entrenched such preferences are with moderns.[140] On the contrary, diversity must be recognized as the necessary resolution of the conflict between the ideal of self-critical individual development and the reality of individual finitude. Through a cooperative endeavor in which each individual pursues some specialized excellence or good, the full range of human potentialities can be developed and each can participate vicariously in this development. A communitarian division of effort in the pursuit of diverse goods provides the closest approximation whereby each mortal individual can achieve the satisfaction—direct and indirect—of complete self-development implicit in the First Amendment.[141] Wilhelm von Humboldt articulated this ideal early in the nineteenth century: ''That which is effected, in the case of the individual, by the union of the past and future with the present, is produced in a society by the material co-operation of its different single members; for, in all the stages of his existence, each individual can exhibit but one of those perfections only, which represent the possible features of human character. It is through such social union, therefore, as is based on the internal wants and capacities of its members, that each is enabled to participate in the rich collective resources of all the others.''[142]

Moreover, each of these three aspects of the First Amendment has an important intergenerational aspect.[143] The progressive diminution of potential experiential heterogeneity narrows not only our own search and dialogue but even more that of our heirs, who are threatened with a double impoverishment. First, our own impoverishment will produce a less rich distillation of experience for us to bequeath to them. Second, the preference-shaping aspect of environmental degradation makes it less likely that steps will be taken to improve or increase the range of environmental quality in the future; as a result our heirs' physical as well as cultural environment will likely be impoverished, perhaps progressively so. The collective search of society over time for the most satisfying ends of existence will be restricted, and the cooperative development of diverse human potentials (which is an endeavor that occurs over time between generations as well as among members of a given society at a given time) will also be impeded.

Historically, the claims for diversity that emerge from this analysis have been satisfied in considerable measure by the combination of the nation's diverse geography, the nonuniform patterns of settlement by immigrants of different cultures, and our federal form of government. However, the opportunities for diversity nourished by our federal system have been undermined by the growth of an industrialized national market economy and associated urbanization. Not only has such growth engendered a greater homogeneity in attitude and outlook, but it has also required states that might otherwise wish to curtail or control the process of economic development to bear a disproportionate share of the costs of promoting diversity. As we have already seen, these "structural defects" are paralleled by political and administrative constraints at the national level that have impaired the ability of the federal government to adopt or implement policies that encourage diversity in the conditions of potential experience. Because environmental quality is a public good, it can be only secured through governmental mechanisms, but the mechanisms available seem to preclude the diversity in environments that we should favor.

Do these considerations justify judicial development of a constitutional right to environmental diversity based on the First Amendment? A First Amendment rationale for judicial intervention is clearly more in accord with our conception of the courts' role than one based on efficient resource allocation. Diversity is the very sort of "structural" principle, ill-served by popular political processes, whose protection requires the interposition of men and women institutionally oriented to a longer view and capable of deflecting the "perhaps unintended or unappreciated bearing [of political decisions] on values we hold to have more general and permanent interest."[144] But recognition of an authority in the federal courts to mandate, on constitutional grounds, governmental policies designed to secure greater diversity in environmental conditions is nonetheless vulnerable to a number of powerful objections.

Objections to a Constitutional Right to Environmental Diversity First, diversity is a nebulous concept ill-suited to judicial elaboration and application. Diversity may be desirable, but how much is enough? What is the particular nature of the diversity that should be afforded and at what sacrifice of other relevant societal goals? It does not appear that reasoned elaboration of any of the principles we have previously discussed could provide persuasive answers of even an approximate sort in given cases posing such questions; the choices in question call for contingent or arbitrary resolutions of the sort that courts are ill-equipped to undertake, particularly in the context of constitutional adjudication.[145]

Second, recognition of a constitutional right to diversity would force upon the courts ultimate responsibility for very large commitments of societal resources. Most First Amendment adjudication simply involves a sanction of nullity to restrain state intrusions upon individual or associational autonomy. But judicial measures to promote environmental diversity could entail substantial resource commitments, either in the form of expenditures to control adverse consequences of development or in the form of foregoing development opportunities in order to preserve diversity. At a time when the polity faces increased concern over resource shortages, judicial abrogation of major resource decisions cannot be easily reconciled with traditional conceptions of the judicial role, particularly where the principles justifying judicial intervention could only be quite contingently related to any particular result that the judges might reach. Recognition of a constitutional right to environmental diversity not only would impair legislative overruling of judicial choices,[146] but it also poses awkward remedial problems. It is one thing for a court to award damages or even issue an injunction against a private firm. It is quite another (as the discussion of transportation control programs below illustrates) to require the executive and the legislative branches to take affirmative steps to provide environmental diversity.[147]

Third, judicial recognition of a constitutional right to environmental diversity leaves the judges too vulnerable to a charge of reading their own value preferences into the Constitution. Whatever the general arguments for diversity, any particular decision sustaining that principle will be perceived, and perceived correctly, as vindicating the interests of environmentalists in opposition to those concerned with economic development. For reasons already canvassed, any such decision is likely to have seriously regressive distributional consequences.[148] The judges are predominantly drawn from the ranks of upper-middle-class professionals who possess the education and wealth to indulge a preference for environmental quality. In such circumstances, judicial decisions seeking to preserve high-quality environments could plausibly be regarded, at worst, as a vindication of personal class-bound values or, at best, as an expression of a paternalism insensitive to sacrifices imposed on those who would form economic development.[149]

A Quasi-Constitutional Rationale for Sierra Club These counterarguments strongly suggest the unwisdom in recognizing a constitutional right to environmental diversity. (A fortiori, they appear to exclude a constitutional right on the part of each individual to a clean and healthy environment.)[150] At the same time, they do not counsel paralyzing judicial diffidence. Given the constitutional affinities in the justifications underlying a principle of environmental diversity and the structural defects in the political process that impede its realization, the courts are warranted, despite the absence of apparent legislative authorization, in taking steps to vindicate that principle in situations in which the force of the counterarguments that we have examined is minimal and the possibility of an alternative disposition by the legislative process is open. This analysis marks out a potential domain of what we have already characterized as "quasi-constitutional" judicial authority.[151] The *Sierra Club* decisions must ultimately be justified as an exercise of such authority.

The *Sierra Club* decisions vindicate powerful interests in diversity that can persuasively claim to transcend contingent judgments about economic efficiency or subjective value preference and secure structural values implicit in the First Amendment and in the federal nature of our political structure. The preservation of pristine areas lies at the very heart of a strategy for environmental diversity. Yet the context in which the controversy arose permitted the court to couch its vindication of the diversity principle in the form of statutory interpretation. Although conventional statutory analysis would not support the courts' rulings, the power and seriousness of the diversity principle justifies dispensation with conventional analysis in the particular circumstances. Moreover, by seizing upon preservation of the status quo in those states with high air quality, the courts minimized the force of potential objections to their action. In utilizing the existing distribution of air quality as the baseline for its decision, in refusing to lay down precise guidelines as to what might constitute "significant deterioration" of existing air quality, and in requiring EPA to implement its decision, the courts were able to avoid the difficulty of directly defining and enforcing permissible differences in air quality. Also, unlike a judicially mandated cleanup effort, preservation of the status quo did not flaunt the potentially massive commitment of resources involved, and it masked the distributional consequences of the court action; development foregone is not so visible nor so obviously regressive in its impact as controlling or shutting down existing sources.

Finally, the court decisions were not ultimately dispositive. Instead, they forced the legislature squarely to confront the problem of diversity in federal policies on air quality and associated controls on development. In addition to provoking congressional attention, the courts' intervention also had an important effect on the ultimate substance of such policies by making a policy of nondegradation the starting point for political adjustments rather than a policy that permits deterioration of air quality to the limit set by national standards. To borrow words developed in a related context by Monaghan, "[The judges' interpostion] triggers an important shift in the political process. The Court, in effect, opens a dialogue with Congress, but one in which the factor of inertia is now on the side of [diversity]."[152]

As a result of the first *Sierra Club* decision, the EPA eventually adopted a zoning approach to the significant deterioration problem, under which states would be divided into differing regions in which different increments of degradation would be permitted.[153] Congress has elaborated on this scheme by proposing legislation that would designate certain undeveloped areas of undisputed national importance (such as national parks) as mandatory Class I areas, in which only minimal departures from existing air quality will be permitted. Other areas, most of which have already experienced a measure of development, would be classified in ways that would permit the states, subject to case-by-case federal approval, to permit further degradation. This legislation was blocked by a Senate filibuster conducted by Utah's two senators on the last day of the 94th Congress, but it is likely that the same or a similar measure will receive congressional endorsement relatively soon.

In the author's view, the mandatory designation of Class I areas is wholly justified because of diversity considerations already adduced, but the potential assumption by the federal government of responsibility for case-by-case decisions regarding further increments of deterioration in areas that are already substantially developed seems unwise. Once the preservation of substantial areas of extremely high environmental quality is assured, the argument for diversity seems too attenuated to impose federal constraints on the decisions of, say, Albuquerque, to choose somewhat greater economic development at the cost of some deterioration of air quality. Moreover, as the unfortunate history of EPA's indirect source program illustrates, such case-by-case decisions have significant local land-use implications that the federal government is ill-equipped to resolve. There may be justification for preventing added emissions in moderately degraded areas because of mounting evidence that adverse health and environmental effects of pollutants such as sulfates may be a function more of total atmospheric loading of the pollutant that ambient concentrations in any one area at any one time. But the need to reduce total loading is not confined to moderately polluted areas. It is even greater in areas that are heavily polluted and calls for measures like an emissions fee, which is specifically designed to achieve a reduction in total loading.

These reservations should not, however, obscure the essential point. The courts intervened to vindicate a principle of environmental diversity that the legislature had not confronted and that would otherwise have gone largely unrecognized. Yet the intervention assumed a form that did not preclude subsequent legislative reevaluation of the court's position. This is quasi-constitutional adjudication working at its best.

EPA Authority to Compel State Enforcement of Federal Transportation Control Plans As this essay has already recounted,[154] the failure of many states to include adequate transportation controls in their implementation plans obliged EPA to promulgate unpopular measures such as gasoline rationing, physical restrictions on automotive use, high parking or access fees, and compulsory provision of improved mass transit facilities. In the face of states' unwillingness to enforce such measures and its own limited enforcement resources, EPA threatened to impose civil and criminal sanctions on state officials who refused to carry out the EPA-drafted controls. Litigation ensued. A majority of the Courts of Appeals that have considered the matter have rejected, in whole or in part, EPA's assertion of the power to require state enforcement of such measures.[155]

The majority decisions have relied upon a combination of statutory and constitutional analysis to sustain their conclusion. They first emphasize the failure of the relevant enforcement section of the Clean Air Act (section 113) to provide explicitly for enforcement against state officials. Because such enforcement would seriously compromise state autonomy, the majority decisions would require a clear statement of congressional purpose to authorize such enforcement.[156] In addition, the decisions advert to constitutional limits on federal power to compel state enforcement action; these limitations were thought either to be inherent in the commerce clause itself (the constitutional foundation for the assertion of federal power over air pollution in the Clean Air Act),[157] or to be imposed by the guarantee clause[158] or the Tenth Amendment.[159]

The judges who have sustained EPA power to require state officials to enforce federally drafted transportation controls have stressed three considerations: the sweeping application of the phrase "any person" in defining the reach of section 113 federal enforcement authority, evidence in the legislative history reflecting an expectation that the act would mandate state enforcement in the area of motor vehicle inspection and maintenance, and the practical necessity for state enforcement of traffic control measures.[160] They have rejected the suggestion of intrinsic or extrinsic constitutional limitations on congressional power to require state enforcement of federally drafted air pollution control measures.[161]

It is not our purpose here to analyze in detail the reasoning of the various decisions or the intricacies of the statutory or constitutional analysis. However, it is appropriate to register two contentions before linking these decisions to the earlier argument in the essay. Congress should not be constitutionally prohibited from requiring state enforcement of measures to reduce pollution from automobiles travelling on state highways.[162] However, a clear statement on the part of the Congress should normally be required in order to conscript state governments into affirmative measures to carry out federal regulatory programs. Because such measures may require substantial resource expenditures by the states and also because the process of conscription will involve considerable interference by the federal judiciary in the normal processes of state government, a clear assumption of responsibility by the Congress should ordinarily be required, not only to provide a visible measure of support for the federal judiciary in the discharge of a difficult task but also to ensure that the "political safeguards of federalism" have had an opportunity to function.[163] Accordingly, normally operative principles of clear statement justify the position of the majority of courts that have denied EPA the enforcement power it claims.

But analysis of the *Sierra Club* decisions raises this question: Why did the majority decisions on transportation controls fail to follow *Sierra Club* and reject normally applicable principles of clear statement in order to sustain federal intervention that would be likely to produce a greater measure of environmental diversity? The transportation control measures proposed by EPA are designed to lessen widespread reliance on automotive travel and promote alternatives—such as mass transit in the short run and "cluster" planning in the long run—that not only would enhance air quality but also might encourage more varied patterns of living and working that would potentially enrich an increasingly homogenized urban living pattern.[164] Nonetheless, the refusal of the majority of courts to require state enforcement of transportation controls can be distinguished from *Sierra Club* and sustained on the basis of contingent and prudential considerations that play a properly important role in the realm of quasi-constitutional law.

The courts would face troublesome remedial problems if such an authority were recognized. To require states to carry out federal transportation control plans could involve the imposition of contempt sanctions on state and local officials who refuse to square their conduct with existing law. It might also require the federal courts to direct the use of local or state funds to provide necessary facilities (such as mass transit facilities) and manpower,[165] and supervise state and local personnel in carrying out transportation control plans. Such difficult ventures—which were avoided to a large extent in *Sierra Club* by imposing on EPA the burden of defining and implementing the nondegradation principle—should not be undertaken in the absence of an explicit congressional mandate or the plain necessity of vindicating a right based squarely on the federal Constitution.

Moreover, federally compelled state enforcement might not prove a workable means of implementing transportation control plans or introducing a greater measure of diversity in environmental conditions. The magnitude of the task to be accomplished and the resistance of local and state officials could render the unaided remedial efforts of the federal judiciary largely ineffectual and perhaps discredit the entire effort to achieve the federal ambient standards through transportation controls. The majority decisions have rather plainly indicated that, in lieu of such an approach, the Congress should impose conditions on federal grants to state and local governments that would require them to undertake transportation control measures in order to receive federal funds.[166] Although the legal limits of such a strategy are unclear, such an approach would minimize the strains on the federal judiciary, and could make local acceptance of the federal requirements more palatable, particularly if the federal grants are geared in part to providing the resources needed to carry out the plans.

The refusal of most courts to sustain coercion of the states with respect to federal transportation controls thus seems wise. However, there may in the future be an increased number of appropriate occasions for judicial intervention to vindicate the principle of diversity in the context of federal environmental programs. Environmental fields other than air pollution control are coming to exhibit leveling uniformities and other distortions characteristic of centralized administration.[167] At the same time, court decisions setting aside administrative decisions that permitted development on public lands contiguous to wilderness areas,[168] threatened endangered species,[169] authorized clearcutting in national forests,[170] and allowed destruction of porpoises by commercial fishermen[171] suggest a broadening judicial willingness to protect environmental diversity against serious erosion by administratively approved commercial development. These decisions, together with *Sierra Club,* indicate that, at a minimum, courts can properly intervene to prevent what is, as a practical matter, potentially irreversible destruction of environmental diversity as a result of administrative policies that do not reflect a widespread and fully considered political consensus in the society. What additional implications the notion of diversity might create for courts is, at present, unclear. Sax's comment on this essay reflects a healthy skepticism and raises questions about the content and judicial implementation of the diversity principle that this essay cannot pretend to have answered adequately.[172] But our common law tradition does not demand a detailed doctrinal blueprint as a prerequisite to judicial testing of new principles. Given the importance of environmental diversity and the structural defects in the political process that inhibit its protection, judges should be encouraged to consider the value of diversity a relevant and important one when they review environmental decision making. The fact that diversity values may conflict with other fundamental concerns—such as equality—should not preclude such encouragement. We value the courts in considerable measure precisely because they provide a forum where conflicts among basic social ideals can be tested under constraints of reasoned decision.

Evaluation

This essay has made scant progress in resolving the analytic imponderables of environmental policy generated by the often conflicting objectives discussed in the beginning section. But it suggests some tentative conclusions whose content is based, ultimately, on the plural character of those objectives.

Both candor and prudence counsel that we should not seek to paper over the conflicts among such objectives and our inability to resolve them by supposing that there is some one calculus or process of decision that can be relied upon to yield correct answers or sound policy. Moreover, the debate over these objectives and their potential reconciliation has value in itself as an important part of the human endeavor to evaluate and shape our own ends, however imperfectly that endeavor is realized. Obscuring the nature of the choices involved by means of oversimplification is ultimately an evasion of responsibility.

These observations have institutional implications. A horizontal and vertical division of decisional responsibility may be necessary if the component elements of choice are to receive sufficient attention in a collective dialogue. In the Clean Air Act and other environmental legislation, Congress has attempted to go quite far in resolving the imponderables in environmental policy with precise statutory mandates. Not only have the particular choices made by Congress neglected important considerations—economic efficiency and diversity, in particular—but the effort (to the extent that it has succeeded) to fence agency and court out of a major decision-making role has had unhappy consequences. At the very least, it has diminished the potentiality for corrective adjustments based on knowledge developed in the course of implementation.[173] Congressional exclusion of other organs of government, including state and local bodies, from a significant role in the choice of environmental policies also tends to muffle salutary dialogue concerning those choices and to diminish the public's commitment to and confidence in the policies that result. Perhaps the most dramatic illustration of this point is the federal preemption of any significant state role in the regulation of nuclear power, a policy that is now paying rich dividends in the form of widespread cynicism and distrust of federal policies with respect to nuclear energy.

It does not follow, of course, that responsibility for decisions should be diffused to the point at which no effective action is possible at all. This essay has provided illustrations of situations, such as nondegradation, in which effective central direction is necessary to the pursuit of environmentally sound policies. But the degree of centralization of decision should not be pressed any further than necessity demands. The rigidities of the Clean Air Act Amendment also may have been historically justified as a form of "shock treatment" calculated to alter settled governmental and private practices and attitudes that reflected an indifference toward environmental degradation. However, the "shock treatment," has largely achieved its intended result, and the time has come for a "second generation" of federal pollution control measures that are less rigid and give greater weight to interests in diversity, on the one hand, and efficient resource use, on the other hand.

Administratively imposed emission fees and other forms of economic incentives, such as marketable pollution permits and mixed fee-regulatory schemes, could play an important role in promoting greater flexibility and efficiency. On the federal level, there appear to be no serious constitutional obstacles to congressional adoption of such incentive systems in lieu of current regulatory strategies.[174] On the state level, the centralized uniformities that characterize the present generation of federal environmental statutes preclude, to a substantial degree, state experimentation with fee systems and other alternatives to traditional regulatory controls. For example, the present Clean Air Act would apparently preclude the use of emission fees in a state's implementation plan to limit air pollution in place of regulatory prohibitions.[175] Accordingly, there is a need to develop realistic modifications of the Clean Air Act and other federal environmental statutes that would permit and encourage experimentation and development of alternatives to regulation at both federal and state levels.[176]

The plural nature of environmental problems and objectives also justifies a substantial role for reviewing courts in checking the parochial tendencies of administrative agencies by attempting to ensure that agencies consider and weigh the full range of considerations that are likely to be implicated in environmental decisions and, if necessary, by straining the relevant statute to enlarge the factors that the agency may take into account. The record with respect to judicial review under the Clean Air Act demonstrates that the courts have discharged this role with considerable success. They have devised novel procedural mechanisms to ensure better agency consideration of all relevant factors without unduly hobbling the agency in pursuit of its responsibilities. Courts have also assumed a demanding but necessary task in reviewing the substance of agency decision to ensure that the agency has rationally weighed all of the relevant considerations that the statute and the practicalities of the situation reasonably admit. This record, which belies the stereotype that the courts (and particularly the District of Columbia Circuit Court of Appeals) are blindly proenvironmental in bias and shows that judges have been solicitous of economic considerations, entitles the courts to take considerable credit for improving the quality of EPA decisions over the past five years.

This essay has also argued that pervasive structural defects in political and administrative processes sometimes lead to systematic neglect of certain environmental values, such as diversity, that can convincingly be characterized as basic and enduring. In such circumstances the courts are justified in assuming a more forward role in advancing such values, so long as prudence is exercised in selecting the occasions for such intervention and the rationale for the intervention leaves open the ultimate possibility of an alternative disposition by the legislature.[177]

In affirming the desirability of a substantial judicial role, we should be careful not to regard its justification as purely instrumental—a means to promoting better policy choices. Environmental issues, as this essay has attempted to suggest, test most acutely conflicting ideals of justice and human excellence in an industrialized society. In the end, the judges neither can nor should dictate those ideals. But the courts provide a forum of government to which proponents of contrasting visions of the good have access as of right.[178] The forum is also one in which the content of contrasting ideals can be tested and, in part, implemented under constraints of reasoned elaboration that legislative and administrative processes do not always assure.[179]

A process of reflective examination and testing cannot promise a definitive resolution of conflicting objectives. In the end its greatest virtue may be in leading us more "securely . . . to think of the desirable ends of social policy as essentially plural and incompatible."[180] That security may be indispensable in addressing the strongest paradox that questions of environmental policy pose. Such questions reveal deep disagreement over the appropriate ends of decision and reiterate the ultimate necessity of pursuing divergent conceptions of value. Yet precisely because environmental quality is so conspicuously a collective good, a measure of community agreement is essential to providing the diversity in environments necessary to nurture "plural and incompatible" conceptions of excellence in thought, perception, and conduct. The paradox of diversity in community, which is deeply rooted in our society, is thus forcefully reflected in questions of environmental policy. The principle of environmental diversity adumbrated in this essay is, after all, nothing more than a restatement of the problem. No one could suppose that judges have some special competence or authority to unravel the ultimate paradox. But no one—not least the Founders—should be surprised if the institution of judicial review supplies a catalytic element that helps further its deeper understanding and affirmation.

Notes

1. The relevant "market failure" can be traced either to the view that environmental amenities are usually public goods or to the view that environmental "bads" such as pollution impose costs external to those that generate such bads. See generally Krier and Stewart, *Environmental Law and Policy,* Ch. 3 (forthcoming 1978) (Bobbs-Merrill).

2. But cf. R. Markovits, "A Basic Structure for Microeconomic Policy Analysis in Our Worse-than-Second-Best World; A Proposal and Related Critique of the Chicago Approach to the Study of Law and Economics," *Wisconsin Law Review* (1975): 950 (Problems of "Second Best").

3. See E. S.Mills and L. J. White, "Government Policies Toward Automotive Emissions Control," infra; M. J. Roberts, "The Political Economy of Implementation: The Clean Air Act and Stationary Sources," infra.

In particular, government regulatory programs have all too often failed to utilize cost-effective measures to achieve given environmental quality objectives. See Roberts and Stewart, "Energy and Environment Interactions: Policy Options and Prospects," in H. Owen and C. Schultze, eds., *Setting National Priorities* (Brookings, 1976): 411–456. The criterion of cost-effectiveness—achievement of an exogenously determined objective at least cost—can be viewed as a constrained version of economic efficiency. It is assumed that in setting the objective the decision maker has matched costs and benefits; unless the objective is attained with minimum resource use, there will be waste, and the objective itself may have to be reconsidered if it was originally set on the assumption of least cost achievement.

4. For additional criticisms, see L. Tribe, "Policy Science: Analysis or Ideology?" *Phil. and Pub. Affairs* 2 (1972): 66; L. Tribe, "Technology Assessment and the Fourth Discontinuity: The Limits of Instrumental Rationality," *So. Cal. Law Review* 46 (1973): 617.

5. The economic literature of the problems generated by preference instability is sparse. See J. Rothenberg, "Welfare Comparison and Changes in Tastes," *Amer. Econ. Rev.* 43 (1953): 885; von Weisacker, "Notes on Endogenous Changes in Tastes," *J. Econ. Theory* 3 (1971): 345.

6. The precise manner in which governmental institutions are structured to make such choices can substantially affect the quantity and nature of the information conveyed and the resulting impact on individual preferences. It was the aim, and in some measure the achievement, of the National Environmental Policy Act to further the dissemination of information about the environmental impact of government policies in order to promote a collective reassessment by individuals of their preferences for environmental quality.

7. Individuals may discount heavily environmental risks that are clearly significant—such as the threat of cardiovascular disease from exposure to elevated levels of pollution resulting from fossil fuel combustion—but that do not impinge on individuals whose identity can be specified in advance and whose etiology is neither dramatic nor clearly differentiated from "natural" causes of morbidity or mortality. On the other hand, individuals frequently appear to accord great importance to avoiding the possible occurrence of dramatic disasters that, if they occurred, would inflict, in a causally obvious way, disease or death on a geographically discrete and identifiable group, such as those living in the vicinity of a nuclear reactor, even though the statistical probability of any such event is extremely low. Compare Charles Fried, *An Anatomy of Values* (Cambridge, Mass.: Harvard University Press, 1970), Chs. 10–12.

8. See *Citizens for Safe Power* v. *NRC,* 8 ERC 1598 (D.C. Cir. 1976).

9. An analogous problem is presented by the satisfaction, through purchase of a given product, of preferences created by advertising for that product. See S. Oster, "The Incidence of Local Water Pollution Abatement Expenditures: A Case Study of the Merrimack River Basin." Ph.D. dissertation, Harvard University, 1974.

10. Cf. John Stuart Mill, "On Liberty," in M. Cohen, ed., *The Philosophy of John Stuart Mill* (Random House, 1961), pp. 241–245.

11. For example, what rule of decision do we adopt if the products selected by persons who have experienced all of the relevant candidates differed depending on the order in which they had historically been experienced?

12. The costs of some environmental measures are distributed through taxation, whose incidence (at least on the federal level) is mildly progressive, but this effect is swamped by the other mechanisms of cost distribution. See Council on Environmental Quality, *Environmental Quality 1973* (Washington, D.C.: U.S. Government

Printing Office), pp. 101–109. See also David Harrison, *Who Pays for Clean Air?* (Cambridge, Mass.: Ballinger, 1975). This regressive effect might be even greater under a system of emission fees that not only caused sources to impose controls on pollution but also required them to pay fees on the remaining pollutants that were emitted (unless the savings obtained by the fee system in securing least cost control outweighs the added burden of fee payments).

13. Harrison, *Who Pays for Clean Air?* pp. 95–107; Freeman, "Distribution of Environmental Quality," in Kneese and Bauer, eds., *Environmental Quality Analysis* (Johns Hopkins Press, 1972), pp. 243, 262–269.

14. Harrison, *Who Pays for Clean Air,* pp. 79–94.

15. See Freeman, op. cit. Compare J. K. Galbraith's sardonic definition of a conservationist as "a man who concerns himself with the beauties of nature in roughly inverse proportion to the number of people who can enjoy them." John Passmore, *Man's Responsibility for Nature* (New York: Charles Scribner's Sons, 1974), p. 105.

16. These factors would explain Rubinfeld's conclusion, based on a study of property values, that persons with higher incomes place a higher value on clean air. D. L. Rubinfeld, "Market Approaches to the Benefit of Air Pollution Abatement," infra.

17. Limited empirical confirmation of these hypotheses is found in S. Oster, op. cit. (note 9); Council on Environmental Quality, op. cit. (note 12), p. 82.

18. This example again raises the question of whether existing preferences should be taken at face value. On the other hand, educating the poor to the benefits of environmental quality may represent a perverse form of benevolence if environmental quality is a luxury good in which only the wealthy can indulge. See Harsanyi, "Welfare Economics of Variable Tastes," *Rev. Econ. Studies* 21 (1953–1954): 204.

19. For development of the notion of minimum distribution of fundamental goods, see F. Michelman, "In Pursuit of Constitutional Welfare Rights: One View of Rawls' Theory of Justice," *U. Pa. Law Rev.* 121 (1973): 962.

20. See infra.

21. For a concise summary of the debate, see O. Herfindahl and A. Kneese, *Economic Theory of Natural Resources* (Columbus, Ohio: Merrill, 1974), pp. 204–221.

22. For discussion of the problem of justice between generations, see John Rawls, *A Theory of Justice* (Cambridge: Harvard University Press, 1970), p. 848; Passmore, *Man's Responsibility for Nature,* pp. 73–100.

23. The argument that natural objects should be recognized as holders of moral and legal rights was first developed in C. Stone, "Should Trees Have Legal Standing?—Toward Legal Rights for Natural Objects," *So. Cal. Law Review* 45 (1972): 450. The notion that humans have obligations to nature that transcend homocentric economic definitions of welfare is sympathetically developed in L. Tribe, "Ways Not to Think About Plastic Trees, New Foundations for Environmental Law," *Yale Law Journal* 83 (1974): 1315. For criticisms, see Passmore, *Man's Responsibility for Nature,* pp. 101–126; M. Sagoff, "On Preserving the Natural Environment," *Yale Law J.* 84 (1975): 205.

24. See, e.g., Endangered Species Conservation Act of 1969, 87 Stat. 903, 16 U.S.C. secs. 668aa et seq.; Marine Mammal Protection Act of 1972, 86 Stat. 1027, 16 U.S.C. sec. 1361 et seq.

25. See M. Sagoff, op. cit. (note 23).

26. By making the costs of environmental control more explicit, a fee system may promote the opportunities for a more explicit consideration of the appropriate levels of environmental quality that we should provide and thereby stimulate the development of information that would improve the quality of such decisions.

27. See L. Tribe, "Technology Assessment and the Fourth Discontinuity: The Limits of Instrumental Rationality," *So. Cal. Law Rev.* 46 (1973): 617.

28. See John Esposito, *Vanishing Air* (New York: Grossman Publishers, 1970), pp. 69–151.

29. See G. Hardin, "The Tragedy of the Commons," *Science* 162 (1968): 1243.

30. The author is obliged to his colleague, Marc Roberts, for this insight.

31. In addition, centralized decision may be preferred by those who advocate government provision of environmental quality at levels higher than those that would be justified by a consideration of citizens' existing preferences (because environmental quality is regarded as a merit good or the experiential ignorance of individuals who have not experienced should be discounted, and so on). Even were it true that decentralized decision in the end more accurately mirrored the preferences of the governed, that would hardly be a virtue for those who advocated a policy of partial disregard of those preferences. To such proponents, the comparative insensitivity of centralized government, its bureaucratic pursuit of policies not fully favored by the citizenry, would be arguments in favor of centralized decision.

32. See infra.

33. This is not to say that market-type mechanisms, such as fees or transferable permits, should not be appropriate ingredients of sound environmental policy but only that decisions as to which environmental policies are sound cannot be reached through the market. The criterion of economic efficiency could be viewed as a surrogate for market processes of decision in determining the appropriate provision of environmental amenities, but the previous discussion indicates that the criterion of economic efficiency alone is not an adequate or satisfactory determinant of environmental choices.

34. See infra.

35. See R. Stewart, "The Reformation of American Administrative Law," *Harvard Law Rev.* 88 (1975): 1577, 1677–1688.

36. Ibid., pp. 1711–1760.

37. Ibid., pp. 1760–1790.

38. E.g., C. Meyers, "An Introduction to Environmental Thought: Some Sources and Some Criticisms," *Ind. Law J.* 50 (1975): 426; A. D. Tarlock, "A Comment on Meyers' Introduction to Environmental Thought," *Ind. Law J.* 50 (1975): 454.

39. Communications Act of 1934, sec. 3307(a), 47 U.S.C. sec. 307(a) (1970).

40. However, as will be shown, the EPA in practice wields substantial discretionary power in air pollution matters, a result that underscores the ultimate limitations of legislative capacity to specify agency power in advance. The discretion enjoyed by state and local agencies in implementation and enforcement is made clear in M. Roberts, "The Political Economy of Implementation: The Clean Air Act and Stationary Sources," infra.

41. The political logic of the 1970 amendments is examined in H. Ingram, "The Political Rationality of Innovation: The Clean Air Amendments of 1970," supra.

42. See, e.g., *Scenic Hudson Preservation Conf.* v. *FPC,* 354 F.2d 608 (2d Cir. 1965); *Office of Communication of United Church of Christ* v. *FCC,* 425 F.2d 543 (D.C. Cir. 1969); see generally R. Stewart, "The Reformation of American Administrative Law," *Harvard Law Rev*. 88 (1975): 1667.

43. See Henry J. Friendly, *The Federal Regulatory Agencies: The Need for a Better Definition of Standards* (Cambridge, Mass.: Harvard University Press, 1962). On the other hand, Kenneth Davis argued that reliance on greater legislative specification was undesirable and unnecessary and advocated judicial measures to force agencies to crystallize their discretion in specific rules and regulations. See Kenneth C. Davis, *Discretionary Justice* (Baton Rouge, La.: Louisiana State University Press, 1969).

44. Ambient standards specify the permissible concentrations of air pollutants in the atmosphere, as opposed to emission limitations, which specify the permissible quantity of pollutants that given sources may discharge to the atmosphere. Emission limitations are obviously required in order to achieve an improved level of ambient air quality.

45. Clean Air Act sec. 109. The six pollutants for which standards were mandated are carbon monoxide, hydrocarbon oxides of nitrogen, photochemical oxidants, particulate matter and sulfur oxides. Sections 108 and 109 of the act, as amended, also require the federal government to establish primary and secondary ambient air quality standards for additional pollutants that (in the judgment of the administrator) have "an adverse effect on public health or welfare" and whose presence in the ambient air "results from numerous adverse mobile or stationary sources."

46. Clean Air Act, section 110.

47. Clean Air Act, section 111.

48. However, "clean" states still enjoy an advantage under the act in accommodating new development; even though new stationary sources must meet the same federal emission limitations regardless of where they locate, states whose current ambient levels are substantially below the federal standards could accommodate new growth without violating those standards, whereas states whose emissions are currently in excess of the federal ambient standards could not accommodate such growth without further reducing emissions from existing sources in order to make room for new sources. This advantage was largely eliminated by court decisions, discussed earlier, which found that this act implicitly precluded any significant deterioration of air quality in areas already in compliance with the federal standards.

49. See infra (note 120).

Preserving higher levels of air quality in presently ''clean'' areas would promote economic efficiency if the benefits of pollution control are higher or the costs of achieving it lower in such states than in presently ''dirty'' states. Although the costs of achieving a given level of air quality in presently ''clean'' states are highly likely to be less than in ''dirty'' states (at least if we take a national perspective and disregard the opportunity costs of the state as a separate unit), the question of benefits is more debatable. On the one hand, health benefits may be far less in ''clean'' and relatively sparsely populated states than in ''dirty'' states that have large populations. On the other hand, aesthetic benefits are likely to be higher in ''clean'' states because a small amount of additional pollution can cause a large loss of aesthetic value in a location that is nearly pristine whereas it will cause a much smaller loss in an area that is already heavily polluted.

Even if lower pollution levels in presently ''clean'' areas are consonant with a crude measure of allocational efficiency, the Clean Air Amendments' structure might be criticized for failing to distinguish among those ''dirty'' areas whose ambient levels are presently in excess of the federal standards. The costs of meeting the ambient standards are likely to be much higher in areas whose ambient levels far exceed the federal standards than in those areas that exceed them slightly (although such cost differences are likely to be accommodated to some degree by the statute's provision for extensions). Moreover, benefits are likely to vary in proportion to population. Yet the act requires the same standards to be met in all such cases. For discussion, see J. Krier, ''The Irrational National Air Quality Standards: Macro- and Micro-Mistakes,'' *U.C.L.A. Law Rev.* 22 (1974): 323–342.

50. Clean Air Act, sec. 112.

51. Clean Air Act, sec. 202.

52. See D. Harrison, *Who Pays for Clean Air*. The overriding aim of the 1970 amendments was achievement of federal ambient standards everywhere. But the degree of automotive emission controls needed to achieve those ambient standards would obviously be far less in relatively ''clean'' areas than in relatively ''dirty'' areas of the country. This fact, coupled with the circumstance that the large number of engine production lines in the domestic automobile industry (over 25) could easily accommodate two or perhaps three different levels of emission limitations, points to the possibility of substantial cost savings in pursuing a ''two-car'' or ''three-car'' strategy in setting new-car emission limitations. Even if there were a single federal emission limitation for all new automobiles, it need not be set at the level of stringency required to guarantee attainment of the federal ambient standards in the most polluted region of the country. An alternative that could also yield substantial cost savings over a completely uniform approach would be to establish a federal emission limitation that would secure compliance with the ambient standards in most areas of the country and to relax simultaneously the preemption feature in order to permit heavily polluted areas to establish more stringent controls on new automobiles as a means of meeting the ambient standards in those regions.

53. There are other respects in which the Clean Air Act fails to promote efficiency in achieving environmental quality goals. The exclusive insistence on regulatory con-

trols to the exclusion of alternative types of incentives, such as emission fees or transferable pollution rights, is an obvious case in point. In addition, the complex structure of the act tends to preclude coordinated abatement measures to ensure achievement of federal ambient standards at least cost. For example, pollutants such as hydrocarbons and oxides of nitrogen are emitted in substantial quantities by existing mobile and stationary sources, new stationary sources, and new mobile and aircraft sources. Cost effectiveness would require that the burden of abatement necessary to meet the federal ambient standard in a given region be allocated to those sources that could most cheaply abate. But the act's division of authority among state and federal authorities, its requirement that federal emission limitations be nationally uniform, and the constraints it imposes on the level of specific limitations (such as those applicable to new automobiles) renders this ideal unattainable.

54. Clean Air Act, sec. 307.

55. These provisions represent, in effect, a codification of prior judicial decisions requiring administrators to take effective action on behalf of the beneficiaries of regulatory programs. See, e.g., *Office of Communication of United Church of Christ* v. *FCC,* 425 F.2d 543 (D.C. Cir. 1969).

56. In the case of state implementation plans, a double backup system was provided. If states failed to enforce the limitations contained in such plans, the federal administrator was empowered to enforce them. Clean Air Act, sec. 113. If the federal administrator also failed to take action, citizen suits could be mobilized to ensure compliance.

57. A much larger number of environmental group actions challenging EPA approvals of state implementation plans have been brought under the general judicial review provision of section 307 of the act and are discussed infra.

58. 475 F.2d 968 (D.C. Cir. 1973). The same issues were involved in *City of Riverside* v. *Ruckelshaus,* 4 ERC 1728 (C.D. Cal. 1972) but were presented solely in the context of California's implementation plan for the Los Angeles region, whereas the *NRDC* litigation challenged EPA's position on a nationwide basis. The court in *City of Riverside* ruled against the EPA, as did the court in *NRDC*.

59. Section 110(e) of the act provides that, at the time a state implementation plan is submitted, the governor of the state may request a two-year delay in the three-year deadline for meeting the federal primary ambient standards if the administrator determines that one or more emissions sources are unable to meet emission limitations required to achieve such standards and that all reasonably available alternative means of promoting compliance (including interim measures of control) have been utilized.

60. EPA subsequently construed the "maintenance" aspect of the court's opinion as requiring regulation of new "indirect" sources, such as parking lots, sports complexes, and highways, that would attract concentrations of automobile traffic whose emissions threaten maintenance of the ambient standards. 38 Fed. Reg. 6279 (8 March 1973). The court, in a subsequent order, issued on 15 February 1974, regarding deadlines for compliance with its decision, apparently agreed with this construction.

61. Whether the court's ruling prohibiting EPA from granting advance extensions of the ambient air quality standards deadline had a serious adverse effect on the longer-run achievement of those standards is far more problematical. The requirement that states first submit implementation plans demonstrating maximum efforts to comply with the 1975 deadline and only thereafter apply for an extension arguably resulted in the initial adoption of ill-conceived or politically unpopular programs that fell short of effective implementation, whereas a plan geared from the beginning to achieve ambient standards by 1977 could have adopted measures with greater long-run administrative and political feasibility. This may have been the case in Massachusetts. See Selig, Padnos, and Bracken, "Transportation Controls Under the Clean Air Act: The Massachusetts Experience," memorandum prepared for the Committee on Environmental Decisionmaking, National Academy of Sciences–National Research Council (1976). On the other hand, it is equally plausible that the difference in procedures followed by the EPA initially and those required under the court's decision were merely formal. See 475 F.2d at 972–973 (J. MacKinnon concurring). More evidence on the development of implementation plans in particular jurisdictions is needed to resolve this question.

62. See infra.

63. Section 510, Agriculture-Environmental and Consumer Protection Appropriation Act of 1975, P.L. 93-563, 88 Stat. 1822.

64. It may be argued that the statute did not mandate indirect source controls. But the requirement that plans contain provisions to maintain ambient standards is plain, and indirect source regulation is, barring a dramatic and unexpected breakthrough in automotive emissions control technology, the most appropriate means of ensuring such maintenance in the case of automotive pollutants where continued compliance with the ambient standards for such pollutants is threatened by new growth patterns.

65. See 475 F.2d 968, 972-3 (MacKinnon, J. concurring).

66. 344 F.Supp. 253 (D.D.C. 1972), aff'd per curiam, 4 ERC 1025, aff'd by an equally divided ct, 412 U.S. 54 (1973).

67. 8 ERC 1695 (S.D.N.Y. 1976), aff'd (2d Cir. 1976).

68. See *Ethyl Corp.* v. *EPA,* 8 ERC 1785 (D.C. Cir. 1976).

69. Section 109 requires that ambient standards be promulgated for pollutants for which the EPA administrator had previously promulgated criteria documents (describing the adverse effects of given pollutants) pursuant to section 108, which, in turn, requires that the EPA administrator "shall" publish criteria documents for each air pollutant "which in his judgment has an adverse effect on public health or welfare" and "the presence of which in the ambient air results from numerous or diverse mobile or stationary sources."

70. In addition to authorizing citizen suits against the administrator, the Clean Air Act provides for citizen enforcement suits against polluters who violate validly adopted emission limitations that have not been enforced by federal or state authorities. There have been few citizen enforcement actions of this sort; in most instances the filing of litigation has provoked action by state or federal officials or has

resulted in settlements that provide for compliance. The small number of citizen enforcement actions in the past may also be due to the fact that national environmental organizations prefer to expend their scarce litigation resources on issues of more general importance than a given source's compliance with a given limitation, and even modest enforcement actions may strain the capabilities of local groups. Because many emissions limitations have only recently come into effect, however, it is likely that more citizen enforcement actions will be filed in the future. On balance, the statutory provision for citizen enforcement suits appears to have been a salutary innovation. There may be occasions where a balancing of considerations would justify some refusal to enforce the letter of the law, but so long as the federal or state governments have the power to revise the underlying emission limitations, it is possible for administrators to provide a safety valve. For this reason, citizen enforcement suits against private polluters are far less likely to raise troubling questions than Congress's utilization of the courts to force the administrator to follow legislatively imposed deadlines and other constraints.

71. Agency decisions to promulgate ambient standards or nonautomotive emission limitations involve a considerable measure of discretion because the relevant statutory criteria are imprecise. For example, ambient standards must "allowing an adequate margin of safety . . . protect the public health" and "protect the public welfare," Clean Air Act, section 109(b)(1), (2); emission limitations must require a level of control "achievable through the best system of emission reduction which (taking into account the cost of achieving such reduction) the Administrator determines has been adequately demonstrated," section 111(a)(1). The statutory provisions in section 211 that authorize EPA to regulate fuel additives for automobiles are similarly imprecise.

One might suppose that EPA's freedom of action was much more tightly constrained with respect to state implementation plans, which are controlled by the overriding statutory requirements of achievement and maintenance of the preestablished federal ambient standards. But even here, as the decisions will confirm, considerable leeway for agency judgment is inevitable. See, e.g., *Texas* v. *EPA,* 499 F.2d 289 (5th Cir. 1974); *Delaware Citizens for Clean Air* v. *Administrator,* 480 F.2d 972 (3d Cir. 1975). The question of whether a given implementation plan for a given region will, in fact, achieve ambient standards involves a host of complex assumptions and predictions, including such matters as the reliability of monitoring data, the appropriate use of diffusion methodologies, the adequacy of state resources to carry out a plan, and the like. The Congress could not dictate the resolution of such matters in advance. The legal discretion implicit in the resolution of technical issues is also apparent in the implementation of the statutory controls on mobile sources. See, e.g., *NRDC* v. *EPA,* 359 F.Supp. 1728 (D.D.C. 1972), sustaining the EPA's action in revising upward, on the basis of data adjustments, the base on which the 90 percent reduction requirement would be calculated. Moreover, in the grant by EPA of extensions of the auto emission deadlines, Congress's apparent effort to tightly constrain the administrator's freedom of choice was substantially undercut by a judicial interpretation that enlarged the effective range of agency choice. See *International Harvester Co.* v. *Ruckelshaus,* 478 F.2d (D.C. Cir. 1973), discussed infra.

72. See Louis L. Jaffe, *Judicial Control of Administrative Action* (Boston: Little,

Brown, 1965), pp. 546–594; R. Stewart, "The Reformation of American Administrative Law," *Harv. Law Rev.* 88 (1975): 1667, 1676–1680.

73. See 5 U.S.C. secs. 554–557. Even where relevant statutes do not provide for a hearing, courts may find that one is required by constitutional due process. See, e.g., *Wong Yang Sung* v. *McGrath,* 339 U.S. 33 (1950).

Trial-type hearings are occasionally required by statute in cases of rule making in which the agency adopts regulations applicable to a whole class of private actors, against whom such regulations may eventually be enforced in individual actions. See R. Hamilton, "Procedures for the Adoption of Rules of General Applicability: The Need for Procedural Innovation in Rulemaking," *Cal. Law Rev.* 60 (1972): 1276.

74. See 5 U.S.C. sec. 553.

75. See Hamilton, op. cit. (note 73), discussing this phenomenon in the context of other forms of economic regulation.

76. See infra (note 84).

77. 462 F.2d 846 (D.C. Cir. 1972).

78. 38 Fed. Reg. 25678 (1973).

79. *International Harvester Co.* v. *Ruckelshaus,* 478 F.2d 615 (D.C. Cir. 1973).

80. E.g., *Texas* v. *EPA,* 6 ERC 1897 (5th Cir. 1974); *NRDC* v. *EPA,* 5 ERC 1879 (1st Cir. 1973).

81. E.g., *Portland Cement Association* v. *Ruckelshaus,* 486 F.2d 375 (D.C. Cir. 1973).

82. See W. Pedersen, "Formal Records and Informal Rulemaking," *Yale Law J.* 85 (1975): 38.

83. In order to obviate the inevitable delays resulting from Freedom of Information Act litigation and in order to avoid the charge of suppressing unfavorable evidence, government lawyers now customarily submit to the court unfavorable as well as favorable documentary evidence considered by the agency in the course of its decision. Ibid.

84. William F. Pedersen in his article "Formal Records and Informal Rulemaking," *Yale Law J.* 85 (1975): 38–88, chronicles the development of EPA "paper-hearing" procedures under the impetus of judicial decisions. He concludes that paper-hearing procedures, combined with the careful judicial review of the record of agency decision that they facilitate, materially improve agency decision making, although he makes a number of suggestions for improvement of paper-hearing procedures in order to develop a more complete record. In questioning the evidence to support this essay's conclusion that the development of paper-hearing procedures at EPA has been an "outstanding success," Sax characterizes Pedersen as a "scholar," implying that his views are merely those of an academic. In fact, Pedersen (who does not hold an academic appointment) was a practicing lawyer in EPA's Office of General Counsel for several years, during which the paper-hearing procedures evolved, and thus is highly qualified to gauge their effect.

In addition, interviews conducted by the Committee on Environmental Decisionmaking (of which the author was a member) disclosed widespread agreement on the desirability of paper-hearing procedures (copies of memoranda summarizing these interviews are on file with the author).

A former General Counsel of EPA concluded that a "trend, prompted by the courts, toward more elaboration of reasons" was "probably a good thing" and stated that he would like to see the procedural provisions in EPA's statutes rewritten to provide for greater use of paper-hearing procedures. One EPA lawyer found that court remands for better-reasoned decisions were "helpful during the Agency's early years," and another stated that "court decisions mandating a fuller articulation of reasons have led to better decision making." Still another EPA lawyer found that the "courts' role has been to some extent constructive, in forcing care in explanation on the part of EPA personnel" and indicated that paper-hearing procedures were superior to trial-type formalities, which "simply give lawyers an opportunity to 'brutalize' technical people and delay the entire process." Coming as it does from EPA advocates who were on the losing end of court decisions remanding EPA decisions because of procedural inadequacies, these remarks add up to a substantiated endorsement of the court's actions. Some of the EPA lawyers expressed concern that paper-hearing procedures invite unduly stringent judicial review and encourage an excessive number of wasteful remands. However, as indicated below, these dangers have not yet materialized.

A leading industry lawyer who concurred in the EPA officials' basically positive assessment, found that a "regularized [paper-hearing record] and a full explanation of the basis for a decision are vital to ensuring fairness." A lawyer for environmental groups was more cautious, in part because he believed it important to preserve the right to cross-examine experts; he concluded, however, that the "court-imposed procedures may have a net beneficial effect."

Based on his own study of EPA decisions during his service over the past year on the Committee on Environmental Decisionmaking, the author is firmly persuaded that the judicial development of paper-hearing procedures has had a substantial beneficial impact on EPA decision making. Because the quality of agency decision making is in some measure subjective, the reader is invited to make his own assessment by examining the adequacy of reasons and supporting evidence in early EPA decisions, such as those involved in *Kennecott Copper* v. *EPA*, 462 F.2d 846 (D.C. Cir. 1972), or *South Terminal* v. *EPA*, 504 F.2d 646 (1st Cir. 1973), compared with those in later decisions, such as *Ethyl Corp.* v. *EPA*, 8 ERC 1785 (D.C. Cir. 1976).

Far more controversial than paper-hearing requirements are occasional court decisions that have gone beyond the requirements of detailed explanation and a paper hearing to require the EPA to grant a limited trial-type hearing on certain more or less narrowly defined issues. For example, *International Harvester* v. *Ruckelshaus*, 478 F.2d 615 (D.C. Cir. 1973), which set aside the EPA's refusal to grant an extension of the statutory deadlines for achievement of certain automotive emission limitations, relied upon inadequacies in EPA's response to manufacturer criticisms of agency methodologies for handling relevant data. In remanding, the court indicated that limited cross-examination would be an appropriate means of dealing with directly disputed technical issues. Similarly, in *Appalachian Power Co.* v. *Ruckelshaus*, 477 F.2d 495 (4th Cir. 1973), the court held that limited cross-examination must be afforded on certain technical and economic issues at some point before emission limitations in a state plan became effective. A review of the actual procedures utilized by the parties upon remand in these and similar decisions indicated that the industries challenging EPA's position did not, in the end, insist upon the use of trial-type procedures; technical issues were instead thrashed out through exchange of documents and informal meetings between technical experts. S. Williams, " 'Hybrid Rulemaking' Under the Administration Procedure Act: A Legal and Empirical Analysis," *U. Chi. Law Rev.* 42 (1975): 401. This experience suggests that cross-examination may have little to contribute to the elucidation of broadly structured technical or economic issues.

A closer procedural question, upon which the federal courts are presently divided is presented where an isolated source challenges the adoption by EPA of emission limitations that apply only to that source and the regulations' validity and reasonableness assertedly turn on isolated factual issues. Compare *Appalachian Power Co.* v. *EPA,* 477 F.2d 495 (4th Cir. 1973), and *Buckeye Power, Inc.* v. *EPA,* 481 F.2d 162 (6th Cir. 1973) with *Anaconda Co.* v. *Ruckelshaus,* 482 F.2d 1301 (10th Cir. 1973). In such circumstances, the greater safeguards of a trial-type hearing process in resolving particularized factual disputes should outweigh the concern that such procedures could lead to undue delays, a danger that could be avoided by requiring the party seeking a hearing to specify with considerable precision the issues on which a hearing is sought and the evidence that would be adduced on such issues. Such threshold requirements have been utilized successfully at the Food and Drug Administration, which has developed a paper-hearing procedure similar to that devised by the courts in the case of the EPA. See, e.g., *Upjohn Co.* v. *Finch,* 422 F.2d 944 (6th Cir. 1970).

85. See infra.

86. 478 F.2d 615 (D.C. Cir. 1973).

87. The relevant statutory provisions, section 202(a)(5)(c), specify that:

Within 60 days after receipt of the application for any such suspension, and after public hearing, the Administrator shall issue a decision granting or refusing such suspension. The Administrator shall grant such suspension only if he determines that (i) such suspension is essential to the public interest of the public health and welfare of the United States, (ii) all good faith efforts have been made to meet the standards established by this subsection, (iii) the applicant has established that effective control technology, processes, operating methods, or other alternatives are not available or have not been available for a sufficient period of time to achieve compliance prior to the effective date of such standards, and (iv) the study and investigation of the National Academy of Sciences conducted pursuant to subsection (c) and other information available to him has not indicated that technology, processes, or other alternatives are available to meet such standards.

88. 38 Fed. Reg. 10317 (1973).

89. However, it might be argued (particularly in view of subsequent legislative and administrative postponements of the original statutory deadlines for achieving 90 percent reduction) that the court acted unwisely in taking both Congress and EPA "off

the hook'' and setting a precedent in favor of postponement. If the court had narrowly adhered to the letter of the statute, might not Congress have been forced eventually to confront the inadequacies of the technology-forcing regulatory-deadline strategy of the act with respect to new-motor-vehicle emissions and to devise a superior strategy, such as the use of emission fees? Such an outcome seems most unlikely given Congress's basic adherence to technology-forcing elsewhere in the Clean Air Act and in the 1972 Federal Water Pollution Control Amendments. Had the court declined to take a measure of responsibility for the sound application of the act's new-motor-vehicle provisions, it is more likely that Congress, under the pressures of ''energy crisis'' would have adhered to the existing regulatory structure but diluted it to an even greater extent than it has been.

90. 486 F.2d 374 (D.C. Cir. 1973).

91. Section 111 of the statute authorizes EPA to require ''the degree of emission limitation applicable through the application of the best system of emission reduction which (taking into account the cost of achieving such reduction) the Administrator determines has been adequately demonstrated.''

92. *Accord, Essex Chemical Co.* v. *Ruckelshaus,* 486 F.2d 427 (D.C. Cir. 1973). In *Portland Cement,* the court, out of evident fear of imposing impossible burdens on the agency, did not require the agency to deal with the contention that the level of controls imposed on the Portland cement plant was excessive in comparison to the controls imposed on other industries with which the Portland cement industry did not compete. Its opinion stated that ''it would be unmanageable if, in reviewing the cement standards, the court should have to consider whether or not there was a mistake in the incinerator standard'' and added that such intraindustry comparisons were proper only ''in the case of industries producing substitute or alternative products.'' 486 F.2d at 389. Under this ruling, EPA is under no judicial obligation to ensure cost-effective achievement of an ambient standard for a given pollutant by allocating the burden of control among the industries that emit that pollutant in accordance with least-cost control considerations. But the EPA is not prohibited from such a course by the court's ruling, which would also insulate such a policy from criticisms by industries that were able to control more cheaply that they were bearing an excessive portion of the abatement burden.

In addition, the court stated that although the EPA must respond to industry cost-benefit analyses it need not produce its own, because of both time constraints and ''the difficulty if not impossibility, of quantifying the benefits to ambient air conditions.'' 486 F.2d at 387.

93. *South Terminal Corp.* v. *EPA,* 504 F.2d 646 (1st Cir. 1973).

94. 499 F.2d 289 (5th Cir. 1974).

95. See, e.g., *NRDC* v. *EPA,* 494 F.2d 519 (2d Cir. 1974), which construes flexibly the statutory requirement that, in granting extension of deadlines to meet ambient standards, EPA must insist on interim controls utilizing ''all reasonably available alternatives.''

96. 421 U.S. 60 (1975).

97. There are two glaring exceptions to this generalization. One is the *Sierra Club* decisions on nondegradation (discussed infra). The other is the Fifth Circuit Court of Appeals' unwise and unwarranted ruling in *NRDC* v. *EPA,* 489 F.2d 390 (1974), that tall stacks or intermittent control systems for stationary sources, which promise substantial economic savings in achieving ambient standards, are not permitted by the act except in special circumstances. Under an intermittent control system, a source operates with a comparatively low level of continuously operating controls that result in a relatively high level of emissions during favorable weather conditions, when emissions are dispersed without causing a violation of the ambient standards. When weather conditions are adverse, the source temporarily utilizes additional measures to reduce emissions, such as switching to cleaner (but more expensive) fuels or curtailing production in order to avoid violation of the ambient standards. Alternatively, tall discharge stacks are utilized to enhance dispersion capacity. Use of tall stacks or intermittent measures to reduce emissions under adverse conditions can result in substantial cost savings as compared to the use of continuously operating controls designed to secure compliance with ambient standards under the most adverse conditions.

Nonetheless, in *NRDC* v. *EPA* the court ruled that the EPA administrator could not approve a state implementation plan that provided for tall stacks or interruptible controls unless there was a showing that it would otherwise be technologically or economically impossible to meet the ambient standards. Section 110 of the statute provides that a state plan must include "emission limitations, schedules, and timetables for compliance with such limitations, and such other measures as may be necessary to insure attainment and maintenance of such primary or secondary standard." The court ruled that a program of interruptible controls was not an "emission limitation" within the meaning of section 110 of the act and therefore could not be utilized unless such measures were "necessary" in the sense of being technologically or economically infeasible.

This reasoning is unpersuasive because it distorts the import of the term "necessary" and because a program of interruptible controls that details procedures for switching to alternative fuels or curtailment of production is as much an "emission limitation" as any other operating procedure, such as continuous utilization of low-sulfur fuel. The objection that interruptible control programs are more difficult to enforce than continuous controls can be met by insisting upon conservative specifications that are enforceable and that will ensure compliance with ambient standards. The real ground for the court's decision is its evident objection to control measures that utilize the natural dispersion processes of the atmosphere rather than continuously operating cleanup technology. This objection appears to reflect a concern that the existing ambient standards are inadequate and a belief that prevention of tall stacks or interruptible controls threatens to degrade the air in other locations. But inadequacies in the existing standards should be attacked directly, either by revising the existing standards or taking additional measures (such as imposition of emission fees) to reduce total atmospheric loadings of pollutants. Nondegradation policies do not justify the court's ruling, because they would preclude installation of measures such as interruptible controls in instances in which no degradation would occur.

Deference to administrative discretion perhaps justifies the decisions in *Kennecott Copper Corp.* v. *Train,* 526 F.2d 1149 (9th Cir. 1975), and *Big Rivers Electric*

Corp. v. *EPA,* 481 F.2d 162 (6th Cir. 1975), which sustained EPA's eventual insistence on continuous controls unless economically infeasible, an insistence based, in part, on administrative difficulties in policing interruptible controls. But there was no justification in *NRDC* v. *EPA* for the court to overrule EPA's position and to impose, on its own responsibility, a requirement of continuous controls when the statute does not clearly preclude use of more cost-effective measures. The court had neither the competence nor the authority to dictate such a conclusion.

98. 96 S.Ct. 2518 (1976).

99. 96 S.Ct. 2518, 2531–32 (1976) (concurring opinion of Mr. Justice Powell, joined by Chief Justice Berger).

Courts have also set aside EPA approval or promulgation of state implementation plans that plainly did not meet the requirements of the act, e.g., *NRDC* v. *EPA,* 478 F.2d 875 (1st Cir. 1973).

100. See *Ethyl Corp.* v. *EPA,* 8 E.R.C. 1785, 1809–11 (D.C. Cir. 1976); id. at 1835–7 (Bazelon, C. J. concurring); id. at 1837–8 (Leventhal, J. concurring); id. at 1839 (Wilkey, J. dissenting); *National Nutritional Foods Ass'n* v. *Weinberger,* 512 F.2d 688 (2d Cir. 1975); id. at 705 (Lombard, J. concurring).

101. See H. Leventhal, "Environmental Decisionmaking and the Role of the Courts," *U. Pa. Law Rev.* 122 (1974): 509.

102. Compare *International Harvester Corp.* v. *Ruckelshaus,* 478 F.2d 615 (D.C. Cir. 1973) with *Delaware Citizens* v. *Administrator,* 5 ERC 1583 (3d Cir. 1973).

In deciding the merits themselves, the judges would bear the salutary and unambiguous discipline of ultimate responsibility, whereas, in reviewing the agency, the judges are put in the ambiguous position of testing the assumptions and predelictions of some other decision maker who has the ultimate burden.

103. See, e.g., *Ethyl Corp.* v. *EPA,* 8 ERC 1785 (D.C. Cir. 1976), where Judge Wright for the majority explored in detail the evidence concerning the health effects of airborne lead, all the while declaiming the inappropriateness of so doing, in order to sustain an EPA decision removing lead from gasoline against the detailed arguments of dissenters that the agency's action lacked evidentiary foundation.

104. See *International Harvester Corp.* v. *Ruckelshaus,* 478 F.2d 615, 650 (Bazelon, J. concurring); *Ethyl Corp.* v. *EPA,* 8 E.R.C. 1785, 1837–39 (D.C. Cir. 1976) (Bazelon, J. concurring). As suggested by Judge Friendly ("Some Kind of Hearing," *U. Pa. Law Rev.* 123 (1975): 1267, 1311–1315), emphasis on procedural safeguards may differ little in practical implication from emphasis on substantive review of discretion, because those judges who emphasize procedural safeguards will tend to insist on greater procedural safeguards in those cases where the substantive issues were not adequately ventilated or explained by the agency.

105. See J. Sax, "The (Unhappy) Truth About NEPA," *Okla. Law Rev.* 26 (1973): 239.

106. For discussion of this problem and proposed remedies, see H. Leventhal, op. cit. (note 101).

107. Concern over such possibilities was expressed by some of the EPA lawyers interviewed by the National Academy of Science Committee on Environmental Decisionmaking (memorandum on file with the author). Although generally approving of judicial development of paper-hearing procedures and detailed judicial review, see supra (note 84), a former general counsel saw "some danger that remands will almost become routine for the courts." Another EPA lawyer saw a "potential danger" that multiplying the issues challengers could raise in court could lead to undue delays and force agency compromises on substantive policies.

108. *Ethyl Corp.* v. *EPA,* 8 E.R.C. 1785, 1805, 1826 (D.C. Cir. 1976). The court itself stated, "There is no inconsistency between the deferential standard of review and the requirement that the reviewing court involve itself in even the most complex technical matters." 8 E.R.C. at 1811.

Other recent decisions sustaining agency policy judgments after detailed judicial review include *NRDC* v. *EPA,* 8 E.R.C. 1913 (5th Cir. 1976) (upholding EPA grant of emission control credit for pollution dispersion attributable to tall stacks built or under construction at the time of a prior decision by the court restricting their use as a pollution control device); *National Asphalt Pavement Ass'n* v. *Train,* 9 ERC 1109 (D.C. Cir. 1976) (sustaining EPA's decision that asphalt cement plants are "significant contributors" to air pollution and its issuance of new source performance standards for such plants).

However, the earlier decision *NRDC* v. *EPA,* 489 F.2d 390 (1974), discussed supra (note 97), does represent improper judicial intrusion into policy questions properly reserved to EPA.

109. For discussion of the favorable impact of judicial review on EPA decision making, see W. F. Pedersen, "Formal Records and Informal Rulemaking," *Yale Law J.* 85:38–89 (1975) and the summaries of interviews reviewed supra (note 84). It is the author's own view, based on more than a year of service on the National Academy of Science Committee on Environmental Decisionmaking, which studied decision making at EPA, that judicial review has been an important element in the improvement of EPA decisions over the past five years.

Sax's request, in his commentary on this essay, for more solid evidence is a difficult one to satisfy. Unlike scientific issues, questions of the quality of EPA decision making and the court's impact on that quality cannot be tested and resolved by replicable empirical experiments. The author does suggest, however, that the reader review the evidentiary record and the agency and court decisions in earlier cases such as *Kennecott Copper Co.* v. *EPA,* 462 F.2d 846 (D.C. Cir. 1972), or *South Terminal Corp.* v. *EPA,* 504 F.2d 646 (1st Cir. 1973) with that in later cases such as *Ethyl Corp.* v. *EPA,* 8 ERC 1785 (D.C. Cir. 1976) and make his or her own assessments.

110. Ibid. at 59–60 (footnotes omitted).

111. 344 F.Supp. 253 (D.D.C. 1972).

112. 4 ERC 1205 (D.C. Cir. 1972), aff'd by an equally divided court sub nom. *Fri* v. *Sierra Club,* 412 U.S. 541 (1973).

113. *Sierra Club* v. *EPA,* 9 ERC 1129 (D.C. Cir. 1976).

114. *Arizona* v. *EPA,* 8 ERC 1238 (9th Cir. 1975); *Brown* v. *EPA,* 521 F.2d 827 (9th Cir. 1975), cert. granted, 44 U.S.L.W. 3682 (1976); *Maryland* v. *EPA,* 530 F.2d 215 (4th Cir. 1975), cert. granted, 44 U.S.L.W. 3682 (1976); *District of Columbia* v. *Train,* 521 F.2d 971 (D.C. Cir. 1975), cert. granted, 44 U.S.L.W. 3682 (1976). These decisions were relied upon in *Friends of the Earth* v. *Carey,* 9 ERC 1007 (S.D.N.Y. 1976), which held that the Clean Air Act does not authorize the maintenance of citizen suits against state and local officials for failure to enforce state implementation plans, regardless of whether the plans had been prepared by the state or EPA.

115. *Pennsylvania* v. *EPA,* 500 F.2d 246 (3d Cir. 1974).

116. See op. cit. (note 114).

117. The pollution controls in the Clean Air Act are founded upon the commerce clause. See, e.g., *Pennsylvania* v. *EPA,* 500 F.2d 246 (3d Cir. 1974).

118. In particular, a nondegradation policy would disadvantage presently "clean" sources in competition for new developments with sources that are just now in compliance with the federal ambient standards, if we assume uniformity in control technologies. For example, assume that the federal ambient standard for a given pollutant is 100 $\mu g/m^3$, that region A's ambient air quality is presently 100, and region B's is 50. Assume further that with the adoption of reasonably available technology, it would be possible to halve existing emissions. Region A can achieve an ambient level of 50 and region B, 25. In competing for new growth, region A has an increment of 50 to offer, whereas region B can offer only 25. Region A can accommodate twice the total industrial development of region B.

119. See, e.g., H. Wechsler, "The Political Safeguards of Federalism," *Columbia Law Review* 54:543; J. Ely, "The Irrepressible Myth of Erie," *Harvard Law Rev.* 87:693 (1975). But see *National League of Cities* v. *Usery,* 96 S.Ct. 2465 (1976), and the decisions cited in note 114, which imply constitutional limits on the exercise of commerce power to require state officials to enforce federal controls on mobile source pollution.

120. The district court in *Sierra Club* relied on four justifications for its conclusion that the 1970 amendments compelled a policy of nondegradation.

First, the court relied upon the act's statement of basic purposes, which included a purpose to "protect and enhance the quality of the Nation's air resources" (section 101(6)). Although the court read the "protect and enhance" language as mandating a policy of nondegradation, it is difficult to construe the general statement of purposes as anything more than hortatory, particularly in view of the operative provisions of the statute. Congressional provision (in section 109 of the act) for geographically uniform federal ambient standards seems plainly inconsistent with a nondegradation requirement that would dictate federally enforced, geographically *varying* ambient standards of greater stringency in those states whose existing air quality is better than that required by the federal standards applicable to other states. Moreover, section 110 of the act lists in detail the requirements that a state implementation plan must meet in order to secure EPA approval; among other matters, the section requires that an implementation plan ensure achievement of the uniform federal ambient standards within relevant time deadlines, but there is no mention of nondegradation. Finally, section 116 provides that nothing in the act (with enumerated exceptions not relevant here) should preclude or deny the right of any state to adopt "any standard or limitation respecting emissions of air pollutants" that is more stringent than that required in a federally approved or promulgated implementation plan and so rather plainly implies that state pursuit of air quality better than that provided under the federal ambient standards is entirely optional. The plain logic that emerges from a consideration of the entire statute is that states must impose emission limitations adequate to secure and maintain the uniform federal ambient standards but need not provide for higher levels of air quality unless they choose to do so.

Second, the *Sierra Club* decision relied upon the fact that, under the Air Quality Act of 1967, EPA's predecessor agency had adopted regulations providing that any air quality standards that "would result in significant deterioration of air quality in any substantial portion of an air quality region" would be unacceptable. In addition, administration witnesses in hearings leading to the 1970 amendments testified that no change in the law was contemplated in this respect. However, the 1967 act provided for state designation of ambient standards subject to federal approval—a scheme that contemplated geographical variations in ambient standards. In addition, the administration proposals on which hearings were held in 1970 proposed uniform national emission limitations that would yield varying ambient standards and curb degradation. But these respective strategies were rejected in the Clean Air Amendments adopted in 1970, which opted instead for a basic strategy of uniform federal ambient standards. In view of the basic shift in strategy, the prior regulations and statements relied upon by the court are of little relevance.

The third consideration on which the *Sierra Club* court rested its decision was a passage in the Senate report accompanying the bill that became the Clean Air Amendments stating that "in areas where current air pollution levels are already equal to or better than the air quality goals, the Secretary shall not approve any implementation plan which does not provide, to the maximum extent practicable, for the continued maintenance of such ambient air quality." S. Rep. No. 1196, 91st Cong., 2d Sess., p. 2 (1970). But this language largely reflects the assumption that the technology-based federal emission limitations for *new* stationary sources established under section 111 would tightly control additional pollutant sources and thus provide "to the maximum extent practicable, for the continued maintenance" of existing air quality. Subsequent experience has eroded the factual underpinnings of this assumption; control technologies, in many cases, are not adequate to prevent new development from substantially degrading air quality, particularly in pristine areas. The excessive optimism in the Senate report regarding the protections that would be afforded by technology-based emission limitations for new sources does not, however, warrant reading into the statute a flat nondegradation requirement for state implementation plans if the explicit logic and other provisions of the statute exclude any such requirement.

Fourth and finally, the court declined to accord weight to EPA's view that the statute did not mandate a nondegradation policy, on the ground that its position was inconsistent with that espoused by the administration in the hearings leading to adoption of the 1970 amendments (a point already dealt with) and because of asserted inconsistency in regulations promulgated by the EPA since the enactment of the 1970 amendments. One such regulation provided that the promulgation of federal standards "shall not be considered in any matter to allow significant deterioration of existing air quality in any portion of any state," yet a regulation dealing specifically with approval of state implementation plans provided that they would be approved so long as they prevented ambient levels from exceeding the federal standards. However, when the background of these regulations is examined, any apparent inconsistency disappears. The first regulation was adopted in response to arguments by industry to states with ambient standards more stringent than the federal standards that such standards represented an expert, considered federal judgment as to the appropriate levels of environmental quality and that states with more stringent standards should therefore abandon them and adopt the more lenient federal standards. The first regulation attempts to state (somewhat elliptically) EPA's position that states should be entirely free to exercise their own judgment in favor of ambient standards more stringent than those adopted by EPA. At the same time, the second regulation makes clear that states are not required to opt for higher levels of environmental quality.

121. See, L. B. Lave and E. P. Seskin, "Air Pollution and Human Health," in R. Dorfman and N. Dorfman, eds., *Economics and the Environment* (New York: W. W. Norton, 1972).

122. See Clean Air Act, section 116.

123. The conventional arguments for federal preemption of state regulatory autonomy center around an asserted need for federal uniformity. See D. Currie, "Motor Vehicle Air Pollution Control: State Authority and Federal Preemption," *Mich. Law Rev.* 68 (1970): 1083. But this justification is wholly inapplicable here, where the effect of federal preemption, through a nondegradation policy, of decisions by "clean" states to opt for more development and lower air quality is to dictate nonuniformity.

124. Cf. *Crandall* v. *Nevada,* 6 Wall. 35 (1868). The state may recoup some indirect benefits through visitor expenditures, but even these may be limited by federal provision of lodging and concessions in federally owned national parks, monuments and forests—federal enclaves whose value depends, in part, on the state's refraining from industrial development that would generate air pollution.

125. The argument for compensating a state forced to adopt a nondegradation policy would be undercut if we analogized such a state to a polluting industrial source. Compensation is normally not made to private firms that are required to maintain a higher level of air quality than they would voluntarily choose, at least when the total social benefits from avoiding increased pollution outweigh the costs to the polluter of providing a higher level of environmental quality. Why should a state be treated any differently? An answer may lie in received political practices and expectations. Many legislative programs are explicitly designed to limit administrative discretion in order to ensure geographical distribution of benefits or burdens on some consistent, evenhanded basis, such as population, determined by the Congress in advance. See Borchardt, "Congressional Use of Administrative Organization and Procedure for Policy Making Purposes: Six Case Studies and Some Conclusions," *Geo. Wash. Law Rev.* 30 (1962): 429. Indeed, the strategy of the Clean Air Act, in providing

uniform ambient standards and uniform emission limitations for new sources, seems to reflect precisely this effort to accommodate conflicting state interests—an effort that militates against a court's subsequent alteration of the balance struck to the disadvantage of a significant group of the parties to it. Another justification for treating states differently than industrial sources is that a state's labor and capital is likely to be less mobile than a firm's and thus less able to avoid the potentially disruptive effects of controls.

126. E.g., *Dandrige* v. *Williams,* 397 U.S. 497 (1970) (welfare assistance payments); *San Antonio Independent School Dist.* v. *Rodrigues,* 411 U.S. 1 (1973) (education).

127. Compare R. Stewart, Foreword, "Lawyers and the Legislative Process," *Harvard Jl. Legis.* 10 (1973): 151, 162–164.

128. See M. Roberts and R. Stewart, Book Review, *Harvard Law Rev.* 88 (1975): 1644.

129. For example, Congress might be unwilling to make explicit compensation arrangements with disadvantaged states for fear of setting a political precedent for compensation whenever regulatory controls had a geographically differential burden.

130. In his *Economic Analysis of Law* (Boston: Little, Brown 1972), Posner argues that judge-made law has been directed primarily at efficient resource allocation, whereas legislative statutes dealing with economic regulation have been directed primarily at redistribution. However, Posner's analysis of judge-made rules occurs primarily in a context in which modifications of liability rules would permit a more efficient working of the market. This analysis has limited applicability to problems such as pollution, which reflect a serious market failure that is beyond correction by judicial liability rules for reasons noted supra. In such cases, it is necessary to rely upon governmental mechanisms to estimate the value to the society of, e.g., clean air. Such valuation presumptively should be undertaken by officials that are politically more accountable to the electorate than the judges.

131. As indicated previously, a policy of environmental diversity might be economically efficient. See also J. Dales, *Pollution, Property, and Prices* (1968), which argues that differences in preferences could mean that a policy of "separate facilities" (areas of low pollution and low economic development and areas of high pollution and high development) would be economically efficient. But the quasi-constitutional argument for environmental diversity developed in the text would claim precedence even if such diversity were not efficient.

132. See, e.g., C. Meyers, op. cit. (note 38); Jonathan Swift, *The Other Eden* (1974).

133. See N. Fabricant, "Economic Growth and the Problem of Environmental Pollution," in K. Boulding, ed., *Economics of Pollution* (1971), pp. 139, 148–149.

134. This reading of the First Amendment assumes a measure of reflective self-development that is profoundly at odds with the economist's (admittedly useful) stereotype of man as the rational maximizer of exogenously determined desires. The quotation from Frank Knight at the beginning of this essay reflects a minority view in

the profession. The following description is more representative: "The neighborhood effect of economic institutions in shaping what we in our culture-bound fashion call human nature is opaque to Pareto-optimality because a man is not likely to be prepared to pay for being made other than he is the amount that it would be worth to him to be a different man, a worth that he could recognize only if he *were* a different man. The pig will not want to pay anything to be made into a Socrates." Alexander, "The Basis of Value Judgments in Economics," in S. Hook, ed., *Human Values and Economic Policy* (N.Y. Univ. Press, 1967), p. 110.

135. Judicial recognition of the value of diversity is found in Fourteenth Amendment due process decisions restraining state efforts to curtail cultural diversity. For example, in *Pierce* v. *Society of Sisters* the Supreme Court invalidated a state statute prohibiting attendance at private schools on the ground that "the fundamental theory of liberty upon which all governments in this union repose excludes any general power of the state to standardize its children," 268 U.S. 510, 535 (1925). See also *Meyer* v. *Nebraska,* 262 U.S. 390 (1923) (prohibition against teaching German below eighth grade); *Board of Education* v. *Barnette,* 319 U.S. 624 (194) (compulsory flag salute); *Wisconsin* v. *Yoder,* 406 U.S. 205 (1972) (compulsory school attendance law as applied to Amish children). In the federal-government context, see *Kent* v. *Dulles,* 357 U.S. 116 (1958) (requiring clear statement by Congress to curtail the right to travel because of its relation to education and self-development).

136. See, e.g., J. Swift, *The Other Eden* (1974); W. Leiss, *The Domination of Nature* (1972).

It may be argued that treatment of environmental amenities as a "product" or "collective good" inevitably lends a commercial and purely instrumental quality to our attitudes toward nature and, in the end, defeats our effort to escape the curious inversion of "artificial environment[s] . . . which are the new harvesting ground for consumers, who stalk the [machine-made products] as the nomad Senaca prowl the forest in quest of natural products." P. Wagner, *The Human Use of the Earth,* 226 (1960). Compare L. Tribe, "Ways Not to Think About Plastic Trees: New Foundations for Environmental Law," *Yale Law J.* 86 (1974): 1315.

137. M. Sagoff, "On Preserving the Natural Environment," *Yale Law J.* 84 (1974): 205.

138. L. Jaffe, "Ecological Goals and the Ways and Means of Achieving Them," *W. Va. Law Rev.* 75 (1972): 1, 2 (quoting Nietzche).

139. J. S. Mill, "Centralization," *Edinburgh Review* 95 (1862): 323, 358.

140. However much we may agree with Mill that "men do not come into the world to fulfill one single end, and there is no single end which if fulfilled even in the most complete manner would make them happy" [J. Mill, "Earlier Letters" 36 (letter to d'Eichthal, 8 October 1829), quoted in G. Himmelfarb, *On Liberty and Liberalism* 337 (1975)], elevation of such sentiments into an intrinsic good would contravene the principles of critical individualism and utilitarianism that we have already surveyed.

141. Implicitly this view may deny the second aspect of the First Amendment already discussed, in that a utilitarian justification for diversity *could* be understood to imply that, over time, collective experience would lead to definitive identification of

those activities that offered humans the greatest satisfaction or at least such an ideal could be approached asymptotically.

142. Wilhelm von Humboldt, *The Spheres and Duties of Government* 12 (J. Coulthard trans. 1954). See also John Rawls, *A Theory of Justice* (Cambridge, Mass.: Harvard University Press, 1971), section 79.

143. The problems of intertemporal change in preferences in both individual and intergenerational contexts are explored in Jerome Rothenberg, *An Approach to the Welfare Analysis of Intertemporal Resource Allocation* (Athens: Center of Planning and Economic Research, 1967). See also L. Tribe, op. cit. (note 4), pp. 652–657.

144. Alexander Bickel, *The Least Dangerous Branch* (1962), p. 25. Compare John Passmore, *Man's Responsibility for Nature* (New York: Charles Scribner's Sons, 1974), pp. 96–97. Speculates that liberal democratic political arrangements are geared to marginal accommodation of competing pressure groups and are therefore unlikely to give adequate weight to environmental values of long-run importance.

145. But cf. H. Monaghan, Foreword, "Constitutional Common Law," *Harvard Law Rev*. 89 (1975): 1.

146. But cf. ibid.

147. One might argue that judicial implementation of an institutional right to environmental diversity in the context of a mixed economy is simply the modern counterpart to judicial development of common-law liability rules with respect to environmental quality in a market economy. Besides the fact that the resources potentially affected would be much greater today, the remedial problems outlined in the text serve to distinguish the two cases.

148. Policies designed to promote cultural diversity also frequently have regressive distributional impacts. See Owen, "Diversity and Television" (White House Office of Telecommunications Policy staff research paper, 1972). Questions the rationale for government regulation to provide "quality" television programming, which attracts comparatively few viewers.

149. Compare A. Wildavsky, "Aesthetic Power or the Triumph of the Sensitive Minority over the Vulgar Mass," in Paulsen and Denhardt, eds., *Pollution and Public Policy* (Dodd Mead, 1973), p. 37.

150. Arguments for a constitutional right to environmental quality have been repeatedly rejected by federal courts. See, e.g., *Tanner* v. *Armco Steel Co.,* 340 F.Supp. 532 (S.D. Tex. 1972).

151. Such situations are analogous to instances of interstate conflict that neither state nor national political processes seem functionally competent to resolve and in which the Court has intervened through the creation of a federal common law. See, e.g., *Hinderlider* v. *LaPlata River Co.,* 304 U.S. 92 (1938); *Texas* v. *New Jersey,* 379 U.S. 674 (1965). The Court has developed federal "common law" in the specific context of interstate pollution. *Illinois* v. *City of Milwaukee,* 406 U.S. 91 (1972). For examination of the notion of a federal common law in a constitutional context, see H. Monaghan, op. cit. (note 145). The notion of quasi-constitutional adjudication seems

implicit in F. Michelman, "In Pursuit of Constitutional Welfare Rights: One View of Rawls' Theory of Justice," *U. Pa. Law Rev.* 121 (1973): 962, 1003–1019.

152. H. Monaghan, Foreword, "Constitutional Common Law," *Harvard Law Rev.* 89 (1975): 1.

Not only did the court's approach permit subsequent revision of its position by Congress, but by refusing to specify in its opinion what might constitute "significant deterioration," the court afforded considerable leeway to the EPA in implementing the basic principle.

153. See *Sierra Club* v. *EPA,* 9 ERC 1129, 1132-4 (D.C. Cir. 1976).

154. See supra.

155. Op. cit. (note 114).

156. See, e.g., *District of Columbia* v. *Train,* 8 ERC 1289, 1296 (D.C. Cir. 1975). Section 113 does authorize enforcement against "any person," and the general definitional section in the statute, section 302, states that "person" includes "an individual, corporation, partnership, association, State, municipality, and political subdivision of a State." However, the majority decisions failed to accord much credit to this generalized definition in view of the fact that section 113, the specific enforcement provision directly relevant, did in contexts other than general enforcement authority differentiate between state governmental units and "persons." See, e.g., *Brown* v. *EPA,* 521 F.2d 827, 834 (9th Cir. 1975), cert. granted, 44 U.S.L. W. 3682 (1976): "Congress would not have intended to take such a step [authorizing enforcement against state governments] in the light of the delicacy with which federal-state relations have always been treated . . . in this obscure manner."

157. *Brown* v. *EPA,* 521 F.2d 827, 837–839 (9th Cir. 1975), cert. granted, 44 U.S.L.W. 3682 (1976).

158. Ibid., at 840.

159. See *District of Columbia* v. *Train,* 521 F.2d 971, 993–994 (D.C. Cir. 1975), cert. granted, 44 U.S.L.W. 3682 (1976), where the Court read dicta in *Fry* v. *United States,* 421 U.S. 542, 547–8 (1975), as support for the view that the Tenth Amendment expresses interests in state autonomy that must be weighed against the federal commerce power.

160. See, e.g., *Pennsylvania* v. *EPA,* 500 F.2d 246, 256–58 (3d Cir. 1974).

161. Id. at 258–262.

162. Existing court decisions on the application of the commerce power to state government activities (see *Maryland* v. *Wirtz,* 392 U.S. 183 (1968); *Parden* v. *Terminal Rwy,* 377 U.S. 184 1964)) rather plainly indicate that states could be required to control pollution in state-owned and operated facilities, even if the pollution were caused in the first instance by private individuals. For example, if the government were to adopt an ambient standard for cigarette smoke, can anyone doubt that a state government could constitutionally be required to enforce such a standard in state-owned buildings, even though the smoke was generated in part by persons who were not government employees? But it seems impossible to distinguish the hypothetical

case from that of a state-constructed and regulated highway system that is utilized by private motor vehicles. In either case, the state has voluntarily undertaken activities that foreseeably result in substantial pollution and, given the requisite effect on interstate commerce, the same principles that justify federal regulation of a state-owned railroad should justify federal requirements that the state take appropriate steps in the management of its public facilities to prevent excessive pollution. There is, in short, no reason in this context not to treat the state as the polluter.

The position here asserted may be at odds with the broader implications of *National League of Cities* v. *Usery,* 96 S.Ct. 2465 (1976), which held that the minimum wage and maximum hours provisions of the federal Fair Labor Standards Act could not constitutionally be applied to states and their political subdivisions under the authority of the commerce power because their effect would invade state sovereignty and therefore transgress limitations on federal power inherent in the federal system. However, Justice Blackmun, a member of the five-man majority in *Usery,* in a concurring opinion explicitly distinguished the case from those involving other issues and asserted that the *Usery* decision "does not outlaw federal power in areas such as environmental protection, where the federal interest is demonstrably greater," 96 S.Ct. at 2476.

163. Op. cit. (note 119).

164. The transportation control cases seem to afford a more favorable terrain than *Sierra Club* for an assertive judicial role to protect interests in diversity, in that the judges would simply be sustaining a position already advanced by the federal EPA, whereas the nondegradation decision was contrary to the position both of the EPA and of some states.

165. For example, the court in *District of Columbia* v. *Train,* required the affected jurisdictions to purchase 475 additional buses for a regional bus fleet.

166. See, e.g., *Maryland* v. *EPA,* 530 F.2d 215, 228 (4th Cir. 1975).

167. The tendency toward centralizing uniformity is seen most clearly in the implementation of the Federal Water Pollution Control Amendments. For example, EPA funding of conventional collection sewers and treatment outfalls in eastern Long Island will lead to depletion of the area's ground-water supply, salt-water intrusion, and harm to shellfishing. Such an approach also neglects alternatives (such as recycling) to source-point technological requirements—alternatives that are both more effective over the long run in meeting specific goals and less environmentally disruptive. The EPA's actions are currently being challenged in court. *EDF* v. *Train,* No. 74-C-1698 (E.D.N.Y.).

168. *Parker* v. *United States,* 448 F.2d 793 (10th Cir. 1971).

169. *National Wildlife Fed'n* v. *Coleman,* 9 ERC 1465 (5th Cir. 1976).

170. *Issac Walton League* v. *Butz,* 522 F.2d 945 (4th Cir. 1975); *Zieske* v. *Butz,* 9 ERC 1061 (D. Alas. 1976); *Sierra Club* v. *Dept. of Interior,* 376 F.Supp. 90 (N.D. Cal. 1974).

171. *Committee for Humane Legislation* v. *Richardson,* 9 ERC 1327 (D.C. Cir. 1976).

172. As pointed out by Joseph Sax in his comment on this essay, the principle of diversity could be utilized to justify development controls in suburbs and towns in ways that would have the effect of excluding new entrants that might threaten existing norms. Courts have acknowledged a legitimate interest in such protection in the context of decisions sustaining such development controls against constitutional attack. See *Village of Belle Terre* v. *Booras,* 416 U.S. 7 (1974); *Construction Industry Ass'n* v. *Petaluma,* 522 F.2d 897 (9th Cir. 1975), cert. denied, 96 S.Ct. 1148 (1976). The decisions do raise the difficult questions of whether an ideal of cultural or income diversity is to be realized *within* each community or *between* separate communities that each assume a distinct, homogenous character. Such issues suggest that the notion of economic and cultural diversity is likely to prove even more intractable than the principle of diversity in natural environments that is the foundation of the discussion in this essay.

173. See the discussion of transportation control plans and indirect source review, supra.

174. Given the absence of past experience with emission fees, a legal analysis of their validity would have to begin by analogizing them to governmental measures that have already received judicial consideration.

Administratively imposed civil money penalties for violations of regulatory measures provide one promising analogy. Such sanctions have long been authorized by various federal statutes, and their use has increased in recent years as federal administrative agencies have discovered their advantages over court-enforced civil or criminal sanctions in speed, cheapness, and credibility. See H. Goldschmid, *An Evaluation of the Present and Potential Use of Civil Money Penalties as a Sanction by Federal Administrative Agencies* (Report to the U.S. Administrative Conference, 1972). There are two potential constitutional obstacles to the use of such sanctions. First, if the courts determine that a given money penalty is "criminal" rather than "civil," the Constitution will require that it be enforced through prosecutions in court with all the procedural safeguards applicable to criminal trials; the judicially developed criteria for determining whether a sanction is "civil" or "criminal" are often ambiguous and confusing. Second, even if the sanction is "civil," due process may require that the person against whom a civil sanction is imposed be afforded a de novo hearing in court on the issue of liability. However, administrative imposition of emission fees would probably be sustained against either of these potential objectives. The purpose of a fee system is clearly regulatory, not punitive. Moreover, once the level of the fee were set (either by the legislature or the agency), the discretion enjoyed by the agency in imposing the fee in any given case would be minimal or nonexistent, an important factor that would mitigate against judicial imposition of additional safeguards. Judicial acceptance of administratively imposed emission fees might be facilitated by providing an upper regulatory limit on emissions from a source, violation of which would trigger traditional sanctions, and by imposing a fee on emission levels below this upper limit. Given the strong policy reasons for administratively enforced emission fees, federal courts would probably not interpose any constitutional objections to their authorization by Congress. See Frederick Anderson et al., *Environmental Changes: Economic, Technical, Legal and Political Aspects,* Chapter 5 (forthcoming). This assessment finds judicial confirmation in *Frank Irey,*

Jr., Inc. v. *Occupational Safety and Health Rev. Comm'n,* 519 F.2d 1200 (3d Cir. 1975), and *Atlas Roofing Co.* v. *Occupational Safety and Health Rev. Comm'n,* 518 F.2d 990 (5th Cir. 1975), which reject constitutional objections (based on the Sixth and Seventh Amendments) to the administrative assessment of penalties by the Occupational Health and Safety Review Commission. These decisions suggest that the position recently advanced by J. S. Abrahams and J. R. Snoden, "Separation of Powers and Administrative Crimes: A Study of Irreconcilables," *So. Ill. Law J.* (1976): 1–150, contending that there are significant constitutional limitations on administrative power to assess penalties, is unlikely to be accepted. However, there might be state law obstacles to the adoption by states of an administratively enforced charge system.

Emission fees might also be framed as a tax, but so doing would obscure their true purpose and raise the possibility that administratively imposed emission fees would offend the principle of nondelegation of legislative power. Cf. *National Cable Television Ass'n, Inc.* v. *United States,* 95 S.Ct. 1146 (1974).

175. The requirement in section 110 of the act that state implementation plans must contain "emission limitations . . . and such other measures as may be necessary to insure attainment and maintenance" of ambient standards was interpreted in *NRDC* v. *EPA,* 489 F.2d 390 (5th Cir. 1974), as requiring continuous controls and precluding the use of tall stacks and other dispersion measures unless they were "necessary" because continuous controls were technically or economically infeasible. See supra (note 97). The court's logic in equating "emission limitations" with continuous controls seems to rule out the possibility that emission fees are an "emission limitation" for purposes of state implementation plan requirements. The regulatory orientation of the Federal Water Pollution Control Amendments similarly tends to preclude or discourage use by states of emission fees and similar incentive systems to control water pollution.

As a result of imaginative state administration, Connecticut has devised a novel system of incentives for complying with air and water pollution control regulations; in effect, polluters who fail to comply with applicable regulations are charged a periodic fee that would tax away all of the economic benefit to the polluter of noncompliance. See *Connecticut Enforcement Project,* U.S. Environmental Protection Agency, I Economic Law Enforcement (Washington, D.C.: EPA, 1975). EPA has been encouraged by the apparent advantages of this alternative to traditional civil and criminal sanctions for noncompliance, which are often slow, costly to enforce, and therefore less credible. This example shows that the opportunities for state experimentation are not totally foreclosed, but it should be emphasized that the Connecticut strategy continues to rely on traditional regulatory controls to reduce pollution; the novelty lies in the sanctions for violation.

176. The requirement that alternatives to present regulatory strategies be *realistic* cannot be overstressed; the implementation of an emission fee strategy or similar incentive system involves numerous administrative and other practical problems that have not always been sufficiently considered by advocates of alternatives to regulation. See M. Roberts and R. Stewart, Book Review, *Harvard Law Rev.* 88 (1975): 1644.

177. In keeping open the possibility of legislative revision, the practice of quasi-

constitutional adjudication bears a certain affinity to Sax's proposal that the courts should invalidate agency action that seriously harms environmental quality whenever such action was not clearly authorized for the legislature. See Joseph Sax, *Defending the Environment: A Strategy for Citizen Action* (1976). It also bears a resemblance to principles of clear statement, which judges have applied to block administrative intrusions on basic liberties when the relevant statute does not explicitly authorize the particular intrusion. See, e.g., *Kent* v. *Dulles,* 357 U.S. 116 (1958). However, the Sax proposal and orthodox clear-statement principles contemplate a negative judicial stance, whereas the practice of quasi-constitutional adjudication may involve judicial affirmation of the court's view of sound policy.

178. See R. Stewart, ''The Reformation of American Administrative Law,'' *Harvard Law Rev*. 88 (1975): 1671, 1711–1759.

179. Ironically, however, it may be in the nature of quasi-constitutional adjudication, at least as it has been practiced thus far, that the court will be reluctant to explore openly the considerations that may ultimately justify its decision, for fear that candor would erode the credibility of the court's decisions as a piece of statutory interpretation. For example, the *Sierra Club* decision contains not a word about the larger considerations of environmental policy broached in this essay. This difficulty, however, should not lead the courts either to abandon the effort or to press on to an explicitly constitutional rationale for decision. What is required is greater judicial confidence, persuasiveness, and ingenuity in devising new techniques of quasi-constitutional adjudication similar to the principle of clear statement utilized in *Kent* v. *Dulles,* 357 U.S. 116 (1958).

180. The quotation is from a recent passage from Stuart Hampshire that could well substitute for the quotation that heads this essay (though it has the disadvantage of not having an economist as its author): ''The problem is not so much that we do not know how to compute social costs in any scientific way, but rather that we have not even securely learned to think of the desirable ends of social policy as being essentially plural and incompatible, and therefore always to look for a balance between them, trading off an advantage on one scale against a disadvantage on another.'' Stuart Hampshire, ''Thinking About Social Costs,'' *The New York Times,* 24 March 1976, section II, p. 35.

Comment James L. Oakes

Continuity and Change in Environmental Law

It is very pleasant for a judge to be asked to comment on a paper that commends the "outstanding success" of the courts in Clean Air Act decisions and that "affirm[s] the desirability of a substantial judicial role." It is doubly nice here because, apart from its praise of the courts, the thrust of Stewart's paper is one with which I essentially agree. Thus, my commentary will be aimed at identifying and expanding upon three related themes that are in varying degrees implicit in Stewart's work. These themes are (1) that much of what judges do in the environmental field is in fact not very different from what they have been doing for many years in related fields; (2) that, to the extent environmental law is truly new, an approach focusing solely on procedural reforms will be ineffective in capturing the essence of this newness; and (3) that the diversity of judges, in both attitudes and backgrounds, and the checks built into our political system together mean that a more substantive role for judges in environmental matters presents little cause for fear and some cause for hope for a better tomorrow.

Continuity with the Past in Environmental Law

In 1973, Chief Judge David Bazelon of the United States Court of Appeals for the District of Columbia Circuit wrote that "environmental litigation represents a 'new era' in administrative law."[1] Although there is unquestionably much that is "new" about environmental law, there are also important links to the past, links that have perhaps been overshadowed in the rush to proclaim a glamorous "new era." In terms of both substantive doctrines and the judge's role, environmental law is merely one part of the continuously evolving structure that we call the common law.

On a substantial level, it is a familiar truism—but one worth emphasizing—that environmental law is derived from a variety of sources, including not only such statutes as the Clean Air Act but also constitutions (state as well as federal) and the law of torts, property, contracts, natural resources, taxation, and international and interstate relations.[2] Each of these sources, of course, has its inherent limitations. None by itself, moreover, enables judges and lawyers to take the necessary broad view of the interrelationships in the environment, a view that is the essence of the word "ecology."[3]

The law of nuisance provides one example of both environmental law's links to the past and the limits of traditional doctrine. When smoke from a nearby factory interferes with an individual's use and enjoyment of his land, he has always been able to sue the factory in tort. But because the court may consider only the harm to the individual and because the factory may be allowed to continue polluting, despite harm to the landowner, if its use of its land is "reasonable,"[4] it is plain that private nuisance suits are a poor vehicle for resolving the question of whether harm to society as a whole from the pollution is outweighed by the benefits to society from the factory's operation.[5] Somewhat more of the necessary balancing can be done by courts in public nuisance suits, in which a public right is vindicated, but here there is a problem of open-endedness—both the harms and the benefits of pollution are so unquantifiable as to make standardless balancing an exercise in pure subjectivity—as well as the more traditional problems of the public nuisance suit's quasi-criminal nature and the requirement that the plaintiff show damages different in kind from those suffered by the general public.[6]

Because of limitations inherent in the very fact of being designed for different problems, traditional substantive doctrines provide only a start in developing environmental law. Far more important are the modes of legal analysis that lawyers and judges have carried over from their traditional legal training into the environmental field. An example with which I am most familiar involves one of the "action-forcing" cases discussed by Stewart, *Natural Resources Defense Council, Inc.* v. *Train.*[7] The district court decision he mentioned required the administrator of the Environmental Protection Agency (EPA) to list lead as a hazardous air pollutant and thereby triggered promulgation of national ambient air quality standards for lead and state implementation of these standards.[8] The decision was recently affirmed by an appellate panel of three judges, including myself.[9] A reader of our unanimous opinion would know that we were dealing with an environmental statute, the Clean Air Act, but would not have any reason to think a "new era" in the law was at hand. As we said in the opinion, "[t]he issue is one of statutory construction,"[10] and we would have taken the same basic approach with regard to any federal regulatory statute. Indeed, any judge, lawyer, or law student would have taken the same approach with regard to virtually any statute, right down to municipal dog-curbing ordinances. Seeking to construe the statute in accordance with its purpose, we looked first at the words of the act, then at the overall structure of the act, then at congressional intent expressed in the legislative history, and finally at judicial decisions on related topics. Whether the result reached was pro- or anti-environment had little or nothing to do with our decision; as Stewart points out in his discussion of the District of Columbia Circuit's decision in *NRDC* v. *EPA*,[11] "[t]he requirements of the statute were plain," so we judges "had no other choice." A very substantial amount of environmental litigation involves statutory construction of this sort, and the traditional tools of legal analysis—leavened, of course, in this area as in all others, by a greater or lesser infusion of judicial (and, it is hoped, judicious) value judgments—provide the principal means of resolving the issues.

Procedural Approaches to Environmental Problems

Despite all the similarities between environmental law and other fields of law, there is much that is decidedly different. Stewart covers many of the differences: the dynamic nature of preferences for environmental quality; the difficulties involved in measuring both those preferences (which are not expressed in dollars in the same way we express a preference for a certain brand of Scotch whiskey) and the harms from pollution; the problem of valuing the interests of living things other than man and of future generations. From the standpoint of the judge, there is the additional problem of the sheer technological complexity of environmental issues, a problem that had led to such proposals as the specialized environmental court[12] and scientific assistants to advise generalist judges.[13] I have expressed my opposition to the former proposal elsewhere.[14] As for the latter, in addition to the fact that it contradicts a normal assumption of our adversary system—that all of the evidence for each side will be presented by that side and the other party has a chance to confront the evidence against it—there is the problem that the expert may simply be wrong or his analysis incomplete, and the judge would have no way of knowing. Although the risk of a mistaken expert is one the judge takes each time he relies on the testimony of an expert witness, at least in the witness situation the other side has had a chance to expose flaws in the expert's analysis via cross-examination.

One judicial response to the problem of technological complexity is well documented by Stewart: the courts necessarily must grant the EPA and other agencies a broad range of discretion but must also require that in exercising that discretion the agencies consider all relevant information and fully explain the reasons for their decisions. An early example of this approach is my court's initial decision in the *Scenic Hudson* case,[15] which involved the Federal Power Commission's approval of a hydroelectric facility on the

Hudson River. The court set aside that approval and remanded to the commission for consideration of alternatives to the project. As Stewart has pointed out elsewhere,[16] however, the remand ultimately changed very little; the FPC, after hearings, reapproved the project with only minor modifications. The court, over my dissent, upheld the FPC.[17]

Scenic Hudson illustrates the limitations inherent in a strictly procedural role for the courts in environmental cases. A court can tell an agency to hold more hearings and consider more evidence, but it cannot, by procedural devices alone, ensure that the agency does anything more than engage in a formalistic ritual, as Stewart's paper recognizes. Of course, the ritual itself has some value, at least if the reviewing court, in an approach taken by my court recently, expresses its awareness of the possibility of a pro forma ritual and admonishes the agency that it must do something more—"at least take a 'hard look' " at the problem—if it wants to win judicial approval.[18] Thus, I do not fully share the pessimism of my fellow commentator today, Joseph Sax, who has written that "the emphasis on the redemptive quality of procedural reform is about nine parts myth and one part coconut oil."[19] On the other hand, given my experience in the *Scenic Hudson* litigation and the problems Stewart discusses with other attempts to take a strictly procedural approach, I cannot share the optimism of those who believe a very limited procedural role for the courts is sufficient. A somewhat more substantive role is necessary if judges are to meet their serious constitutional obligation to check unwarranted abuse of executive discretion.

Judicial Competence and Substantive Review

Stewart adequately documents and defends the increasing willingness of courts in environmental matters "to review in detail the bases for agency judgment but to acknowledge in the end that the judgment is the agency's." The problems with judges delving into the substance of agency policy, particularly in complex technical areas, are familiar and need only be outlined here. Litigation is not regulation; the case-by-case approach courts necessarily must take may lead them to ignore the larger picture or the difficulties involved in continuous involvement with a problem. Judges, moreover, are generalists, and they lack the supporting expert resources that the head of an administrative agency can command.[20] Finally, it is almost a cliché in any field of law that judges are not elected; consequently, in environmental law, judges lack the legitimacy and responsiveness necessary to make the difficult, ultimately political (in the best sense of the term) trade-offs between environmental quality and economic growth.[21] A corollary to the latter point is the concern, expressed by Stewart elsewhere, as it was by Thayer,

Frankfurter, and Bickel before him, that judicial custody over genuinely political issues could lead to "atrophy of political consciousness and responsibility."[22]

I believe, however, that all of these quite valid concerns about a substantive judicial role have been magnified by courts and commentators out of proportion to their importance. To take the first point mentioned above, litigation does not force judges to wear blinders, and no conscientious judge will do so. Judges are informed about the larger context within which individual decisions are made by other litigation, outside study, and the mass media, as well as by the parties to the particular case. Each side, of course, has its own view of what is important within that larger context, but the essence of judging resides precisely in this sort of informed choice. As for being unaware of the difficulties encountered in the day-to-day process of ongoing regulation, no one who has read a brief on behalf of the EPA or heard the government argue in support of the actions of the FPC could think judges unaware of the problems these agencies face. If, in our decisions, we sometimes appear less sympathetic to these problems than the agencies might desire, it is perhaps because experience suggests that some skepticism about their claims is warranted, especially if the agency involved has a dual role, one part of which is promotion of the very business it is regulating.

Far more fundamental is the criticism that judges are legal generalists who cannot intelligently analyze technical, nonlegal issues. Both assumptions underlying this criticism—that judges are generalists and that generalist review is necessarily unintelligent—should be examined. Many judges come from specialized practices or public service in a specialized field, and on the appellate bench they tend to get opinion-writing assignments in their areas of interest. In my own case, I have written a fair number of environmental and energy-related opinions, in part, perhaps, because my background includes service as special attorney to the Vermont Public Service Board in connection with energy imports from Niagara and St. Lawrence; as a lobbyist in favor of the importation of hydroelectric power from Quebec as an alternative to building a nuclear power plant; and as the person responsibile, during my term as state attorney general, for establishing the state's policy toward Vermont's only nuclear plant with regard to water and air considerations.

Concerning the second assumption, I believe that some generalist oversight of the experts' decisions promotes decision making in the public interest. When experts talk only to each other, they tend to forget for whom the decisions are being made; when they are forced to explain their decisions in terms that even a judge can understand, it is more likely that the public will also understand. Relatively minor procedural adjustments, moreover, can aid court comprehension of technical matters without sacrificing our adversarial tradition; the Fifth Circuit, for example, has experimented with informal pre- and post-argument conferences at which the judges and the parties discuss the issues in some depth.[23] The most important decisions, finally, those involving trade-offs between conflicting goals, are ultimately not technical in nature,[24] although an expert may help clarify the likely consequences of alternative choices. Arguably, these decisions should be made by a more politically responsive branch of government than the judiciary, but this allocation has nothing to do with expertise.

On the question of the judiciary's political responsiveness, it is good to remember first that, as mentioned above, judges do not rise to the bench out of a vacuum but come from a rich array of diverse backgrounds. Many judges have had some connection with public service prior to donning the robes, and a very large number, including myself, have had political experience in the legislative or executive branches, or both, of state and federal government. We do not suddenly forget whence we came, and many of us make diligent efforts to keep in touch with people from all walks of life, as well as with a certain amount of economic and scientific thinking. There are sufficient checks within the judicial system, moreover, to prevent any one view from dominating court decisions. For example, when lower courts started freely awarding attorneys' fees to public interest litigants, the Supreme Court, taking note of Dawson's law review criticisms, put the brakes on in its *Alyeska Pipeline* decision.[25] To take another example, although I sometimes share Stewart's concern about "upper-middle-class" judges "indulg[ing] a [regressive] preference for environmental quality," I have met many judges whose earlier legal careers were spent defending large industries and whose preferences for environmental quality have somehow been sublimated. The main charge against judges for the last twenty-five years has not been that they are too elitist but that they are too egalitarian. Most of us, most of the time, are, in any event, simply applying the law, as in the NRDC case I mentioned earlier.[26] When we do have subjective value choices to make, the diversity of our backgrounds and current attitudes ensures that no one side will win too often, and the intervening give-and-take is beneficial for us and, I hope, for society.

The final assurance that the judiciary will be politically responsive comes from the political process itself. The turnover in judicial personnel ensures that each president and each senator (senators dictate most federal court appointments) will have a chance to appoint those who, within certain parameters, think the way he and his supporters do. There are limits on this type of accountability, of course—President Eisenhower is reported to have rued the day he appointed Earl Warren chief justice—but, in general, the process works, as it apparently did for President Nixon, who shaped significantly the current Supreme Court. Once courts have made a decision, moreover, the political branches can reverse it, as Stewart discusses, except on constitutional grounds. An example is the current legislative expansion of attorneys' fees provisions in civil rights cases in the wake of *Alyeska Pipeline*.

Finally, of course, as was said by Finley Peter Dunne through Chicago bartender Mr. Dooley, "no matter whether th' constitution follows th' flag or not, th' supreme coort foolows th' illiction returns."[27] Given the recent election returns, it seems an appropriate note on which to close. Judges add something to environmental litigation that may ultimately be undefinable but that in no sense poses dangers to the health of the body politic. Stewart and I share a belief that the court's role has been a positive one in the past. With a little more creativity, as exemplified in *Sierra Club* v. *Ruckelshaus*,[28] which Stewart cleverly rationalizes as a "quasi-constitutional" promotion of environmental diversity, and a little less reticence, the court's role may be an even more positive one in the future.

Notes

I would like to express my appreciation to my law clerk, Phillip Spector, for his assistance in editing and footnoting this paper.

1. *International Harvester Co.* v. *Ruckelshaus,* 478 F.2d 615, 651 (D.C. Cir. 1973) (concurring opinion), quoting *Environmental Defense Fund, Inc.* v. *Ruckelshaus,* 439 F.2d 584, 597 (D.C. Cir. 1971).

2. See Atkeson, Introduction, *Federal Environmental Law* (E. Dolgin and T. Guilbert, eds., 1974), p. 5; Oakes, "Environmental Litigation: Current Developments and Suggestions for the Future," *Conn. Law Rev.* 5 (1973): 531.

3. See Atkeson, op. cit. (note 2), pp. 4–5.

4. W. Prosser, *Law of Torts* (4th ed., 1971) section 89, pp. 593, 596–602.

5. Oakes, op. cit. (note 2), p. 547. See generally Porter, "The Role of Private Nuisance Law in the Control of Air Pollution," *Ariz. Law Rev.* 10 (1968): 107; Schuck, "Air Pollution as a Private Nuisance," *Nat. Resources Law* 3 (1970): 475.

6. Oakes, op. cit. (note 2), pp. 547–549. Some of the traditional problems would be alleviated were courts to adopt the more liberal public nuisance standards proposed in *Restatement (Second) of Torts,* section 821B (Tent. Draft No. 17, 1971). See generally Bryson and Macbeth, "Public Nuisance, the Restatement (Second) of Torts, and Environmental Law," *Ecology L.Q.* 2 (1972): 241.

7. 8 Envir. Rep. Cases 1695 (S.D.N.Y.), aff'd, No. 76-6075 (2d Cir. 10 November 1976), slip op. 485.

8. See 42 U.S.C., section 1857c-3 to c-5 (1970).

9. *Natural Resources Defense Council, Inc.* v. *Train,* No. 76-6075 (2d Cir. 10 November 1976), slip op. 485.

10. Id., p. 491.

11. 475 F.2d 968 (D.C. Cir. 1973).

12. See, e.g., Whitney, "The Case for Creating a Special Environmental Court System," *Wm. & Mary Law Rev.* 14 (1972); 473.

13. See, e.g., Leventhal, "Environmental Decisionmaking and the Role of the Courts, *U. Pa. Law Rev.* 122 (1974): 509, 550.

14. Oakes, op. cit (note 2), pp. 553–555. See also Thompson, "The Role of the Courts," *Federal Environmental Law,* op. cit. (note 2), pp. 233–235.

15. *Scenic Hudson Preservation Conf.* v. *FPC,* 354 F.2d 608 (2d Cir. 1965), cert. denied, 384 U.S. 941 (1966).

16. Stewart, "The Reformation of Administrative Law," *Harvard Law Rev.* 88 (1975): 1667, 1777–1780.

17. *Scenic Hudson Preservation Conf.* v. *FPC,* 453 F.2d 463 (2d Cir. 1971), cert. denied, 407 U.S. 926 (1972). The *Scenic Hudson* litigation was continued in *Hudson River Fishermen's Ass'n* v. *FPC,* 498 F.2d 827 (2d Cir. 1974), in which the commission was ordered to reopen hearings on the proposed plant.

18. In ordering the Postal Service to prepare an environmental impact statement (EIS) on a facility it wanted very much to build, the court wrote:

Since the Postal Service is evidently committed in a very real sense to the movement of its facilities and the transfer of its employees to [the planned facility], there is danger that an after-the-fact EIS such as we are ordering could result in pro forma, automatic reconfirmation of the presently contemplated plan. Our order is intended, however, to ensure that the agency does more than give the present plan rubber-stamp approval. The agency must at least take a "hard look" at the environmental consequences and alternatives, *Kleppe* v. *Sierra Club,* ______ U.S. ______, 96 S.Ct. 2718, 49 L.Ed.2d ______ n.21 (1976); it must not act arbitrarily or capriciously in its decisionmaking process.

City of Rochester v. *United States Postal Serv.,* 541 F.2d 967, 987–79 (2d Cir. 1976).

19. Sax, "The (Unhappy) Truth About NEPA," *Okla. Law Rev.* 26 (1973): 239.

20. These resources, it should be noted, are far more extensive than those that would be available to a judge under the proposal for an expert assistant, discussed in text at notes 13 and 14 supra.

21. These considerations are more extensively discussed in Oakes, op. cit. (note 2), pp. 552–553.

22. Stewart, op. cit. (note 16), p. 1803.

23. See *State of Texas* v. *EPA,* 499 F.2d 289, 297, and n.8 (5th Cir. 1974).

24. See Stewart, op. cit. (note 16), p. 1684.

25. *Alyeska Pipeline Serv. Co.* v. *Wilderness Soc'y,* 421 U.S. 240 (1975).

26. See text at notes 7–11 supra.

27. *Mr. Dooley on Choice of Law* (E. Bander, ed., 1963), p. 52.

28. 344 F. Supp. 253 (D.D.C. 1972), aff'd, (D.C. Cir.) (memorandum opinion), aff'd, 412 U.S. 541 (1973) (4-4 per curiam).

Comment

Joseph L. Sax

The word puzzlement formed in my mind as I read Richard Stewart's paper. I am not at all sure of the impression he aspires to leave upon his readers. Like much legal writing, the paper oscillates between the descriptive and the prescriptive. As a description and classification of judicial review of air pollution cases, the paper is both unexceptionable and unexciting. As a brief in support of increased judicial review of administrative decision making, it is not particularly original. As an effort to prove the conclusions he seems to draw, it is poorly substantiated. Perhaps most important, it seems to draw conclusions about some of the most profound dilemmas in administrative law without probing them very deeply. Yet one responds to the paper with some diffidence, for, like many law review articles, Stewart's has an "on the one hand, on the other hand" quality that frequently leaves the reader uncertain whether the author is urging a solution or merely suggesting a solution that might be urged, while noting carefully all the problems that it and every other approach present.

For example, on first reading Stewart's discussion of judicial review of the merits, I thought he was claiming that the judges had found a way significantly to improve environmental decisions without turning courts into ad hoc legislatures or administrative agencies. If that were so, we would be some distance along toward the resolution of an intensely difficult legal problem. On second reading, I noted how carefully Stewart speaks of the "design" of the new standard of judicial review rather than of its outcome and how cautiously he observes that the record he has reviewed simply "*suggests* that the court exercise of such a role has enhanced the quality of ultimate decisions." Perhaps his paper will suggest that conclusion to some readers. I hunger for evidence rather than suggestion. In this respect, the article disappoints.

Where Stewart is more explicit—saying that procedural review and the development of the "paper-record" technique "has been an outstanding success"—his evidence is less than overwhelming. He cites another scholar's article and interviews conducted with participants in air pollution cases. The interview evidence Stewart has provided is considerably less decisive than his paper suggests,[1] but on the matter of the paper-record technique, it is not so much the thinness of proof that troubles me as the failure to recognize that he is engaging one of the most puzzling problems in administrative law.

The question of the best technique for administrative decision making, traditionally conceived as a choice between rule making and trial-type hearings, involves a problem of true profundity—the breadth with which one conceives an issue as to which a particular decision has to be made. Rule making has been appealing because it permits an agency to look very broadly at the problems that impinge upon a particular decision—at issues of public health, jobs, and income redistribution and at prospects for future technology—and to bring that knowledge and judgment (though necessarily vague in many respects) to bear upon the resolution of a particular issue. Certainly there is merit to encouraging this broadness of vision, just as there is the demerit that making decisions this way impedes detailed review and verification by a court. The trial-type approach to decision making has precisely the opposite merits and demerits. The more one moves toward the trial model, the more a reviewing court can examine information and suppositions rigorously but at the expense of breadth. One can only look deeply at a very few things. All of this was explicated in a slightly different context in a 1932 *Harper*'s article by Harold J. Laski entitled "The Limitations of the Expert." It is an article worth volumes of administrative law scholarship. I pluck only a single, evocative sentence from Laski's article: "The expert," he said, "sacrifices the insight of common sense to the intensity of his experience."

That is the fundamental problem presented by administrative decision making and judicial review. And it is not a problem that is solved by fastening on the compromise process that Stewart calls the "paper hearing." The paper hearing has its merits, in my opinion. It permits the courts to examine administrative decision making somewhat more closely than has been the case in the past, without all the burdens of a traditional trial record. It serves as a legitimating device for judicial decision on the merits of cases under circumstances in which courts are reluctant to admit (or even believe) that they are involving themselves deeply in the merits. But to imply, as Stewart seems to do, that the paper hearing even gets close to the roots of the fundamental problem presented by the conflict between rule making and adjudication is to devalue the difficulty of the problems with which legal scholarship has to deal.

Finally, let me turn to Stewart's discussion of what he calls "quasi-constitutional" decisions. I find this part of his paper the most troublesome of all. The strategy is essentially the same as that found earlier in the paper; it is an attempt to deal with a very difficult problem by suggesting a lawyer's compromise. Stewart sees a role for the courts in providing some leadership on very difficult social decisions about the allocation of resources, of which the nondegradation standard is but a single example. Yet he is, conventionally, unwilling to give the courts the final word on the subject. So he seeks a middle ground, which he calls "quasi-constitutional," or structural. The notion that the courts have a useful role to play in articulating important social policies, leaving it to the legislature to consider and perhaps reject those policies subsequently, is a decision-making strategy that I find congenial, and, as Stewart notes, in this respect his approach has an affinity to views that I have expressed.

What is troublesome for me is the content of the social policy he suggests, to which he gives the name "diversity," and which he describes as a nonconstitutional spin-off of notions emanating from the First Amendment. I submit that the idea, as Stewart presents it, will not withstand careful analysis. One need only read Stewart's own excellent catalog of the issues involved in making environmental policy, set out in the beginning section of his article, to see the difficulty. Diversity is simply too all-embracing an idea to be helpful. It is a justification for any interest that cannot garner majority support. Shall we have judicially created wilderness areas? It can be done in the name of diversity in a society that can be said to be rushing toward urbanization of its agricultural land and commercialization of its forests. Shall courts permit or require some communities to remain semirural outposts of the upper middle class, where high-density apartments, mobile homes, and urban crime can be walled out? It can be done in the name of diversity. Shall they permit some states to prohibit the building of nuclear plants or the storing of nuclear wastes or the prerogative of towns to veto new jetports within their boundaries? To do so might promote diversity.

All these and a hundred more common contemporary problems could be stuffed into the bag called diversity. And I would be the last person to suggest that these claims are without merit. The problem is that they simply reflect one of many important values. The pressure to open attractive suburban communities to urban development reflects a powerful need for housing and for escape routes from urban decay. Nondegradation raises the question whether and how the burden of air pollution and industrial development should be shared among the regions of America.

Stewart recognizes some of these problems, and yet he seems to think that a nonconstitutional judicial advancement of one of the many conflicting values (namely, the value he calls diversity) is to be encouraged. It is this conclusion that I find mystifying, at best. To be sure, the judicial decisions will not be final, as would a constitutional holding. But they will nonetheless have the substantial effect of judicial determination of the legislative agenda. And they impose upon the courts an enormous burden of social judgment, which, although it is legislatively rejectable, is just the sort of social judgment, I think, that Stewart earlier in his paper describes as straining judicial competency.

There may be a role for the courts in setting legislative priorities, though it is not obvious why this should be so, and Stewart certainly deals summarily with this far-reaching suggestion. More important, the vague notion of diversity leaves quite uncertain the scope of such judicial leadership. Stewart seems to think that what he calls diversity is a value that can stand alone as self-evidently appropriate because it is like the First Amendment. Yet, as he recognizes, diversity intrudes upon a number of other very important values, such as pressures for equality, which in some settings are also "ill-served by popular political processes." Why should not the courts take similar leadership in promoting other, perhaps competing values? The answer is not obvious from the case Stewart makes.

I would like to think that there is a case to be made for some of the ideas that Stewart has in mind when he talks of diversity and of an expanded role for the courts in promoting that interest. But the case has not been made in this paper. If Stewart undertakes that difficult task, he will have produced an invaluable contribution to the literature.

Note

1. The following excerpts are taken from interview summaries provided by Stewart upon my request.

Question: Has the judicial role with regard to procedural formalities been a constructive one? Has the court's insistence . . . on full explanations of the reasons for decisions actually improved the EPA's decision-making process?

Answers:

Robert Zener, EPA General Counsel: . . . There is a government-wide trend, prompted by the courts, toward more elaboration of reasons, and this is probably a good thing, although there is some danger that remands will become almost routine for the courts. . . .

Richard Denney, EPA Assoc. General Counsel for Pesticides: . . . The court remands for better-reasoned decisions were helpful during the Agency's early years, but now there is a potential danger: the more that is explained, the more issues and problems challengers will be able to raise in court. Many of these challengers want delay for its own sake, not rational agency decisions, and their ability to raise procedural issues in court can at times "intimidate" the agency into modifying substantive provisions.

Ray McDevitt, EPA Assoc. General Counsel for Water: The court's role has to some extent been constructive, in forcing more attention to detail and greater care in explanation on the part of EPA personnel. . . .

Bruce Terris, attorney in private practice: . . . It is important to get experts on the witness stand; conflicting written statements are of little use to the decision maker. . . . The court-imposed procedures may have a net beneficial effect.

4 The Political Economy of Implementation: The Clean Air Act and Stationary Sources

Marc J. Roberts and
Susan O. Farrell

There is no point in having good ideas if they cannot be carried out.
Jeffrey Pressman and Aaron Wildavsky

Introduction

In this paper we will examine recent experience in the implementation of the Clean Air Act Amendments of 1970 with regard to stationary sources (that is, excluding motor vehicles). Under that legislation all regions of the country were supposed to meet so-called "primary" air quality standards by the end of 1975. Yet today, in 1977, although the air is generally cleaner than it was in 1970, many areas still regularly violate those standards,[1] and progress toward cleanup has been quite uneven among states and for given states over time. Such a review should suggest how the program might be improved and help us to assess the value of the regulatory approach to environmental problems generally.[2]

The argument we will make is straightforward. The initial task given to the state agencies by the amendments was to devise a state implementation plan to achieve air quality standards. This assignment was technically and politically very difficult, especially if the plan were to be economically efficient, consider natural uncertainties, and include allowances for long-run economic growth. With limited resources at their disposal, agencies generally developed very simple plans using simple methods. The problems of economic efficiency, uncertainty, and growth have been generally ignored. Furthermore, given the technical uncertainties and ambiguities, there has been room for a substantial amount of political influence in the specification of those state plans. The combined result of simplicity and politics is to make it uncertain whether state plans will achieve air quality goals, even if they are effectively implemented and enforced.

In fact, implementation and enforcement have been and will continue to be highly uneven. State agencies have substantial discretion in deciding what a specific source must do to comply with the implementation plan. They also face substantial procedural difficulties and manpower costs in compelling a reluctant source to undertake control measures. As a result, specifying cleanup requirements for specific sources often becomes a bargaining process. Once the needed equipment is installed, it is often both technically difficult and expensive to monitor emissions in order to ensure that the source is properly operating and maintaining its facilities.

Under the circumstances, a substantial amount of voluntary compliance is essential for successful implementation of the act. Such compliance, in turn, seems likely to depend both on public concern with air quality and on the extent to which waste sources believe that what is required of them is both fair and socially rational. The recent shift in public opinion away from environmental concerns and the economic and technical limitations of current implementation plans are therefore likely to cause continuing problems because they will tend to diminish voluntary compliance.

Past experience included a number of instances in which politics, leadership, organization and management all played significant roles in determining what actually occurred at the state level. Because of their political and economic circumstances, those agencies often have neither the means nor the incentives to be "tough" on polluters. Yet, although political developments and their likely impact on agency budgets are not promising, there are a number of steps that might be taken to improve the functioning of the current program. Many of those steps can be taken only at the state level. Hence, whatever its intentions, Congress cannot eliminate the substantial discretion enjoyed by the states as long as the latter are actually responsible for implementing a federal program. This conclusion is one way of answering our initial question as to why the goals set forth in the 1972 amendments in fact have not been achieved.

A premise in all that follows is the view that waste sources have obvious financial incentives to delay and minimize control efforts. Cleaning up air pollution generally costs money: the less required, the lower the cost to the firm. And insofar as one firm can do less and incur lower costs than those of its competitors, its relative profit position would clearly tend to improve. Delay also means that the capital that would be required to comply with the regulations can be used for other purposes. In the meantime, requirements might be relaxed, either for a particular firm or for the industry generally. Faced with such incentives, the public relations and money costs of a protracted legal battle to delay or reduce compliance costs may seem a reasonable investment, especially for a source with high control expenses, such as a large firm in a basic material processing business (which typically generates lots of waste products).

The paper that follows develops these arguments on the basis of our studies of the implementation of the Clean Air Act in five states, with some exploration of four others. (For a list of those surveyed see Appendix A.) For these states, both the development and the enforcement of the state implementation plans (SIPs) required under the act are reviewed. The conclusion deals with ways to strengthen enforcement and voluntary compliance efforts and considers some of the implications of the experience for environmental and related regulatory problems generally.

Background and Context

The Clean Air Act Amendments of 1970 gave the federal Environmental Protection Agency (EPA) thirty days to establish "primary" and "secondary" air quality standards. The primary standards were to protect public health. The stricter secondary standards were to protect public welfare; that is, they can take account of material and vegetation damage. Once proclaimed, the states had nine months to submit written plans for limiting air pollution emissions from various sources (the SIPs) for EPA approval.[3] Once they had been wholly or partially approved, the states had to enforce the plans, which were supposed to be designed to achieve the primary standards by 1975. The EPA also has power to monitor and to enforce state plans through the federal courts. And it can alter state plans and take over all of a state's enforcement efforts if either is deemed unsatisfactory.[4] Furthermore, the EPA is to proclaim nationally uniform emissions limits for new sources.

This program can be seen as a logical extension of the 1967 legislation. That act directed the department of health, education, and welfare to designate selected areas as "Air Quality Control Regions," and states were supposed to set both ambient standards and emission control plans for those regions. However, the department was slow to designate Air Quality Control Regions, and the states were slow to submit implementation plans for those regions. In addition, HEW was slow to review those plans that were submitted. Only twenty-one plans were submitted and none had been approved prior to passage of the 1970 legislation.[5] Under the current act, the EPA sets nationally uniform ambient standards. In effect, every region of the country is part of some Air Quality Control Region.[6]

The 1970 legislation also streamlined cumbersome federal enforcement procedures. Formerly, the EPA could only act in emergencies and cases of interstate pollution and then only via a series of conferences, hearings, and negotiations. Now it has the power to act directly against violators of state plans either by issuing a "cease and desist" order or by referring the cases directly to the regional U.S. Attorney for court action. Unlike the Water Act, however, the 1970 legislation does not permit the administrator himself to directly impose civil penalties on violators.[7]

In sum, states theoretically must approve their own enforcement efforts or have the job taken away. They were given initial responsibility for what is politically the potentially most vexatious responsibility, namely, deciding what control burdens were to be imposed on various existing stationary sources. But again, failure to perform acceptably means federal preemption. In all, tight time limits and, if necessary, extreme measures characterized the program. As the Senate committee report put it, "The Committee determined that existing sources of pollutants either should meet the standard of the law or be closed down."[8]

What is the actual state of air quality as a result of this program? The most recent Council for Environmental Quality report,[9] which relies on 1974 data, suggests that 111 of 198 ACQRs studied showed at least one monitoring site at which the primary standard for annual levels of "total suspended particulates" was violated. In 1974, about a third of the nation's population lived in regions where the standard was violated. Furthermore, at 16 sites, two each in eight representative metropolitan areas, the twenty-four-hour standard was violated anywhere from 3 to 116 days a year. Eleven ACQRs violated the ambient standard for annual SO_2 violations; daily violations were less frequent (varying from 2 to 54). Finally, in 1975 the CEQ forecast that a number of the nation's major cities—notably, Los Angeles, Philadelphia, Chicago, St. Louis, Louisville, Cincinnati, Providence, Detroit, Minneapolis, and Milwaukee—would not meet either of these two critical standards by 1976.[10]

These results do represent an improvement in air quality, and some of the change is no doubt due to the act. However, much of the recent improvement occurred before the state implementation plans were approved by the EPA in May 1972, largely in response to changing relative fuel prices, which brought about a decline in the use of coal for urban heating beginning in 1968.[11] More recently, rising fuel prices and continuing slow economic growth have also helped to lower emissions. Thus, regardless of where one judges the efforts on the spectrum from "success" to "failure," progress has not been as great as those who drafted the 1970 amendments intended. This is the outcome we will try to explain.

Formulating State Implementation Plans

Given the time, resources, information, and expertise available to the state agencies and the technical difficulties of the problem, there was no way for state agencies to develop cost-minimizing emissions limitations programs carefully calibrated to achieve the primary standards. This situation has had several consequences. First, the plans that have been adopted by the states in fact might not have achieved the ambient standards, even if effectively implemented. Second, the weak scientific rationale for these plans has increased the ability of political actors to shape their initial content or to compel their revision. Finally, this same weakness has tended to undermine the willingness of waste sources to voluntarily comply with plan requirements. In this section, these points are illustrated from the experience of the states that were studied.

The critical step in developing an economically efficient implementation plan is to link air pollution emissions to the ambient concentrations of pollutants they will produce. Unfortunately, the ways in which pollution emissions diffuse and become transformed in the atmosphere are not all that well understood. Although there are mathematical models[12] that simulate these processes, they reproduce actual events imperfectly making their use as much art as science. The reason, in part, is that all such models must make some simplifying assumptions about the meteorology and topography of a region. These assumptions depart in varying degrees from actual conditions at any one time.

Such a model, however necessary, is not sufficient to develop a plan for meeting ambient standards at least cost. A state must also acquire information about the cost to each source of reducing emissions by varying amounts. The problem of combining this data to develop an economically efficient plan is not particularly difficult conceptually, but it can be extremely difficult in practice. Actually performing such an analysis requires a great deal of meteorological and economic data and substantial expertise in the use of machine computing.

State implementation plans were supposed to be filed with the EPA just *nine months* after ambient air quality standards were promulgated. Devising an optimum implementation scheme in that amount of time would have taxed a large, well-prepared, and technically first-rate organization. Most state agencies, in contrast, were neither large nor technically very expert. Nor were the relevant data easily available. Much of the needed meteorological information had never been collected, and much of the data on control costs could come only from engineering studies that had not been performed. As a result, one estimate is that only ten states made use of formal modeling methods in creating their plans.[13] Even that number may be an overestimation, because some states where it was supposedly used (for example, Massachusetts) simply used the model as a rough check on plans developed by other techniques.[14] Indeed, even today, some states still have no sophisticated diffusion modeling capacity, and others are just beginning to undertake such analysis.[15]

Lacking a convincing analytical basis for imposing different control requirements on different sources, state agencies found it very tempting to pick a single emission standard for *all* sources in a state or region, or, at most, to introduce a few distinctions based on process and size differences. As a basis for such limits, the "linear rollback" technique was often employed to develop SIPs. To use this technique, it is necessary to determine the ratio of observed pollution levels in the worst area of a state or region to the levels specified in the standards. A uniform emissions limit that will reduce total emissions by the same ratio is then imposed.[16]

Such techniques are both crude and arbitrary. Because the weather can be more or less unfavorable, how bad a contingency should be factored into the calculations? What allowance should be made for probable future economic growth? What assumptions should be made about natural and/or otherwise uncontrolled background sources of particulate emissions such as blowing dust from farmland and construction sites? All these decisions and many more had to be made. Often they were resolved without any clear rationale or indeed any conscious recognition of the question. For example, a given pattern of emissions will produce varying ambient conditions under different meteorological circumstances. This reality was not dealt with in an analytically sophisticated manner. Instead, some simple approach was used. In some cases, *average* conditions were assumed, and some ad hoc margin of safety was added into the calculations. In other cases, so called "worst case" conditions were assumed—although that concept itself is ambiguous. And in most cases, there was little or no analysis of how much margin was prudent to make room for new industry by insisting on more than otherwise necessary cleanup of current emissions. Instead, requirements were typically expressed in allowable emissions rates per unit of output, so emissions can be expected to rise and air quality to deteriorate with further economic growth.

For agencies under severe time pressure and with limited skilled manpower, there was no time for such complexities—even if they understood them and they did not in all cases. Furthermore, raising such issues was only likely to increase political pressures when it was already difficult to avoid having efforts at developing a plan overturned by the governor, the legislature, the courts, or the EPA. Successfully maintaining one's bureaucratic position under such circumstances required a rather careful balancing of competing political forces. In 1971–72, when the SIPs were being developed, the potential electoral impact of environmental groups pushed political leaders in several states to intervene in standard-setting on the side of tightening those standards. (See following discussion of experience in Florida, Massachusetts, and Los Angeles.) However, knowledge that this had occurred only served to compound industry feelings (given the crudeness of the analysis just noted) that the standards were not fully legitimate. This attitude in turn both exacerbated compliance difficulties and reinforced industry determination to relax SIP requirements. And, more recently, as the balance of political forces has shifted, they often have been able to do just that.

Unfortunately, the resulting implementation plans are not always very cost-effective; often (for example, the SO_2 regulations in Connecticut) the resulting rules were uniform across the state.[17] But, is it really true that equal degrees of cleanup are appropriate in the very different circumstances that often obtain in different areas of a single state? In other instances, such as Massachusetts and New Jersey, more stringent limits were imposed in urban areas.[18] But the arbitrary boundary lines of the areas, and the weakness of the underlying analysis, still make the specific provisions of the plan hard to justify as being economically rational. Similarly, some states set similar requirements for emissions rates regardless of a source's fuel.[19] Other states did require varying control levels for different fuels or for sources of different ages, process types, and sizes.[20] Clearly, both approaches cannot be economically efficient. And the relative stringency of regulations in various cases was seldom determined on the basis of careful analysis.

Another result of the balance between political pressure and the lack of expertise was the writing of unenforceable regulations. State personnel often felt a need to provide for the control of certain pollutants without knowing how to specify either the manner or the level of control. For example, Georgia regulations[21] require that "All persons responsible for any operation, process handling, transportation or a storage facility which may result in fugitive dust shall take all reasonable precautions to prevent such dust from becoming airborne."[22] The North Carolina regulations specify that plants cannot be permitted to operate "without employing *suitable* measures for the control of odorous emissions including wet scrubbers, incinerators, or such other devices as may be approved by the commission." (emphasis added)

Overall, the process produced SIPs more notable for their simplicity and their political sensitivity than for the cost-minimizing properties. One of the most striking examples of both of these tendencies comes from Los Angeles, where, until recently, the county Air Pollution Control District imposed flat quantity restrictions on the emissions of nitrogen oxide from any source, regardless of the size of the facility.[23] Obviously, it is much more difficult for a large facility to meet such a limit than for a small one. And surely, compliance with such a rule will entail very different marginal control costs for different size sources. But the rule was adopted soon after several county supervisors had lost reelection bids because of the failure to "do something" about smog and at a time when the public concern was increasing over the possible development of large fossil-fuel power plants in the Los Angeles Basin. In the midst of an acrimonious conflict with the Los Angeles Department of Water and Power over such a facility (the Scattergood #3 unit then under construction), the political pressures for decisive action were very high.[24]

The same circumstances that allowed political pressures to be reflected in SIP formulation facilitated later amendment of these plans as popular attitudes and the balance of political forces shifted over time. Last summer, for example, the Los Angeles Air Pollution Control District rewrote and relaxed the regulation just discussed.[25] Similarly, conditions have also changed in Florida, where the key lobbyists for stringent air pollution control traveled from hearing to hearing with state authorities when the state's proposed implementation plan was first developed.[26] The individuals on the public body that approved the plan had been appointed by the new governor, who assured them of their freedom from industrial pressure.[27] As a result, a relatively "tough" plan was adopted. More recently, under continued attack from increasingly well-organized utility interests, the state agency has substantially postponed dates for compliance with key control provisions.[28]

In Massachusetts, the original state implementation plan was made stricter by the department of public health under intense pressure from a governor who was in the midst of a reelection campaign against an opponent who was attempting to make environmental protection an issue.[29] However, in 1974 the state legislature passed a law requiring that state standards be relaxed to the extent feasible and still in line with federal specifications.[30] It also authorized sources to proceed with the experimental implementation of schemes to burn higher sulfur content fuel during more favorable meteorological conditions. The bill, written by a utility industry lobbyist, passed almost unnoticed on a voice vote after little debate.[31]

In Georgia and South Carolina, state legislatures passed similar laws, although the Georgia bill was vetoed.[32] In North Carolina, textile mills and food processors were responsible for the changes in the SIP. They focused primarily on requirements for self-monitoring rather than on emissions requirements.[33] The industry version passed one house of the legislature but was ultimately amended in line with state agency suggestions on the grounds that the proposed requirements (to prevent the state's laws from being more stringent than those of any other state) were unnecessary, unwise, and unmanageable.[34]

Relaxation of an SIP implies that someone has made a mistake somewhere. Either the initial program was stricter than was needed to meet standards or the new one is too lenient. In either case, relaxation undermines the creditability of the requirements. It also suggests the limitations of federal review procedures when federal agencies too lack a scientific basis for opposing local political trends. Insofar as plans become "too lenient," air quality goals will not be attained, even with effective implementation; this situation is likely to become increasingly troublesome over time, given the failure to allow explicitly for economic growth.

Implementation of State Implementation Plans

Implementing an SIP involves four steps.

1. The agency must decide what each source must do to comply with the plan. Typically, a permit issued with the agreement of the source specifies these requirements.
2. Needed capital investments must be made and any changes in operating procedures instituted.
3. Day-to-day operations of processes and control equipment must be monitored to check on continued compliance.
4. If a source refuses to cooperate at any stage of the process, the agency may have to undertake "enforcement actions," that is, impose on the offender sanctions designed to discourage such behavior.

State experience with these four steps reveals that resource limits, political pressures, and variations in managerial capacity play a large role in the process. The technical difficulties of monitoring and the cumbersome nature of enforcement procedures imply that state agencies have substantial discretion as to how they proceed.

State agencies appear to have the least discretion in the first stage: determining what is required of each source. In actuality, here the state agency often has the most discretion. In order to determine source-by-source limitations, most states require that every source have a permit to operate. Once it has been identified, each source is asked to submit information concerning outputs, processes, production rates, hours of operation, fuel characteristics, and pollution control equipment. From this information and EPA-published "emissions factors,"[35] the agency can determine the likely hourly emissions from each source. (The emissions factors are listings of the amount of various pollutants produced by a given process when performed at a given rate.) Calculated emission rates are compared with the emissions standards in the SIP in order to determine the amount of control required. Typically, a permit also carries with it a "compliance schedule," a timetable that specifies when various steps will be taken.[36]

Although this procedure sounds straightforward, in actuality it is not. Many firms operate more than one process, and different processes often vary significantly in the amount of pollution they produce. Also, operating rates often vary in the course of the year in response to seasonal demand variations. And obviously the greater the rate, the more pollution. In addition, the sulfur and ash content of the fuel will vary from time to time—especially if coal is used. This factor also alters the amount of pollution generated. Finally, pollution control equipment varies significantly in achieving its designed effectiveness, depending upon process rates, fuel characteristics, and how it is operated and maintained. For example, an electrostatic precipitator designed to remove 95 percent of the particulates from the stack gasses of a coal-fired power plant could easily function at 70 to 80 percent with inadequate maintenance, high operating rates, and fuel that was lower in sulfur content than allowed for in the design.[37]

Thus, the agency has significant discretion in deciding how conservative to be in specifying control requirements for a particular source. Should the equipment be designed to cope with the worst case with regard to operating rates, fuel composition, and so on or to cope with average conditions or somewhere in between? Part of the problem facing the agency is that, because of the limitations on monitoring capability—reviewed below—the state typically cannot simply specify allowable emissions rates, leave all control decisions to the source, and then just keep track of the result. Instead, formally or informally, it becomes involved in judging the adequacy of the proposed control measures.

An example of this type of discretion is the case of two identical coke ovens in Ohio that applied for permits, one under the jurisdiction of the Toledo air pollution agency, the other under the jurisdiction of one of the state's district offices. The district office denied a permit, whereas the Toledo air pollution agency granted one. In theory, the same SIP requirements applied to both.[38]

Not only does the agency exercise discretion, but, implicitly or explicitly, the source must agree to the permit conditions and the compliance schedule or the state is faced with an enforcement problem. The result is that the permit process becomes a bargaining situation in which both sides face steadily increasing costs the longer it takes to reach an agreement, especially if court action ensues. Yet, the possible gains from delay and/or reduced requirements mean that some firms may choose to be noncooperative. In deciding how to respond, the agency must trade the time and manpower costs of a tough stance and the gains in increased cleanup that toughness might achieve, against the value of a quick agreement that frees its limited resources for other tasks.

For example, we were told of a source that insisted that its permit requirements not be based on the process that polluted the most but rather the process that was in use the greatest proportion of the time (40 percent). Rather than confront a lengthy and costly legal battle, which it might loose, and delay all compliance efforts in the meantime, the agency agreed.[39]

Furthermore, the quality of the permit process also depends on the accuracy of the information available to the agency and the expertise with which it is used. A lack of qualified personnel means that much of the information supplied by the industries cannot be verified. Instead, only standardized data available from the literature can be used to check industry claims. In addition, the agency must rely on the source's statements concerning its operations. This dependence can lead to serious inaccuracies, given that apparently similar sources vary significantly in their emissions as a function of design details and operating practices.

In fact, there are substantial differences in staff size and quality both within states over time and across states. These differences often reflect more fundamental patterns of public attitude and political balance. Some states, such as Massachusetts, have long-standing political traditions that support a civil service system based on patronage and job creation. This tradition leads to low salaries and low technical qualifications for most state agencies, including air pollution. In contrast to most states, where inspectors have engineering degrees, in Massachusetts the majority are only high school graduates. A total of eight engineers are responsible for inspecting and processing permits for some 10,000 sources. Georgia, which has one-third the number of sources Massachusetts has, has twice as many engineers assigned to such tasks.

Indeed, agencies are sometimes organized so that top management does not fully control the permit process. In Ohio, for example, the central office only reviews permit demands. Thus it has the power to loosen but not tighten field office requirements.[40]

The permit process is critical because it can significantly effect the actual level of cleanup. For example, a change from 99 to 98 percent in the efficiency of control equipment may seem small—and is easily the kind of variation that might occur through different permit processing—yet the change represents a doubling of emissions. Thus, for all the court action surrounding the question of when variances in SIPs can be granted, permit conditions and the timetables attached to them as compliance schedules represent a potential source of significant temporary or permanent variances from the SIP not approved by the EPA. If ad hoc variances are given, the SIPs will prove inadequate and violations of the air quality standards will occur.

Experience in North Carolina offers some evidence that this is not a purely theoretical concern. On paper, the emissions limitations supposedly placed on sources should have improved air quality by a far greater margin than has actually occurred. Although this lack of success could be due in part to the use of inaccurate assumptions in the formulation of the plan, it may also be due to inaccurate permit policies.[41] Similarly, but less seriously, widespread slowness in coming into full compliance with the standards, even though sources are obeying implementation schedules, could have contributed to the failure to meet the deadlines of the 1970 legislation.

The second stage of the implementation process—seeing that the firm makes the required investments or changes in its operating procedures—is relatively easy. Typically, the compliance schedule attached to the permit specifies a series of intermediate tasks—such as conducting studies and placing orders. Because such events leave a "paper trail" of correspondence, contracts, vouchers, and so on, and ultimately some obvious physical results, it is often easy to verify whether or not they have been accomplished. In part because sources know that the state could check up if it chose to, many states rely heavily on self-reporting for monitoring at this stage. That is, waste sources have to file regular reports on what they have done and whether they are keeping to the schedule.[42]

The third step in implementation, determining whether or not the source's emissions are within prescribed limits, is the most difficult. Operating rates, fuel characteristics, maintenance levels, and so on all influence actual emissions. Ideally, one would like to monitor continuously what goes up the smokestack. In practice, it has seldom been done because such monitoring devices are expensive or, as in the case of sulfur compounds, simply unavailable.[43] Also, the devices themselves tend to be unreliable and in need of regular adjustment. In addition, a fair amount of pollution arises because of "fugitive emissions," or leakages from process units and storage tanks; monitoring these is generally impossible, regardless of the pollutant.

In the absence of continuous monitoring, many states either perform or require a large source to perform "stack tests," in which emissions are monitored for a few hours by special equipment brought in for that purpose. But these tests can cost $2000 or more each for each stack (a large source may likely have five stacks).[44] And they do not provide any check on day-to-day performance—especially when announced in advance to the sources because notice allows them to prepare their equipment to function as well as possible for the brief period of the test. (In Georgia, they try to get around this problem by forbidding such maintenance in the several weeks prior to the test.)[45] Furthermore, such tests must be closely supervised by (or even performed by) agency personnel to be meaningful because the results are quite sensitive to test methods, operating conditions, and so forth.[46]

Requiring a stack test can at least help get the source into compliance—which is one of the rationales used in North Carolina for asking that they be performed.[47] However, the value of such a device depends on how rapidly performance can deteriorate after an acceptable test. Unfortunately, in many situations changes in operating practices or fuel characteristics that significantly effect emissions can be made almost immediately, once a test is over. And poor maintenance could begin to have an effect in a matter of days or weeks.

Stack tests are especially open to being unrepresentative in the case of SO_2 emissions because few sources have control devices for SO_2 and fuel sulfur content, which determines emissions, is easily changed by changing fuel. Thus, states typically require large sources to report fuel sulfur content regularly. Where oil is used, such reports are not difficult to verify because oil is a manufactured commodity, sold with explicit chemical specifications, and its movement leaves a paper trail of bills, invoices, refinery records, and so on. Coal, in contrast, varies significantly in sulfur content, even as it comes from any one mine. Many large users mix coal from a number of mines, whose coal could be very different. Although accurate sulfur content sampling can be done (by sampling the crushed coal just before it is burned), usual state procedures rely on fuel contract specifications, source reports, or infrequent random samples, all of which are much less reliable.[48]

The most common inspection procedure involves simply having an inspector visit the plant without doing any actual monitoring.[49] The only emissions violations he is able to discover are obvious ones, such as dark smoke plumes (gross violations of the particulate standards), obvious odors, and so on. Violations of SO_2 and NO_x standards will not be revealed by such inspections. Under such circumstances, the focus of the inspection has to be on whether or not the source is complying with easily observable permit requirements, such as whether or not required control equipment is in place and apparently operating. Determining whether emissions are within specified limits is difficult to do without monitoring because examining fuel composition, operating rates and practices, equipment condition, and so on only gives an indirect and unreliable indication of what is going up the stack. Furthermore, few actual inspections can be that rigorous, and even if they were, they do not reveal whether the firm is altering its behavior and its emissions while it is not being inspected.

The superficiality of most actual inspections stems, in part, from a general lack of resources. States have anywhere from 2,000 to 40,000 sources of air pollution, excluding small furnaces. Of these, between 200 and 1,500 are major sources: they have the potential to emit over 100 tons yearly of any single pollutant. The agencies have between 15 and 200 inspectors responsible for permitting and inspecting these sources. Consequently, inspectors would have to perform anywhere from three to thirty inspections daily just to inspect each source annually. Such a schedule is clearly impossible, given travel time, the need to write reports, and the time entailed by even the most cursory review. In many cases, the inspectors themselves do not have the technical background to perform such a detailed review.

One result of this manpower shortage is that agencies must set up inspection priorities. In Illinois, noncomplying sources in areas that do not meet air quality standards are inspected four times a year, whereas apparently complying sources in areas that meet standards are inspected less than once per year.[50] In other instances, agencies have tended to pay attention to those violations that cause complaints and/or are likely to be perceived and understood by laymen. In Massachusetts, inspectors on high buildings, with a clear view of a large number of sources, look for dark plumes. Such a visual indication or a citizen's complaint leads the agency to contact the firm to see what the problem is.[51] Hence, smoky (or smelly) sources in urban areas are under the closest control. These sources are not necessarily those that do the most damage to the environment. This approach protects the agency against public complaint because it acts on those violations that the public is most likely to notice on its own. And given the political difficulties of the enforcement process, discussed next, this strategy also means giving priority to cases in which popular concern and support are likely to be most forthcoming.

The account thus far has described the problems an agency has even with a cooperative source. If a source refused to agree to permit conditions or fails to keep to its implementation schedule or violates emissions limits, the fourth stage of the process, enforcement, must be put into motion. Enforcement procedures are apparently quite similar from state to state. Usually, a phone call is made to the firm and a notice of violation is sent requesting that the firm amend the situation. If no response is forthcoming, from one to six more notices are sent. Then the company is ordered to an informal conference. If no agreement is reached, there will probably then be a formal administrative hearing at which the firm will be ordered to take action. Depending on the state, this order may include a fine, a compliance schedule, or both. The judgment usually can be appealed through several administrative levels in the state, and thence to the courts, each stage takes six months to two years. Once a compliance schedule, or permit, has been established, by negotiation or by the court, it can still be disobeyed and the entire process repeated. Thus, the time and resources costs to the agency to get compliance from an uncooperative source can be very substantial.

In most states, the agency must also have the cooperation of both the state attorney general and the state court judges before it can impose sanctions on uncooperative firms. These officials typically are sensitive to political pressures, especially if they are directly elected. For example, a state official in Massachusetts reported that a former attorney general would not prosecute polluters before checking with local political figures.[52] This problem can be difficult to overcome where the cases are tried in local courts. In one case in a southern state, the judge repeatedly delayed or did not appear for hearings in a pollution case against a local firm.[53]

Judges often prefer that money be spent on cleanup rather than on fines. And they have a hard time knowing what level of penalty is fair or appropriate, so they might try to avoid the issue. In addition, few state agencies have the power to directly impose administrative civil penalties. As a result, there are almost no ''first-instance sanctions'' in air pollution enforcement. That is, almost no source is penalized the first time it is caught in violation. Only repeated, flagrant violations and a failure to make or honor subsequent agreements result in the assessment of fines. Thus a rational source has little incentive to comply until it is caught or to make difficult and expensive efforts to remain in compliance—especially because inspectors are spread so thin that many violations will not be noticed.

Given limited agency resources and procedural and monitoring problems, successful enforcement requires that most sources comply ''voluntarily''—before being discovered to be in violation—so that enforcement against them is not necessary. Hence the agency must strike a delicate balance to be tough yet reasonable. The rationality of the SIP itself and of the agency's approach to its implementation are critical in this regard.

One state where this balance has not been achieved is Ohio, where the agency refused to discuss SIP provisions on SO_2 with industry. Industry, in turn, found its provisions unfeasible and greeted them with hostility. There ensued a well-financed and successful court challenge, prolonged state agency paralysis, and finally an EPA-proclaimed plan for SO_2 emmission limitations.[54] Some observers also suggest that it is not unusual for firms in Ohio to submit obviously unacceptable applications with every intention of delaying any resolution by engaging in repeated appeals of any decision.[55]

Similarly, the original administrator of the Illinois EPA intentionally set out to limit the ability of his agency to compromise with waste sources. He organized separate groups for each functional task, such as permitting or monitoring, and limited communication both among them and with the outside world—he even limited the number of their telephone lines. In the adversary situation that resulted, noncompliance was widespread. A subsequent director decided to be more conciliatory toward waste sources and altered internal procedures in order to encourage some level of compromise.[56] Some officials in the state believe that this trend could go too far, but these same people admit that not only has cooperation and compliance improved, but also the legislature has ceased making large cuts in the agency's budget.[57]

Yet another way to deal with the same problem is used in North Carolina, where inspectors assist the sources in filling out permit applications and coming into compliance, as well as in handling equipment breakdowns. The agency claims that this involvement increases the accuracy and efficiency of the permitting process and leads industries to promptly notify the agency when something goes wrong.[58] Such cooperation increases the capacity of the agency to assure compliance with the SIP.

Because the central office seldom has the resources to thoroughly check and recheck its own staff, the beliefs and quality of the staff also effects the degree to which a state achieves compliance with the SIP. If inspectors do deal with the same firms over and over again—to increase communication and compliance—they can come to face significant interpersonal pressures to be reasonable and understanding. They might find it difficult to report violations or impose formal sanctions if the source appears to be acting in good faith, especially if they themselves are not fully convinced of the reasonableness of the requirements. Thus, in Georgia when a power plant's collecting device malfunctions in such a manner as to require a plant shutdown in order to repair the unit, the inspector will usually allow the plant to wait until its regularly scheduled plant maintenance to fix the unit.[59] Thus it may be a matter of months before the unit is functioning properly again.

Indeed, in many states key positions are held by long-service employees whose perspectives were shaped in the preenvironmental era. These older engineers tend to emphasize cooperation and persuasion, which were the accepted professional norms in their business and are not very comfortable with confrontation, court action, and sanctions. The enforcement actions of the Massachusetts attorney general have been stymied by this problem: some regional offices refused to refer their cases to his office; they preferred instead to "talk to" the sources, rather than to subject them to court action.[60] Thus, although we have argued that some openness to sources can be helpful, it is clearly possible to give up too much in the process.

In such a context, staff expertise can play a critical role in determining actual outcomes. In Massachusetts, for example, the new attorney general noticeably increased the size and skill level of his environmental staff. Even though the laws and regulations, as well as the political framework of the state, seem to make pollution control more difficult than formerly, there has been a substantial increase in both the number and success of enforcement actions by the state.

Similarly, managers vary greatly in their ability to use the media to publicly embarrass an uncooperative source by calling attention to its behavior. (This tactic can be especially helpful in regard to a large producer of brand-name consumer products.) For example, the somewhat outspoken and colorful former director of the Georgia Environmental Protection Division successfully employed the media to prevent sources from availing themselves of the complicated administrative appeal process. The current director does not use the press in this way,[61] however, and perhaps coincidentally, for the past eighteen months a large source has been using the appeals process to contest a substantial fine.

Because these agencies do have discretion and are connected to the larger political process, it is to be expected that the process will have a significant impact on enforcement. For example, the legislature can decide that the agency is being "too tough" and reduce its funding, as happened in Illinois a few years ago.[62] Individual staff members, too, can feel such political pressures. One staffer in Massachusetts who tried to take vigorous action against a polluter was promptly transferred to another job because a state legislator disapproved of his action.[63]

In sum, resource constraints, technical difficulties, political pressures, the costs of confrontation, limited administrative controls, and accepted attitudes all combine to make implementation of the Clean Air Act anything but a routine and mechanical process. Instead, the agency and its members have substantial discretion. And the Congress, the courts, and the state legislatures (not to mention the agency heads themselves) have only limited control over how that discretion is exercised. When we couple this "slack" with regard to implementation and the technical and economic limitations of the SIPs themselves, it is almost surprising that air quality has improved as much as it has. The fact that most sources do accept some level of control as legitimate, together with the relatively easy enforcement of sulfur content standards for fuel oil, no doubt accounts for part of this change. But if present achievements are not good enough, what else can or should we do?

Conclusions and Recommendations

Our survey of state experience under the Clean Air Act has led us to the following conclusions.

1. Because State Implementation Plans are based on technically inadequate analysis and have been revised in light of political pressures, we cannot be sure that they will achieve primary, still less secondary, air quality standards, even if well implemented.
2. These same factors have led to economic inefficiency in the plans.
3. The form of the regulations in the SIPs makes failure noticeably more likely in the face of continued economic growth and unfavorable atmospheric conditions.
4. Given the substantial discretion enjoyed by state agencies, the permits issued to individual sources and the compliance schedules that govern their implementation are a potentially significant source of implicit variances from the SIP with respect to both the timing and the extent of cleanup required of individual waste sources—typically not subject to the same stringent formal limits that have been imposed on the variance process.
5. In most states, imposing penalties on reluctant sources at any stage of the process is uncertain, cumbersome, and time-consuming, and hence, enforcement authorities are involved in a continuous process of negotiation and informal accommodation with waste sources.
6. Given these characteristics of the process, there is no way for congressional action to tightly constrain the implementation of the 1970 amendments. Instead, variations in local politics, managerial capacity, and technical expertise will continue to have a major impact on actual air quality.
7. Because the existing monitoring and enforcement apparatus cannot effectively coerce a generally recalcitrant set of waste sources, a substantial amount of voluntary compliance is needed if clean air goals are to be attained.

Given all these characteristics of existing implementation activities, the fact that the act's air quality goals have not been achieved on schedule is hardly surprising. Achieving more effective implementation under these circumstances will not be easy. But several obvious morals stand out.

1. Given the need for voluntary compliance, the rationale for current levels of pollution control must be more persuasively developed. Thus the EPA needs to do additional studies and develop more information concerning the effects of emissions on health, well-being, property, and the ecosystem.
2. The states should make a serious effort to develop enough atmosphere modeling capacity so that they can better relate source-by-source emissions limitations to ambient goals and use that capacity to produce economically efficient SIPs. In that way the state could help save socially valuable resources, protect themselves against court challenges to their SIPs, and, perhaps most important, make the regulations more defensible and reasonable in the eyes of waste sources—and possibly more deserving of compliance. The recently proposed SO_2 compliance plan for Ohio, which the EPA developed, seems like a step in the right direction in this regard.[64]
3. As part of such revisions, we do need to take explicit account of uncertain atmospheric conditions and make conscious decisions about how much we wish to spend as "insurance" in the form of normally unnecessary control equipment to lower the probability that we will violate the standards under unfavorable circumstances. The current rules of "one violation per year" does not adequately cope with this problem because that apparent margin of flexibility simply leads to a restatement of the question as "What probability should we accept of achieving only one violation?"
4. If we are to maintain ambient standards, new economic growth can occur only insofar as some current sources reduce emissions rates. To prevent a steady decrease in air quality with continued economic growth, we should begin to express emissions regulations not as percentage rates but in the form of total allowable physical quantities of various substances specifically allocated to various sources.[65] However, it is far from clear that society is willing to delegate to its air pollution control bureaucracy the extensive and detailed control over economic activity and future development that allocating and reallocating such quotas would involve. In this context, the use of transferable quotas (that is, marketable effluent permits) has much to recommend it.[66] The recently proposed regulations on new sources in so-called "nonattainment areas" apparently will come close to creating such a system. Per-

fecting those regulations and extending the scheme to other areas has much to recommend it. Unless we adopt such a scheme or else assess fees even for allowable discharges, no source has an incentive to do more cleanup than necessary and thus make room for added emissions from others. The alternative is to steadily tighten the SIPs and require backfitting of further control devices in previously acceptable facilities.

5. To provide added incentives to waste sources to first agree with and then adhere to permit requirements and compliance schedules, some simplified administrative procedure is needed for imposing fines on those who delay at either step. In this context, the two-part Connecticut scheme for imposing civil sanctions on delaying sources is quite promising. Before agreeing to a schedule, a source is liable to a fine equal to the interest on the capital value of the control devices it is required to install for the period since the SIPs enactment. Upon agreement, the previously accumulated fine is waived, but a firm becomes liable for similar fines if it delays meeting its agreed-to schedule. Equally important are provisions for immediate whole or partial payments of fines into escrow accounts pending successful completion of the project and a series of smaller administrative civil penalties for failing to meet reporting requirements. Although rationalized as a way of taking away the gains of delay, the real scheme is especially attractive for its procedural simplicity, its apparent rationality in the eyes of businesses and the courts, and its ability to prod recalcitrants with procedural penalties or requirements for escrow payments.[67]
6. To improve our capacity to control actual emissions, improved monitoring and data processing capacity is urgently required. Recent EPA regulations that would require automatic monitoring on large sources by 1977 are clearly an important step forward.[68] Rapid completion of companion regulations on fuel sampling, especially for coal-fired sources, is also essential. And we need a better understanding of the relationship of what monitors measure (particulate emissions) and ways of improving the reliability of the monitors themselves. Otherwise, we just transform the problem of constraining source operation and maintenance behavior from control devices and operations to the monitoring equipment itself.
7. Until such monitoring capacity is developed, there are some arguments for trying to regulate sources in ways that do not require the unavailable capabilities. In particular, the use of low-sulfur fuels appears to be a more enforceable basis for regulation than the use of flue gas cleaning devices, which present acute operating, maintenance, and monitoring problems.

8. The EPA "emissions factors" methodologies should be more carefully defined to deal with process-rate and process-mix problems. And once defined, the use of approved methods should be made part of the condition for the acceptance of state implementation plans. Doing so would improve overall compliance with the SIP and would lessen the extent to which state agencies are subject to source-by-source pressures for favorable interpretations, because the states would then have to (and be able to) reply, "I'd like to help you, but I have no choice." The successful adoption of such restrictions, however, may well depend on also acting on the previous recommendation, so that the "tough" requirements also can be characterized as "rational" in the eyes of waste sources. No matter how accurate and carefully defined the emissions factors are, however, they will never in themselves allow the agency to control the actual day-to-day emissions from a source.
9. Even the existence of the capacity to discover operating violations, however, will not produce energetic compliance unless there are penalties for those who do not keep their processes and equipment within bounds. If monitoring can be made inexpensive and reliable enough, there would be a strong argument for a scheme in which fines were routinely assessed on the basis of the duration and magnitude of emissions violations. If analyzing monitoring outputs and fine computation can all be done automatically by computer, so much the better. Notice that this procedure implies that, unless sources install enough control equipment to satisfy the "worst case" with regard to process mix and operating rates, they will pay such fines at least occasionally. Such a scheme, by the way, would impose smaller dollar penalties on smaller sources than it would on larger sources that committed the same violations, in *percentage* terms. This property might let us set the fine levels high enough to be noticed by large sources without seeming to be an unfair imposition on smaller ones.
10. Until a computerized system of fine imposition can be implemented, states need a way of swiftly imposing fines. At the very least, EPA approval of the SIP should be conditional upon the power of an agency to impose civil penalties with few procedural delays on those who do violate emissions requirements.
11. Even with such changes, much of the success of the Clean Air Act enforcement will necessarily depend on the personal qualities and management skills and bureaucratic powers of the state agencies. Better review of permit decisions, the recruitment of young and committed personnel who want to make the program work, and effective rallying of

local public support for the program are all potentially significant contributions to what actually occurs.[69] These items will become even more important if previous recommendations are adopted because enforcement would become potentially costly to waste sources and the management and corruption-preventing tasks of state agency leadership would become that much more difficult.

12. In part because state politics will inevitably continue to play a role in state agency actions, the EPA needs to aggressively pursue its own enforcement activities. In at least one state we studied (Ohio), the EPA appears to have taken on substantial responsibilities for widely backstopping a state agency. In Massachusetts, the EPA proceeded against the two incinerators that the state legislature had tried to exempt, on the grounds that such action did not constitute an acceptable modification of the SIP.[70] Better cooperation between the EPA and state agencies in "dividing the work" where appropriate could be quite helpful.[71] However, again if tougher EPA enforcement is not to create enormous opposition, they must develop some broad belief in the rationality and reasonableness of the regulations being enforced.[72]

From a broader perspective, it is worth noting that any market or fee scheme, which economists have often urged be used in environmental regulation, would encounter many of the same monitoring and enforcement problems as the regulatory approach currently being pursued. Insofar as cheap and effective continuous monitors are not available, effluent fees, in particular, become difficult to use because their whole rationale is the imposition of small fees for each small variation in emissions. And in any case, violations of permits or deceptions with regard to fee payments will still have to be discovered and perhaps time-consuming court action undertaken to ensure compliance. So far, enforcement efforts have been most successful in areas in which source actions leave a paper trail (a key feature of income tax enforcement and compliance, too). Unfortunately, these are still only a minority of instances.

If we do choose to pursue fiscal incentives, one advantage of rules plus graduated fines or marketable permits as opposed to effluent fees is that they offer the agency some direct control over the *quantity* of emissions. This apparently greater degree of control over outcomes may enable the fiscal incentives approach to succeed despite the general hostility to such measures in the Congress in the past. And such control also does provide some hope for dealing more effectively with the long-run growth problem. However we proceed, the tighter controls that economic growth will require mean that a scientifically and economically defensible basis for imposing such controls (that is, technically sophisticatated SIPs and strong scientific support for ambient standards) and a well-oiled machine with appropriate incentives at its disposal for seeing that they are expeditiously carried out, will be essential to public acceptance and successful implementation.

Finally, an obvious moral from the experience of the Clean Air Act is that Congress operated on too short a time horizon. Society simply cannot mobilize the political-economic-legal system to do something as dramatic and complex as achieve air pollution control in a three-year period. The technical problems, the need to negotiate and build consensus, and the safeguards built into our system of judicial review simply prevent such rapid progress. It takes years for agencies to get organized, to set up and improve procedures, to acquire expertise, and to become effective at their tasks. In general, the state agencies we have studied are doing steadily better jobs—a hopeful sign in light of the fact that the problems of implementing the act will, in general, get more difficult over time as sources move from the easily monitored stage of making investments to the more-difficult-to-supervise situation of operating that equipment to achieve compliance. If we really do mean to have clean air under those circumstances, then some of the changes recommended here or other proposals designed to deal with the difficulties we have identified need serious consideration.

Notes

1. Council on Environmental Quality, *Environmental Quality,* sixth annual report, December 1975, pp. 310–326.

2. For a more general overview of policy options, see M. J. Roberts, ''Environmental Protection: The Complexities of Real Policy Choice'' in Irving K. Fox and Neil A. Swainson, eds., *Water Quality Management: The Design of Institutions* (University of British Columbia Press, 1975); Allen V. Kneese and Charles L. Schultze, *Pollution, Prices and Public Policy* (Brookings Institution, 1975). For an analysis of some early water pollution control programs that reaches many similar conclusions to our own, see Mathew Holden, Jr., *Pollution Control as a Bargaining Process. An Essay on Regulatory Decision-Making,* Cornell University Water Resources Center, publication #9, October, 1966.

3. The amendments were actually signed 31 December 1970. Although 30 days were given to proclaim ambient standards, they were not actually published in final form for about 120 days (30 April 1971); see 35 F.R. 8186, et seq. Guidelines for submitting state implementation plans were issued on 14 August 1971 (36 F.R. 15486 et seq.), and the plans themselves were due on 31 January 1972, nine months after the ambient standards were announced. Thus the tight schedule announced in the legislation slipped by three months at the very first step.

4. For example, because Ohio was unable to establish a satisfactory SO_2 emission standard, the USEPA promulgated a sulfur regulation for Ohio on 27 August 1976; see 41 F.R. 36324, et seq.

5. U.S. Comptroller General, *Assessment of Federal and State Enforcement Efforts to Control Air Pollution from Stationary Sources* (U.S. General Accounting Office, 1973).

6. Although Air Quality Control Regions established under the 1967 legislation tended to be logically constructed (e.g., to include a metropolitan area), many of those created under the 1970 amendments are mere administrative conveniences, such as "Nebraska except Omaha."

7. J. Clarence Davies III and Barbara S. Davies, *The Politics of Pollution,* second ed. (Bobbs-Merrill, 1975), p. 205. The authors later argue, however, that procedural limitations have led the federal government largely to ignore this statutory authority; see ibid., pp. 207–210.

8. See Senate Report No. 91-1196, 91st Congress, 2nd Session, p. 3, (1970).

9. Note 1, supra.

10. Ibid. pp. 317, 319, Table 4, and Table 5. These tables do not tell us by how much or for what time periods or exact areas each city can be expected to violate those standards.

11. Ibid. Figure 13 (p. 333) and Figure 15 (p. 335). The former figure records a decline in winter seasonal sulfur dioxide peaks beginning in 1968–1969 and much smaller summertime changes. This difference appears to be related to the decline in retail coal deliveries noted in the latter figure because the distillate and residual fuel oil that replaced much of the coal for price reasons is typically noticeably lower in sulfur content.

12. Turner, *Workbook on Atmospheric Dispersion* (U.S. Environmental Protection Agency, 1970).

13. Davies and Davies, op. cit., p. 187.

14. In Massachusetts, only two zones were established and uniform rules established for all of the state outside of the Boston area. This effort hardly constitutes a cost-minimizing plan.

15. For example, North Carolina only recently has developed facilities more complex than a workbook in atmospheric dispersion for modeling the impact of new sources. Interview with Brock Nicholson, head of technical support section, Division of Environmental Management, 29 August 1976, Raleigh, North Carolina.

16. Davies and Davies, op. cit., pp. 184–185.

17. Interview with Ron Brubaker, former assistant state attorney general, 11 August 1976 in Columbus, Ohio.

18. In New Jersey, for example, the sulfur content regulations on liquid fuels exempt the state's southern region from restrictions on residual fuel oil tighter than 1

percent, whereas the northern regions have a 0.3 percent restriction. See New Jersey Department of Environmental Protection, New Jersey Administrative Code, Title 7, Chapter 27, subchapters 9.2 (a) and 9.4 (b).

19. North Carolina Administrative Code, Title 17 Department of Natural Resources, Chapter 2, Environmental Management, subchapter 2D, sections .0503 and .0516 and Alabama Air Pollution Control Commission, Rules and Regulations.

20. For example, the Georgia regulations allow a new source that is over 250×10^6 BTU/hr to emit 0.7 lb. $NO_x/10^6$ BTU when firing coal, 0.3 lb. $NO_x/10^6$ BTU when firing oil and 0.2 lb. $NO_x/10^6$ BTU when firing with natural gas. Old sources and smaller sources are not given any emission limitation. Georgia rules and regulations for Air Quality Control, Chapter 391-3-1-.02.

21. Rules and Regulations for Air Quality Control, note 20 supra.

22. North Carolina Administrative Code, Title 17, Department of Natural Resources, Chapter 2, Environmental Management, subchapter 2D, sections .0512 and .0522.

23. See Los Angeles Air Pollution Control District, Rule 67, adopted 1969.

24. This account is based on M. J. Roberts and J. Bluhm, *The Evolution of Environmental Decisions* (forthcoming).

25. Telephone interview with Duane Bordvick, variance section, Los Angeles Air Pollution Control District, 28 September 1976.

26. Roberts and Bluhm, op. cit.

27. Telephone interview with David Levin, former Chairman of Florida Pollution Control Board, 8 June 1975.

28. In May 1975, the state's interim standards, which allowed Gulf Power Company to burn 3.5 percent sulfur coal, were extended to December 1977 to postpone a requirement for approximately 1 percent coal. Cole interview, note 22 supra.

29. See Harvard Business School, case 9-371-076, BSI 110: "Boston Edison," pp. 25–32.

30. The bill was introduced by State Senator Daniel J. Foley of Worcester as Senate No. 1594. It was signed by Governor Sargent on 9 July 1974.

31. Interview with Massachusetts state representative who prefers to remain anonymous, September 1974.

32. Interview with Moses McCall, Head of the Land Protection Branch of the Georgia Environmental Protection Agency in Atlanta, Georgia, on 27 July 1976.

33. Interviews with various members of the North Carolina Environmental Management Division, 29 and 30 July 1976 in Raleigh, North Carolina. Interview with Marshall Racklee, director of the Central Field Office of the N.C. DEM in Raleigh, North Carolina. Telephone interview with Milton Heath of the Institute of Government of U.N.C., Chapel Hill on 17 September 1976.

34. Telephone interview with Milton Heath, note 33 supra.

35. United States Environmental Protection Agency, AP-42.

36. Interviews with Ed Cave, permitting engineer of the Illinois Environmental Protection Agency, in Springfield, Illinois, on 14 July 1976, and John Mitchell, director of the Air Permit Compliance Program of the Georgia Environmental Protection Division, on 27 July 1976. Racklee 37 supra.

In order for a permit system to work, the state must have the appropriate information and that information must be accurate. If a state asks only for the average SO on ash content of the coal, as in Georgia and North Carolina, it will never be able to predict the worst concentration of emissions a source can put out. If a state never checks to see whether or not the information is true, again the emissions predictions will not be correct. Thus, without careful information gathering and scrutiny, a permit can serve as a license to pollute.

37. A.B. Walker, *Experience with Hot Electrostatic Precipitators for Fly Ash Collection in Electric Utilities* (Research-Gottrell, Inc.; presented at the American Power Conference, 1974).

38. Interview with Charles Taylor, director of Division of Program Development and Review of the OEPA, in Columbus, Ohio on 10 August 1976.

39. Interview with state source that chose to remain anonymous.

40. Interview with Tom Crepeau, head of the Air Permit Records Section of the OEPA on 10 August 1976 in Columbus, Ohio, and telephone interview with Clyde Watkins, director of the Division of Field Operations who supplied information on state practices on 22 September 1976. The interpretation that this could tend to encourage lax application of the rule (in order to avoid later review) is our own.

41. Nicholson, note 15 supra. He believes the discrepancy is due to an underestimation by the agency of the contribution of fugitive dust to ambient air quality when the SIP was written.

42. This point was widely reported by state officials, including Cave, note 36 supra; Mitchell, note 36 supra; and Racklee, note 33 supra.

43. 40 *Federal Register:* 46240 et seq.

44. Telephone interview with Michael Fogel of the stack test section of the Georgia Environmental Protection Division on 24 September 1976.

45. Ibid.

46. Telephone interview with Jerry Levi, the air enforcement inspector for Massachusetts in Region I, U.S. Environmental Protection Agency, on 23 September 1976. The EPA ranks the procedures used to evaluate a source's emissions according to their validity. A government-performed stack test is considered to be the most accurate assessment of a source's emissions. A stack test performed without any government observers present is the least acceptable procedure. The use of emissions factors is considered to provide a more accurate representation than an unobserved stack test.

47. Telephone interview with Marshall Racklee, 17 September 1976.

48. In Georgia, Victor Belba, inspector for the Georgia EPA, said in an interview in Atlanta on 28 July 1976 that Georgia samples the coal from power companies quarterly according to a careful averaging procedure, but Georgia is the only state that has a regular program. John Desmond, director of the Massachusetts Department of Air Pollution Control, on 22 September 1976 stated that coal is sampled occasionally but there is no formal sampling plan; John Hagaman, head of the monitoring section of the North Carolina DEM, and John Romans, head of the field operating section, said the same thing in an interview on 30 July 1976 in Raleigh, North Carolina. Many power plants do sample the coal for their own purposes however.

49. Interviews with Victor Belba, note 48 supra; Miles Zamco, head of the field operations section of the Illinois Environmental Protection Agency in Springfield, Illinois, on 14 July 1976; Joe Darling (regional inspector), Steve Capone, and Steve Dennis (regional engineers) of the Metropolitan Air Pollution Control District in Boston, Massachusetts, on 22 September 1976.

50. Zamco, note 49 supra.

51. Darling, Capone, and Dennis, note 49 supra.

52. Interview with Massachusetts state official who choose to remain anonymous.

53. McCall, note 32 supra. In September 1976, during a telephone interview, Brubaker (note 17 supra) also recounted an incident when he preferred to settle a case out of court rather than face a hostile local judge.

54. Brubaker, note 17 supra; and Crepeau, note 40 supra.

55. Brubaker, note 17 supra; and telephone interview with Tom Boltaglio, assistant director of the air enforcement section, region V office of the U.S. EPA.

56. Interview with Russell Eggert, an assistant attorney general for the state of Illinois, in Springfield, Illinois, on 15 July 1976; and John Palinsar, head of Enforcement Services, Illinois Environmental Protection Agency on 13 July 1976; also Davies and Davies, op. cit., pp. 126–166, 212–216; and Richard Briceland, director of the Illinois EPA on 15 July 1976.

57. Palinsar, note 56 supra; and Eggert, note 56 supra.

58. Racklee, note 33 supra; and Romans, note 48 supra.

59. Belba, note 48 supra.

60. Interview with Charles Corkin, assistant attorney general for the commonwealth of Massachusetts, in Boston on 22 September 1976.

61. Interview with a number of Georgian sources who prefer to remain anonymous.

62. These pressures can also affect the types of enforcement cases an agency brings. For example, the consolidated environmental management program plan for the transition quarter FY 1976 and initial draft FY 1977 of the North Carolina DEM shows that .12 man-years are used to take enforcement action against permit and special orders and .12 man-years are spent on open burning regulations. Note that these violations are easily discovered, proven, and stopped.

In Illinois, where there has been a concerted effort to take action against major sources rather than minor sources, Palinsar (note 56 supra) mentioned that he wishes that the Illinois Environmental Protection Agency would reemphasize nuisance actions because they are easy to prosecute and because they generate public support.

63. A source within the agency who prefer anonymity, relates how one inspector in Massachusetts apparently cited a source for violations of the air pollution regulations after being told not to by the legislator in whose district the source was located. Within a week, the inspector had been transferred to the monitoring section of the agency. Similarly, a former official of the state-run Metropolitan District Commission of Massachusetts said, ''every time we need a new pump for the water supply system, we have to bargain in the legislature and promote someone from lieutenant to captain in the park police.'' On Massachusetts politics more generally, see E. Litt, *The Political Culture of Massachusetts* (MIT Press, 1965).

64. Those regulations impose very different sulfur content limitations on different sources—derived from diffusion modeling and taking account of source size, other sources in the area, topography, etc. See 41 *Federal Register,* 36324 et seq.

65. Note that the worsening of the ambient situation with economic growth is only *possible,* not *necessary.* It is at least conceivable that changing technology and factor prices will induce enough of a shift toward cleaner processes to allow us to grow without more stringent external controls, at least as selected parameters. The decline in urban coal use during the 1960s and the associated improvements in ambient quality (see notes supra and associated text) exemplify such a phenomenon. The steady switch to nuclear power for electricity in place of higher-cost oil may well provide another. Of course, as the text suggests, we can hardly count on such favorable developments to solve our problems.

66. See the discussion in Roberts, op. cit.; M. J. Roberts and M. Spence, ''Effluent Fees and Marketable Licenses for Pollution Control,'' *Journal of Public Economics.*

67. Connecticut Enforcement Project, *Economic Law Enforcement,* vol. 1 (Department of Environmental Protection, 1975).

68. 40 *Federal Register* 46240 et seq.

69. Many people observed that there is a generation gap in the state agencies. The people at the top levels tend to be water pollution engineers left over from the public health days. The people in the middle management and below tend to be young, idealistic engineers who came to the agency when it was first formed, often straight out of college. This generation gap has led to frustrations over enforcement techniques. The ''old guard'' still tends to have the attitude that negotiation will achieve compliance, whereas the younger people would like to see more stringent action taken against the polluters. One can hypothesize that one of the reasons that the Clean Air Act has improved air quality is that federal money forced states to hire people, and this ''new blood'' has enabled the agencies to have even moderately forceful programs.

70. Interview with Levi, note 46 supra.

71. Corkin, note 60 supra. Also see the letter sent by Charles Corkin II, Morton Goldfein, and Troy Webb for the National Association of Attorneys General to John Quarles deputy administrator of the EPA, on 29 July 1976.

72. The federal personnel are not themselves immune to political pressures. One state agency source, who prefers to remain anonymous, alleged that there was little point in referring cases to the U.S. Attorney in his area, because such cases were almost always put into the lowest priority category and, hence, not prosecuted in a timely manner. Clearly, various regional offices differ slightly in this regard.

Appendix A (as of 1/77)

	Georgia
Total facilities	3,000
Major facilities: (potential 100 T pollution/year)	276
not in compliance	15
on schedules	12
# federal schedules	4[b]
# not following schedules	1
Total federal NOV's	17
Total state fines/amount	~15/$20,000
Inspectors: engineers/salary	16/$13,700[h]
technicians/salary	4/$12,510[i]
Stack tests in state/year	200
Fuel samples performed by state/year	50
Monitors	130
sites	30
% source specific	30%

[a] our facilities are of unknown compliance status
[b] Also on state compliance schedules.
[c] It was not known whether one additional facility was in compliance with its schedule
[d] It was not known whether thirteen additional facilities were in compliance with their schedules.
[e] FY 1976 only

Illinois	Mass.	N.C.	Ohio
40,000	10,000	10,000	20,000
1,441	600	885	950
45[a]	70	83	240
43	45	57	205
30	20	2	30
2[c]	0	12	26[d]
47	39[e]	54	50
340/[f]	0/0[g]	11/$10,000	0/0
30/$16,000[i]	8/$14,000[h]	19/$15,000[h]	160/$12,000[h]
0	16/$7,000[i]	21/$11,000[i]	
200	65	101	160
0	465	277	0
378	132	140	607
—	57	—	—
0	<1%	25%	0

[f] As of May 1976 Illinois had assessed a total of $1,649,579. 10 in fines against 550 sources of air, land, water, and noise pollution.
[g] One oil company was fined $100,000 for selling fuel with an excess concentration of sulfer.
[h] responsible for permitting, inspecting, plume monitoring.
[i] responsible for inspecting and plume monitoring.

Comment

Henry Beal

The authors characterize as the chief deficiencies of state implementation plans (SIP) their simplicity of formulation and inconsistency of application. Because SIPs are the chief means through which the Clean Air Act is implemented, such deficiencies are said to explain the failure of the act to produce clean air in a timely manner.

Simplicity might be considered a virtue in most government regulations, but in this instance it is manifested by the formulation of uniform emissions limits legally effective over widespread geographic areas and applicable uniformly to air pollution sources that experience diverse marginal costs of pollution control. ''Irrational'' is how the authors label this simple approach. The result has been pervasive misallocation of the pollution control burden, proved by the widespread failure to attain clean air standards, and unsound cost distribution. Perceiving these faults, say the authors, businessmen have declined to comply voluntarily with laws they regard as arbitrary, and politicians have sought to weaken environmental controls that struggling business doesn't like and that state air pollution control agencies are unable technically to defend. Simplicity is, in this case, irrational, and irrationality has made the attainment of clean air standards *uncertain*.

Inconsistency of enforcement, whatever the emission limits may be, can make the attainment of clean air more difficult. The authors' study of several state air programs allows them to identify a series of inconsistencies that I would like to classify as resulting from either *structural* or *operational* difficulties. The chief structural impediment to consistent enforcement is the cumbersome, ineffective mechanism in most states for applying sanctions against unlawful behavior. The chief operational inconsistencies result from the inefficient and unpredictable exercise by enforcement officials of their substantial discretion during reviews of new source permit applications and inspections of existing sources for compliance. Inconsistencies exist between individual states and, in each state, between individual decisions. Inconsistency of enforcement reinforces the uncertainty that clean air goals can be attained.

Simplicity These descriptions of program flaws seem largely accurate to me. Of course from the perspective of the Clean Air Act, which demonstrates little concern for the methods of state enforcement and even less for economic rationality, such characteristics are flaws because they produce uncertainty in the attainment of standards. Had SIPs produced instead clean air, the authors' paper and this discussion would be largely unnecessary. To improve the rationality of the SIPs, the authors recommend a detailed analysis of the pollution reduction burden assigned to individual pollution sources involving computer models of pollutant dispersion. Conceptually, this approach is more rational in terms of air pollution control than the uniform ''rollback'' approach used in the development of most SIPs. Diffusion models are certainly capable of greater analytic precision than rollback models. Practically, there are a few drawbacks worth considering.

First, the accuracy of diffusion models depends on the accuracy of the data input and such data is notoriously suspect. Although inventories of annual emissions are improving slowly, data about shorter-term emissions can probably never be gathered with the timeliness required for the models to simulate real-world conditions. As a result, simplifying assumptions are incorporated in the analyses and much of the theoretical accuracy of the models is washed out. If among the assumptions there is a safety factor accounting for inherent flaws in the model, such as inability to accurately account for terrain features, and for the inadequacies in the data, the outcome of a source-by-source analysis may well be source-by-source overcontrol. This notion does not bother me, but it may annoy economists.

Second, as a percentage of control agency budgets, the cost of the analysis will be spectacular. This notion may not bother economists, but it annoys me. A decision to allocate to a single source any particular allowable emission rate will affect, in different degrees, the emission rates allowable to every other source in the area impacted by pollution from the first source. The number of possible scenarios thus approaches the factorial of all sources being considered. To model a single scenario can cost up to $5000 in a state the size of Connecticut, although each analysis need not be done on a statewide basis. All sorts of simplifications might be made to reduce the costs of analysis by treating all small sources as a group or applying common emission factors to a whole industrial category. To the extent this is done, one saves money but moves closer to the rollback approach. If there were any prospect that an appropriate scenario would allocate pollution control burdens in a cost-minimizing fashion, perhaps those who hoped to benefit could be induced to pay for the search. However, cost of control is not a variable considered by diffusion models and that fact raises the third drawback.

Third, no particular emissions limit allocation will necessarily recommend itself as "most rational" in air pollution terms. If it is possible to find a technologically achievable allocation of pollution control burdens that will make the air clean enough to meet act standards, it is all too likely that many variations of such allocation could be found. Choosing among allocations is emphatically not a function for an air pollution control official. Who should choose? Use of some market mechanism (such as emission fees) would result only by chance in "selection" of one of the control burden allocations found acceptable through diffusion modeling, and the chance, in my opinion, is of a low order of probability. Instead, perhaps a cadre of economists could try to estimate marginal control costs for all the affected sources, then apply such costs as a factor to each scenario, and find the cost-minimizing choice. I think it would be an edifying experience to try to get such confidential data from the business community, but the attempt is unlikely to succeed. Perhaps the choice could be made in terms of size of source by adopting the scenario that favors the largest or smallest sources or in terms of the political muscle of the sources or the control agency's estimate of the enforceability of each scenario. In sum, the use of diffusion models may lead to rational allocation of control burdens, but rationality is not automatic. What is rational in terms of some values (such as pollution reduction) may not be too rational in terms of other values (such as politics or economics), and apriori it is rather difficult to say in what sense a more complex and sophisticated pollution analysis will result in greater rationality.

The fourth category of drawbacks I shall call "miscellaneous." There is growing evidence that, for a couple pollutants in some parts of the country, our ability to link through diffusion models the causes and effects of pollution levels is either (1) very inadequate or (2) being applied on the wrong scale (that is, on a state scale rather than a multistate scale). With pollutants such as hydrocarbons both problems exist. Near-term remedies are unlikely, and for these pollutants the rationale for control is likely to become less sophisticated rather than more. Indeed, "less rather than more," a classic conservationist credo, may be the only near-term rationale. A policy based on enforced conservation of resources and minimization of waste strikes me as acceptable although its justification may require more acts of faith than acts of government. Finally, there is the issue of growth margins and their adequacy, an issue intensified by the federal policy of "prevention of significant deterioration." A rational distribution of the pollution control burden is highly dependent on perceptions about the amount of clean air resource to be left unused for new development. In most states these perceptions are ill-defined and it is rare to find them articulated in any form.

Are these drawbacks so overwhelming that a state air pollution control agency ought to ignore the author's proposal? Definitely not. The proposal is really quite good. Its merit, however, lies in its capacity to make more certain whether or not an existing SIP can ever hope to achieve clean air. Diffusion model analysis for appropriate pollutants in appropriate areas performed at a pace scaled to the resources of the pollution control agency can result in a more definitive statement concerning the expected effects of the control program and the need for additional control. Such analysis can serve a valuable public information function. In the course of attempting to change any SIP in the wake of the analysis, a host of political and economic issues will be explicitly raised for consideration. The outcome is unlikely to be "rational," but it can aspire to greater certainty.

Inconsistency Inconsistent application of any SIP control program makes attainment of clean air standards more uncertain. Flawed control-agency structure and operating procedures both can contribute to inconsistent enforcement. In general, the authors find that structural complexity tends to interfere with the effective punishment of malefactors and recommend a simplified, automatic sanction system. Operational inconsistencies, say the authors, result from the erratic exercise of discretion by control agency officials in cases in which compliance is technically uncertain and sources can be given the benefit of any doubt. The recommendations in both problem areas are really quite good; in those instances in which they are currently impractical, they at least suggest a sound direction for change.

In regard to the authors' recommended operational changes, the notion of uniform guidelines for engineering assessments of source compliance is an excellent one. The selection of emission factors, the use of worst-case or average-case analysis, the choice of diffusion models and the definition of significance of impact shown by such models are not matters that should be subject to substantial variations among states. Failure to achieve uniformity leads to a sort of pollution "forum shopping" by the business community and encourages political pressure on control agencies to compete in the laxity of analytic criteria. However, although the more debilitating differences can be eliminated in the fashion recommended, there is no prospect whatever that engineering judgment and thus discretion can be eliminated in the application of the guideline to individual circumstances.

Also valuable is the recommendation that actual emissions of regulated pollutants be measured by devices installed at the source. Widespread use of such devices, perhaps, as suggested, telemetering their data to a computer, would take us close to a regulatory nirvana that would allow comprehensive management of the air resource. With or without telemetry, such devices are critical to any emission fee program that purports to relate the amount of the fee to the amount of excess pollution discharged. Alas, there are two problems. First, the technology may not be good enough to make such programs possible and those programs that are possible will be quite costly. For example, an instrument capable of measuring actual emissions of sulfur oxides with some assurance of quality and hope of reliability costs about $10,000. Its annual operating costs may approach that amount. Telemetry equipment is equally costly. If the source burns fuel, one may want to monitor, at a capital cost of several thousand each, particulate, hydrocarbon, and nitrogen oxides emissions, as well as sulfur oxides. With something simple, such as an asphalt batching plant, one might settle for measurement of particulates. With any measurement, the question arises whether the air sample pulled from a stack (which may have a diameter of several feet) through a "probe" (which may have a diameter of an inch or less) can be said to measure total emissions with sufficient accuracy. By comparison, during a so-called "stack test" as many as forty or more samples are taken from different locations in the stack in an effort to insure a representative sample. If fines, fees, and legal action are to be based on measurements from these devices, their accuracy must be provably good. For certain sources, however, I believe these problems can be overcome; for some others, an approach based on measurement of some proxy variable, such as "pressure drop," may be satisfactory.

Second, reliance on these devices simply redirects from the pollution control equipment to the monitoring device the judgment exercised by the enforcement official: is the monitor working properly; is it properly calibrated; is the sample representative? In some cases, especially if a simple proxy variable is used to measure compliance, the judgment may be easier to make and more certain. In no case will the exercise of judgment be eliminated.

Regarding the authors' recommended structural changes, from the perspective of management efficiency and enforcement effectiveness, it makes sense to move the point of decision to sanction to the lowest practical level in the bureaucracy and to set decision-making criteria that propose automatic sanction assessment once a violation is determined to exist. It is this facet of the Connecticut Penalty Plan, as much as the clarity of its penalty calculation formula, that makes it a valuable model. Where the decision-making process has been allowed to function within this structure, enforcement effectiveness in Connecticut has been remarkably improved. Needless to say, the more effective a tool is in its ability to motivate behavior, the more unlikely it will be left to function automatically. Due process safeguards and political jitters can formally and informally result in a shift of the decision-making point in individual cases to commissioners (political appointees), hearing officers, or judges and cause applications of sanctions to become less certain. The threat of occasional rearrangement of the structure does not argue against its basic soundness. Automatic sanctions are a good idea, but exercise of considerable judgment will characterize much of the decision making in this area.

In my opinion, then, the authors are moving in the right direction, but for the wrong reasons. My concern for motives is not idle quibbling. The authors are apparently motivated to their recommendations in part because of their opinion of state air pollution program officials. By implication and sometimes overtly, they indicate that state officials are inept, ignorant, incapable, and have been saved from a life of white-collar crime solely because their enforcement programs are so ineffective that businessmen consider the offer of a bribe to avoid enforcement a redundancy. Astonishingly, federal officials are portrayed as comparative competents prevented from taking immediate enforcement action by the Clean Air Act's delegation of power to the states. Hogwash—on all counts.

In view of this bleak picture of the human resources available to solve the nation's air pollution problems, it is no surprise that the authors, who have already abandoned hope that the invisible hand might save the day, recommend reliance on the invisible inspector. Uniform guidelines are proposed as a means of making the inspector's job as ministerial as possible; monitoring devices are proposed to eliminate engineering judgment from determinations of compliance; and the Connecticut civil penalty program and excess emissions fees are proposed to remove all human elements from the decision to sanction.

Well, the uncertainty of attaining national air quality standards is not going to be eliminated by creating an impersonal mechanics of pollution control that renders inconsequential the role of the fallible enforcement official. One's sense of priorities comes out wrong, in my view, when otherwise good recommendations are offered as a means of displacing people from decision-making roles rather than as a means to make the decision making more effective. I know that I am exaggerating the authors' position in this matter. They do devote some space to urging that state air pollution agency officials be better recruited, better trained, and better motivated. This sentiment, however, is a commonplace in analyses of government and in this paper tends to lose whatever weight such urgings usually carry because it has been preceded by a series of recommendations that seem to claim credit for making enforcement official sentience unnecessary. Improvements in analytic tools and procedures should be made in the awareness that their success is dependent on the character, training, and morale of the persons who implement them. Investments in human capital should accompany investments in new procedures and devices to analyze or measure pollution, and probably should precede them. With this modest reorientation of purpose I think the recommendations offered in the paper can be of value to persons responsible for implementing the Clean Air Act.

Comment

Charles Corkin II

Roberts and Farrell argue that the strategy of the 1970 amendments to the Clean Air Act is inflexible and cannot be implemented within our political and legal system.[1] They conclude that simplified enforcement procedures and stricter penalties are required. Their treatment of enforcement procedures implies that the states make it difficult for the federal Environmental Protection Agency (EPA) to carry out the purposes of the act.

My thesis here is straightforward: existing law is adequate to ensure compliance with air quality standards. Existing enforcement procedures are simple and adequate, and the states can obtain compliance. The Clean Air Act requires that they be assisted in such tasks by EPA, which does not have the capacity to assume more than an ancillary role in enforcement.

The reason for failure in enforcement is unrelated to the law or available enforcement procedures; rather, it is a result of the public perception of the importance of clean air and the fear that the cost of pollution control is greater than the rewards. For example, the threat that a manufacturer will move to another state rather than face enforcement action may make elected and appointed officials hesitant to take a firm stand on environmental protection. As a result of EPA's reluctance to show its enforcement power nationwide, pollution havens still exist wherever local officials choose to allow them.

New law and procedure are not necessary. The horse has already been led to the water; the task is getting it to drink.

The Act

The 1970 amendments to the Clean Air Act "reflect a genuine effort to develop an elaborate form of cooperative federalism in which both the federal government and the states are given vital roles."[2] In enacting this statute, Congress determined that "the prevention and control of air pollution at its source is the primary responsibility of states and local governments."[3] EPA is essentially "relegated by the act to a secondary role in the process of determining and enforcing the specific source-by-source emission limitations which are necessary''[4] to meet air quality standards. No other approach could work in our federal system.

Stated simply, the Clean Air Act requires that:

1. EPA establish national uniform ambient standards for several pollutants.[5]
2. Each state submit for federal approval a written plan for the implementation, maintenance, and enforcement of air quality standards within each state.[6] This plan is referred to as a state implementation plan, or SIP.
3. Each state enforce its SIP or suffer the indignity of standing by while the federal government attempts to do it.

In principle, this scheme preserves the states' primary role in enforcement and insures national uniformity. According to the act, officials of each state can draw on the resources of EPA in order to pursue a vigorous enforcement policy, secure in the knowledge that polluters will not move their tax dollars and jobs to another state because the other state or EPA will be there with an equally vigorous enforcement program. State officials, the theory goes, must enforce the SIP in order to prevent federal preemption of one of the state's primary responsibilities.

In practice, state officials do not act according to the theory. Instead they balance tax dollars, jobs, and other political considerations against the necessity or desirability of clean air. EPA also fails to follow its statutory mandate in some instances, in part because a large bureaucracy cannot move with the speed required by the act and in part because some EPA officials are simply not interested in assisting state enforcement officials. My analysis of EPA's failure follows.

The Majestic Simplicity of Enforcement

Roberts and Farrell state that the enforcement and implementation processes involve substantial discretion by state agency officials and that EPA is unable to control the abuse of that discretion. Accordingly, their argument continues, ''We have not had a tough program and current efforts for the control of stationary source air pollution are unlikely to attain ambient goals.''

This conclusion is erroneous. It is true that state agencies have traditionally favored a nonadversary, ''jaw-boning'' approach to compliance. In Massachusetts, however, the Department of Environmental Quality Engineering has selected litigation as the most effective method of persuading sources to comply with the SIP. In Massachusetts, implementation schedules, lengthy meetings, correspondence and memoranda ad nauseum are associated only with the federal bureaucracy. When the names of sources out of compliance are forwarded to the attorney general for action, a complaint is filed within two to three weeks in the state superior court asking for immediate closure and for a receiver to take over the company or municipality so that it can be run to comply with the law. The results of court action are wonderfully staggering: almost every source requests an opportunity to discuss the matter and the average case is disposed of in from one to three months. And the disposition includes rigorous terms. The defendant normally agrees to an implementation schedule that state engineers have determined to be the most stringent possible and agrees to pay a fine (usually $1000 per day) if that schedule is not met. The agreement is embodied in a document called a consent judgment, which is approved by the court. Frequently the polluter pays a fine for his violation of the July 1975 attainment date.

The threat of fines, however, is minor in comparison to the threat of imminent closure and the appointment of a receiver. Contrary to Roberts's and Farrell's conclusion that delays of up to two years are not unlikely if court action is pursued, a motion for preliminary injunction (seeking closure of the source) may be heard in less than two weeks. Because the law is clear that sources may not operate out of compliance after 31 July 1975, a source would rather "make a deal," however tough, than to face a judge, who might close it immediately.

It might appear that actions against municipalities would be the most difficult because there is currently a fair amount of sympathy for the beleagured American city. In July of 1975, however, the Suffolk Superior Court ordered the Boston Incinerator closed after a two-week trial. In one of two written opinions in the case the judge indicated that he did not believe the court had the authority to permit the incinerator to operate in violation of the source emission limitations set forth in the SIP. In a subsequent case, a one-hour argument on the commonwealth's motion for a preliminary injunction resulted in an order closing the city of Lowell's incinerator. Still other consent judgments have been obtained closing the incinerators of Holyoke, Fall River, and Framingham.

Litigation is so successful an enforcement tool that when I recently asked the EPA regional enforcement director[7] for the names and addresses of sources the commonwealth hadn't yet sued, she responded that EPA had none.[8]

The Unsatisfactory State-Federal Relationship

The 1970 Clean Air Act amendments were enacted largely because no state would be able to enforce the costly requirements of air pollution control if one of the polluter's options were to move across a state line for more lenient treatment.

The most important reason for incomplete fulfillment of the act's objectives is EPA's failure to establish a strong, effective, and uniform national enforcement policy. Without EPA's policy rhetoric, there would not be even a semblance of a functioning enforcement program. For example, in Massachusetts, which has the largest population and the largest number of air polluters in the New England region, EPA has filed only one legal action against an air source since the inception of the program. The defendant was the town of Weymouth, whose incinerator was exempted from state enforcement. The complaint was filed in court only after agreement was reached allowing the incinerator to violate source emission limitations for ten months beyond the date specified in the SIP approved by EPA.[9] Winchester, another of three towns exempted from state enforcement, operated its incinerator for a year and a half beyond the statutory date; EPA brought no court action.

This inaction is extraordinary in the light of the fact that the EPA enforcement division employs a staff of more than fifty, operates its own computer, has a personnel budget that exceeds a million dollars annually, and has access to a separate and comparably well-staffed division responsible for surveillance and analysis of air pollution.

One reason for the government's failure to enforce is EPA's organizational structure, which is not conducive to action. The agency has its headquarters in Washington, D.C., and ten regional offices located throughout the United States.[10]

The hierarchy of headquarters personnel is illustrated in figure 1. Each regional office is an approximate replica of headquarters, as is illustrated in figure 2. Although it is said that headquarters sets policy and the regions execute it,no one knows where the responsibility for a particular decision lies. As a result, there is a great amount of friction, sometimes hostility, between headquarters and regional personnel. This friction is compounded by the state's involvement.

Consider the EPA decision-making process associated with the question of possible enforcement action against a pollution source. It may be helpful to refer to figures 1 and 2 while tracing the following steps in EPA's processing of a typical case:

1. A computer produces a printout that recites that Badair Corporation has not complied with the 1973 state order requiring the installation of pollu-

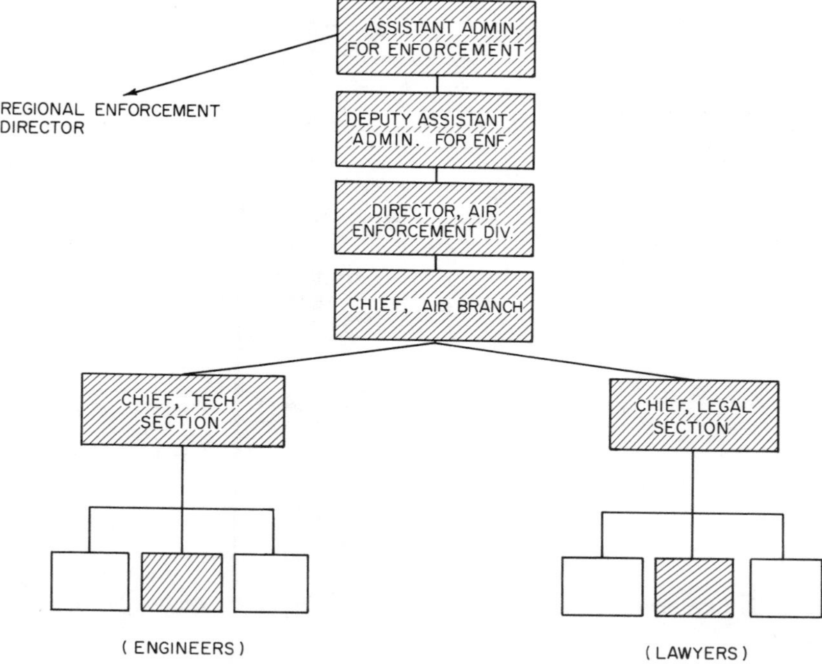

Figure 1. EPA headquarters enforcement office

tion control equipment necessary to meet the appropriate emission limitations under the EPA-approved state implementation plan.

2. This information moves through several unknown steps to an engineer and a lawyer (see bottom line in figure 2).
3. The engineer and lawyer consult and agree that something should be done.
4. The lawyer refers the matter to the legal review section chief for his concurrence and the engineer refers the matter to the technical review section chief for his.
5. The two section chiefs, the engineer, and the lawyer meet to determine whether concurrence still exists. By this time as many as six memoranda have been passed among the participants.[11]
6. The meeting noted in step 5 and the exchange of memoranda normally result in a meeting with the enforcement branch chief, after which more memoranda are written.
7. The enforcement director holds a meeting attended by the branch chief,

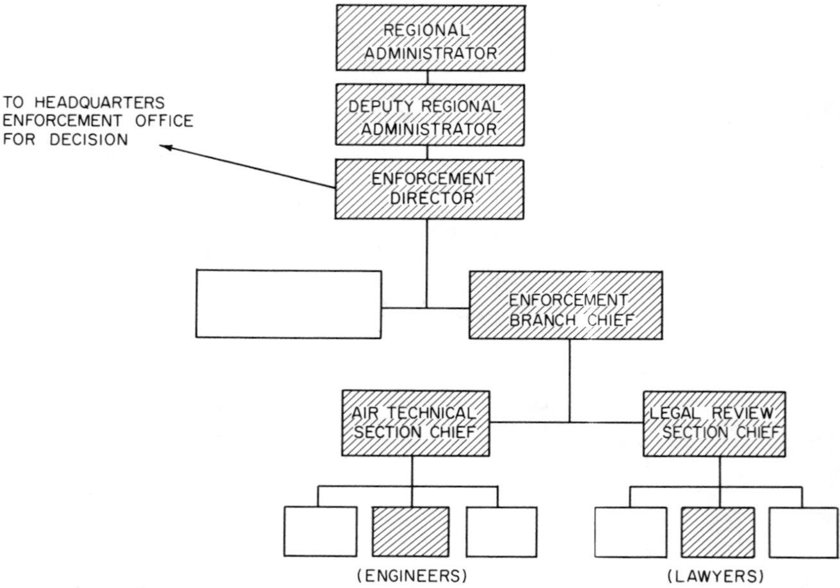

Figure 2. EPA regional enforcement office. A flow chart of the decision makers

the two section chiefs, the engineer, and the lawyer. It is reiterated that Badair Corporation is in violation of the Clean Air Act and that something should be done. The question now is what to do.

Because of agency policy, EPA has never brought a criminal action in Region I;[12] the enforcement alternatives are: (a) to refer the matter to the state attorney general to handle on behalf of the state agency; (b) to refer the matter to the U.S. Attorney; or (c) to issue an administrative order. The first is unacceptable to EPA, because it would reduce the involvement of federal officials. The second alternative, which transfers the action and the control from EPA to the justice department and a U.S. Attorney's Office, is cumbersome and unreliable. The third alternative, issuance of an administrative order, is an illegal "revision" of the state implementation plan, because it does not comply with the requirements of section 110 of the act. Nevertheless, it is the most common enforcement practice because it increases the involvement of the federal bureaucracy and because each issued order counts as a "bean" under the agency's credit system.[13] For example, a regional office gets credit for every order issued to a noncomplying source, even if each of the orders merely extends the compliance schedule of the last. The regional office gets no credit if the matter is referred to the state attorney general or if information and assistance is given to him so that an assistant attorney general and a state engineer can solve the problem by persuading a court to order the source into compliance or closed.

8. Accordingly, the enforcement director tentatively[14] decides that the regional administrator should issue an administrative order. A new series of meetings brings together the polluter, the deputy regional administrator, and the regional administrator in various combinations and at various times. Many memoranda are written. Agreement with the polluter regarding a comfortable compliance schedule is zealously pursued.
9. Further advice and concurrence are sought from headquarters. Before a decision is made, the headquarters enforcement official who was contacted by the regional office consults with each of the other headquarters officials who are symbolized by shaded boxes in figure 1. They hold meetings and write memoranda. The regional enforcement director is ultimately notified of the decision she is to make.
10. After EPA and the polluter have agreed on an implementation schedule,[15] regional personnel prepare a draft order, which is forwarded to Badair Corporation for the signature of a person authorized to indicate the corporation's consent. Many copies are made.
11. In some cases, the regional administrator calls a press conference, at which he announces the order and elaborates on how EPA is cleaning up the environment.

If, on the other hand, the enforcement director decides at some point in the process that court action is desirable, additional reviews are necessary before it can begin, and the hierarchies of the justice department and the U.S. Attorney are called into action, if ''action'' is the appropriate word.

It is not clear who has the authority to make a decision on any particular matter associated with litigation. The justice department or the U.S. Attorney, depending on the particular judicial district in which the polluter is located, attempts to control the handling of the case. No one person in the complex of hierarchies in EPA has authority to make any decision regarding procedure or settlement, and frequently no one person in EPA has the authority to recommend a settlement position.[16]

EPA is not on the list of favored clients of the justice department, and environmental cases are not processed with loving care. This situation contrasts with the treatment of environmental cases by most state attorneys general, who have separate offices that deal exclusively with environmental litigation.

As I suggested earlier, the result of the cumbersome EPA procedure is a practice of nonenforcement. Although the agency is on a public relations campaign to demonstrate its enforcement potency, most state officials are not misled. Those who do not want enforcement in their states do not enforce and are not concerned about preemption.

A further consequence of the way in which EPA handles a case is that its personnel become frustrated by the agency's ineffectiveness and by their own inability to accomplish specific results associated with the protection of the environment. This frustration is often translated into hostility toward the state officials who actually perform enforcement functions. Too much of EPA's energy is spent contesting with the states. Little cooperation exists, and true support of the states, as required by the Clean Air Act,[17] is meager.

A Solution to the Problem

The theoretical solution to the problem is a system that assures that EPA will take enforcement action if the state does not and directs EPA's energy to assisting the states rather than squabbling with them. Enforcement and therefore compliance would soon be a reality if states were assured that their enforcement actions would not drive industry to neighboring states (because the federal government would preempt any state that was recalcitrant) and if states knew that the resources of the federal government would be available to them.

The practical answer is more difficult. No one has been able to cut a federal bureaucracy down to a workable size. Nor is it easy to encourage EPA personnel to give meaningful assistance to the states.

Nevertheless, there are at least three changes that EPA should make to enhance enforcement under the Clean Air Act:

1. The most important change must originate in headquarters. The agency must recognize the value of state enforcement action and give credit to the regional office for successes in state courts. Such recognition would make it more palatable for federal officials to assist the states rather than withhold information and compete for the action.
2. There should be a drastic reduction in the number of EPA enforcement personnel and a redistribution of skilled personnel so that the states, which have primary enforcement responsibility under the act, have increased resources. The reduction in EPA personnel would result in more significant work for the remaining federal officials and a concomitant reduction in malaise and frustration. This redistribution could be effectuated by the assignment of federal personnel to the states pursuant to the Intergovernmental Personnel Act.[18]

3. There should be direct grants to the state attorneys general to insure adequate legal support for state enforcement programs.

The adoption of these measures would be a proper response to the congressional policy statements set forth in the Clean Air Act.

Notes

1. M. J. Roberts and S. O. Farrell, "The Political Economy of Implementation: The Clean Air Act and Stationary Sources."

2. *Brown* v. *EPA,* 521 F.2d 827, 835 (9th Cir. 1975).

3. Section 101(a)3.

4. *Train* v. *NRDC,* 421 U.S. 60, 79 (1975).

5. Total suspended particulates, sulfur oxides, carbon monoxide, nitrogen oxides, photochemical oxidants, and hydrocarbons.

6. Section 110(a)(1).

7. Leslie Carothers, EPA, Region I.

8. EPA tracks only major sources, according to Lawrence Goldman in the Region I office.

9. The incinerator was closed by the town in May 1976 in accordance with a consent judgment that was prepared before the action was filed.

10. The regional offices are located in Boston (I), New York (II), Philadelphia (III), Atlanta (IV), Chicago (V), Dallas (VI), Kansas City (VII), Denver (VIII), San Francisco (IX), and Seattle (X).

11. EPA, like other federal bureaucracies, memorializes results of most meetings in writing. Because these memos are not subject to acquisition under the Freedom of Information Act, no one can be certain of their actual number or their content.

12. Conversation with Stephen Schroeder, EPA, Region I.

13. Conversations with Lawrence Goldman and Leslie Carothers, EPA, Region I.

14. Virtually all decisions made at this level are tentative. The comfort and security of federal bureaucrats requires that they can "keep my options open" and it is important not to make binding commitments. The words "tentative" and "tentatively" are frequently used in government service.

15. No fine or penalty is collected.

16. Such decisions are memorialized and communicated in letters or memoranda that are generally signed by one person. In order to avoid the responsibility of making the decision, however, the signatory insists on concurring signatures on the file copy of the letter.

17. Section 101(b) states that one of the purposes of the act is to "provide technical

and financial assistance to State and local governments in connection with the development and execution of their air pollution and control programs.''

18. 5 U.S.C. 1371 et seq.

5 Regulatory Strategies for Pollution Control

A. Michael Spence and
Martin L. Weitzman

Introduction

The debate over regulatory policies to control pollution has been dominated by advocates of both effluent standards and effluent charges. This paper argues for a combination of the two that achieves the objectives of each but is more finely attuned than either alone to the practical problems that are often encountered in regulation. The conventional case for effluent charges makes some good sense, but it is based on an unrealistic assessment of the amount of information actually available to regulators, information about the sources, extent, and costs of pollution. It therefore misrepresents the practical options actually open to government regulatory bodies. On the face of it, the case for effluent standards might appear to be stronger. Yet, as we shall argue, the strategy for setting standards can also be improved.

The regulatory problem, in general terms, is to constrain sectors of the economy to achieve the ''right'' reductions in pollution levels. At the same time, however, the regulatory authority must hedge against the major sources of uncertainty that arise when limited information about polluters and effluents is available. Dealing effectively with this uncertainty involves building into the regulatory process sufficient flexibility to avoid the worst consequences of mistakes. This paper argues that a combination of effluent charges and standards will be the most effective policy.

The problem of uncertainty is not inconsequential. Pollution entails substantial adverse effects for society. But the costs of cleanup or the costs to the industrial sector of equipment to reduce effluents at their source are also substantial. Any mistake about these relative costs and the implied appropriate regulatory policy can cost the economy as a whole many millions of dollars. Because of the magnitude of resources involved, it is important to devote a sufficient amount of effort to ensure that the incentives faced by the regulated sector are consistent with social objectives.

Economists tend to think of pollution as a problem that results from a market failure. That failure is the complete absence of a market for the effects that polluters have on those adversely affected by pollution. (A market, in this context, should be thought of as an arrangement in which people pay for the things they do that affect others.) The fact that markets in the effects of effluents do not and are not likely to exist implies that the costs of pollution are *not* necessarily a factor in the decisions of the polluters. That implication, in turn, tends to result in excessive effluents.

Economists do not just identify the problem as one of a missing market. With a certain rationality, we also incline toward fixing the problem with something that looks like and acts like a market, that is, a system in which prices are attached to effluents. It is usually called an effluent charge or an effluent fee system. The effluent charges are imposed. The regulated sector is then allowed to adjust to them by some combination of paying the charges and reducing effluents.

To most people, the identification of the pollution problem as one of a missing market or a market failure seems odd, to say the least. There never was a market in pollution, and there is not likely to be one in the near future. Pollution, for many people, is a problem like that of preserving wilderness areas or keeping the streets clean. It has little to do with markets, potential or actual. Opponents of prostitution do not react favorably to proposals to control that industry by imposing (perhaps sizeable) excise taxes that approximate the social and moral costs of the activity. There has been a similar feeling about effluent charges as a response to the degradation of the atmosphere. Recently, opposition to effluent fees by regulators and environmental groups seems to have declined somewhat.

To the extent that the presumption in favor of effluent charges is based upon the view that the underlying problem is a missing market, the case has not been convincing to the majority of the public or to noneconomist professionals such as engineers and lawyers. Moreover, public acceptance is not entirely an academic matter. Many of the major regulatory efforts in the United States have been fashioned and modified in the Congress. The general form of the regulations does not usually emanate from a technically oriented agency like the EPA.

Those who mistrust the effluent charge approach typically favor effluent standards. These are maximum levels of effluents that are deemed acceptable and consistent with the maintenance of the quality of the environment in which people live. They are set by the political process and met by the regulated sector. This approach to the regulation of pollution corresponds much more closely to most people's perception of the problem. There is a collective decision about what is and is not acceptable conduct, and the government's task is simply to enforce the laws. The setting of standards is the major perceived alternative to effluent charges. Thus far, standards have been winning the battle in the political arena.

The economists' case for effluent charges is not entirely based upon the missing market hypothesis. To suggest that it is would be to do an injustice to its proponents. The case is often amplified and buttressed by other arguments. Standards are rigid, at least in principle, and therefore are insensitive to costs. Of course, if costs turn out to be much higher than expected, the standards can be relaxed. But the effluent charge advocate would argue that this type of relaxation occurs automatically and in a controlled way with an effluent charge. Moreover, the knowledge that a standard may be relaxed can create an incentive for the regulated to create the impression that costs are or will be high.

It is argued further that standards distribute the cleanup among polluters in ways that are potentially inefficient. What one would like is a system that causes the sources with the lowest cleanup costs to do most of the cleaning up. Effluent charges do distribute cleanup activity among sources efficiently.[1] Whatever the level of effluents actually achieved, it is achieved at least cost.

There is a final argument. It is that the government should take the position of standing in for the public, whose interest is not represented in the absence of regulation. The public interest is properly represented when the additional benefits of reduced effluents are commensurate with the additional costs that result from effluent reduction. The public, after all, pays these costs in the end, in the form of higher prices, displacement of jobs, and so on. Effluent fees, it is sometimes argued, are a reasonable way of putting the public benefit into the equation. The fees "represent" the benefits to the public of reduced effluents or, equivalently, the costs of pollution to the public.[2] By contrast, it is said, rigid standards seem to imply that the social costs of exceeding the standard are high enough to make it unreasonable to contemplate emissions in excess of the standards. They implicitly misrepresent the damages from pollution. No one really believes that the social cost of exceeding the standard is infinite. Then, the argument goes, one is left wondering about the rationale for adopting what must inevitably be seen as a somewhat arbitrary standard in the first place.

There are two basic problems in regulating pollution, setting aside enforcement problems for the moment. One is distributing cleanup among effluent sources in an efficient manner. The second is trading off costs and benefits and adjusting effluent levels until costs and benefits are commensurate. Effluent charges accomplish the first objective. If the initial effluent charge is not the correct one, it can be adjusted and, if necessary, readjusted until the appropriate trade-off between costs and benefits is achieved. Thus, the second objective is achieved by a process of trial and error. In the course of this process, the regulators need only adjust one number, the effluent charge. It is adjusted when the incremental benefits of effluent reduction are perceived to differ significantly from the effluent charge, for it is the effluent charge that is supposed to approximate these benefits. The idea, then, is that the regulators go through a cycle: setting charges, obtaining a response from the regulated sector, and resetting the charges. A similar argument could be made for effluent standards, but the economist would argue that it is more difficult to iterate in this case because there are more variables to control.

The underlying presumption is that the regulated sector can and will effortlessly, costlessly, and frictionlessly adjust to changes in the regulations until the hypothetical optimum is reached. This is not a very accurate description of our world. The image of a frictionless and responsive regulated sector is misleading for several reasons. First, the limited information about costs and cleanup technologies that the government possesses is also a problem for effluent sources. The real expenditures required to meet particular standards or to respond to particular regulations cannot be taken back. They are sunk costs. Each time the regulations are changed there are additional costs.

Second, the cleanup technology is usually highly capital intensive. The investments required to respond to a particular set of standards cannot easily be reversed. In fact, they may not be reversible at all.

Third, it takes time to mobilize any organization to engage in a new activity. Once in motion, most business organizations do not easily change direction. Frequent changes in regulations may create serious implementation problems for a well-intentioned business management.

Fourth, the organizational inertia just described applies also to the regulatory organization involved in implementation and enforcement. Learning on the job is rendered significantly complicated by frequent changes in the rules that are being enforced.

Fifth, if the regulated sector anticipates regulatory changes that, in turn, are responses to costs in the regulated sector, then the simple model of the regulated sector responding myopically to each new regulation, be it a standard or a charge, is not realistic.[3] If the regulated sector's behavior affects the rules and the polluters know it, they are unlikely to take each new set of rules at face value.

These factors conspire to make any regulatory process that involves repeated adjustment to new information costly, time-consuming, and perhaps infeasible. These remarks apply to any fine-tuning regulatory process and not just to the effluent charge approach. The same comments would apply to a system of continuously adjusted standards.

If it is costly or impossible to deal with uncertainty by adjusting until the relevant aspects of costs and benefits are known, then it is important to think of the regulatory problem as one of imposing rules based on the best available information, however limited. The combination of the rules and the regulated sector's responses will produce results (effluent levels, cleanup costs, and price changes for many products) that must be endured for some extended period of time. This is not to say that that the rules are immutable but rather that the initial rules are important because their effects will last for an extended period.

One cannot help feeling that the occasional hostility of the debate between proponents of standards and effluent charges is in part the result of a failure of communication among the interested parties about what the practical constraints on regulatory activity may be. Perhaps the two sides have been operating with essentially different models. Economists, using the previously described frictionless model, have labored with some success to explain the merits of the price system and appear to have difficulty understanding why there are so many recalcitrants among the policy makers. The proponents of standards have felt that their approach is safer and more practical, although the case for standards as a control strategy has been less effectively defended on formal grounds.

In the literature and in discussions of control problems with those who have had to frame, implement, and respond to regulation, certain facts emerge upon which most would agree. They are facts that can act as guides in developing regulatory strategies.

First and most important, both the benefits and costs of effluent reduction are uncertain at the time the regulations are imposed. The uncertainty can be reduced through the expenditure of time and resources, but it cannot be eliminated.[4]

Second, adjustments in levels of effluents cannot be made easily or costlessly, for the reasons cited earlier.

Third, any control system requires some form of monitoring. The costs of and available technologies for monitoring vary from one pollution problem to the next.[5]

Fourth, the link between ambient air quality standards and emission levels at effluent sources is imperfectly understood. The diffusion models required to predict ambient air quality from emissions are complicated. They have not been available long, and those that are available are not necessarily understandable to state enforcement agencies.

Fifth, regulation not only affects the regulated sector's incentives to reduce emissions, but it also affects the incentives for research and development in the area of new technologies for cleanup, a subject to which we shall return.

Regulation places restrictions and imposes costs on the regulated sector. There is a large variety of different kinds of restrictions that can be imposed. Perhaps unfortunately, two have attracted almost all of the attention. They are standards and effluent charges. After looking at the important characteristics of costs and benefits in the second section, in the third we examine standards and charges. The reason is essentially two-fold. First is to argue that, between the two, the preferred option depends on some important features of pollution damages and cleanup costs, especially the uncertainty about costs. Second, an examination of the relative merits of each alternative as a response to limited information is essential to understand what we think may be a better practical alternative: combining standards with effluent penalties for emissions in excess of the standards.

Therefore, in the fourth section we outline a regulatory strategy based on standards supplemented by penalties that look and act like effluent charges. This strategy combines the better features of both of the currently debated alternatives. We believe it will appeal to the practically oriented as a reasonable and useful modification of current control systems based on standards.

The Nature of Benefits, Costs, and Uncertainty

Although we can and must quantify the benefits of clean air to make intelligent decisions about desirable levels of air quality, it would be a mistake to think that it is easy to determine the damages from pollution. Ideally, we would like what the economist calls a damages function—a dollar measure of the harm or disutility caused by various levels of pollution. But such numbers are hard to come by. For one thing, there is a large psychic component—polluted air is undesirable, in part, because it is unpleasant. Attaching numbers to people's preferences is always difficult, especially when they disagree. And even when less subjective elements are involved, many of the more tangible damages from pollution, including health effects, are hard to measure.

So a crucial aspect of pollution damages is uncertainty. This acknowledgment does not mean we know nothing about the benefits of clean air. We may know upper and lower limits but be somewhat fuzzy about the area in between.

The uncertainty in benefits does not really affect pollution control strategy per se (the choice between standards and fees, for example). Whatever course is actually tried out may not make us less uncertain about benefits. There is a slight qualification because changes in ambient air quality attributable to regulation provide new data that can be used to improve estimates of the health effects and other impacts of pollution. But the "value" of cleaner air would still remain largely uncertain. In contrast, the way that the regulated sector responds to a particular set of regulations will, over time, tend to reduce our uncertainty about cleanup costs. Uncertainty in benefits can be narrowed only by more research, carried out presumably by some branch of the public sector. Under the circumstances, the economist tends to suppress the uncertainty in benefits by working with the reasonable compromise of an expected damages function. The expected damages function may be higher or lower depending on whether damages are anticipated to be higher or lower. It represents our best single estimate of damages at the time when a regulatory decision must be reached. Henceforth when we speak of damages, we will implicitly be speaking of expected damages.

There remains a problem of translating the various consequences of pollution into an index that is commensurable with cleanup or abatement costs. In discussing this problem with a variety of people who have been involved in the framing and implementation of the clean air act, we have discovered a way of phrasing the issue that seems to have some appeal. Imagine that pollutants are currently at some fixed levels. One can ask what maximum additional abatement costs would be tolerated to achieve a further 10 percent reduction in pollutants. The answer to this question translates into a statement about marginal damages measured in dollars. But it seems easier to confront the issue by comparing the consequences of effluent reduction and abatement costs directly in this way than to attempt to attach a dollar value to effluent reduction abstracted from abatement costs.

Perhaps the most important single property of pollution damage is that the extra damages of an additional unit of effluent often increase (or at least do not decrease) with the overall level of pollution. This is sometimes called the "principle of increasing marginal damages." Although it is not universal, the principle seems to have general validity. When the air is fairly clean, an extra unit of effluent does less damage than when the air is already heavily polluted.

As we shall see presently, the form of an optimal pollution strategy very much depends on the shape or curvature of the damages function. Two extreme cases merit special attention.

A relatively straight damages function means that marginal damages do not increase very much with pollution levels. This function would be characteristic of a situation in which increased pollution leads to a steady, even deterioration without any dramatic changes.

When the damages function is highly curved at some level of pollution, marginal damages are increasing rapidly around that point. This might be a fair description of a "threshold effect" in pollution. In such situations, marginal pollution damages rise precipitously as pollution starts to become dangerous or uncomfortable.

The principle of nonconstant marginal damages means that it is difficult to place a single, unambiguous price tag on pollution. Unfortunately, marginal damages depend on the level of pollution. To price effluents correctly, we would need to know what the level of pollution is or will be. The fact that we do not know the pollution level in advance makes it difficult for a fee system to function well. Naturally this problem is going to be more acute when marginal damages are changing rapidly than when they are relatively constant.

Turning now to costs, our starting point is the cleanup cost function. This is simply a schedule giving the dollar outlays necessary to obtain a certain reduction in emissions.

Most analysts believe that the incremental cost required to eliminate an extra unit of pollution goes up as the effluent level declines.[6] That is, it is less costly to eliminate the first 5 percent of effluents than it is to eliminate the second 5 percent, and so forth. Economists call this phenomenon the "principle of increasing marginal costs."

The shape of the cleanup cost function will have some bearing on the form of an optimal pollution strategy, just as does the shape of the damages function. Sharply increasing marginal costs give rise to more highly curved cost functions, whereas slightly increasing marginal costs are associated with relatively straighter cost functions. There are no general principles for determining the curvature of cost functions. It depends on the situation and varies from case to case. In short, it is an empirical matter.

It seems to be a fact of life that the regulators don't know cleanup costs to a high degree of accuracy at the time the regulations are imposed. It is especially true when a new or unproven technology is involved, as with auto emission controls. There is no way of knowing beforehand exactly what it will cost to achieve a certain cleanup level. Estimates can be made, but the final costs will not be known until mass-produced equipment is in place, if then.

The uncertainty in cleanup costs is essentially due to lack of experience. It can be reduced by research but not altogether eliminated. No one knows precisely what it will cost to achieve some cleanup level because it has never before been tried. Once a full-scale effort has been launched, the relevant costs will eventually become known but not before.

There is another important feature of cleanup costs that goes along with the uncertainty. Not only are costs unknown, but it is also difficult and expensive to find out what they are. Sometimes economists and others share a tendency to conceptualize regulation as a process of continual fine-tuning. A certain strategy is adopted, and marginal costs and marginal benefits are observed; if they are not equal, the fees, standards, or other parameters are smoothly adjusted until an optimum is obtained.

As argued earlier, this may be an inappropriate way of viewing the problem. In order to have a chance to work, a regulatory strategy must be left in place for an extended period after it has been adopted. As we see it, analysis of regulatory strategy should start from the following point of departure: *the regulators are forced to make decisions in an uncertain environment and they must live with the consequences for some time*. Among these consequences is the possibility that costs will turn out to be higher or lower than was expected. The above principle will provide a framework for analyzing certain important issues that are outside the scope of the fine-tuning model. With it in mind, we turn to the relative merits and demerits of effluent fees and standards.

Fees vs. Standards

The two best-known regulatory strategies for controlling pollution are effluent *standards* and effluent *fees*. They are easily comprehended and are frequently contrasted. In this section we propose to analyze carefully the comparative advantage of each of these control strategies.[7]

It is useful to analyze standards and fees for at least two reasons. For one thing, this issue is of interest in itself because there is a long-standing policy debate about the comparative merits of these two control modes. For this reason alone it is important to understand how fees and standards work and to be able to identify situations in which each one is likely to outperform the other. A second motivating factor is our own interest in promoting a mixed standard-fee system, which we, and others, feel may be superior to either standards or fees alone. To understand how the proposed mixed system works and just exactly why it is better requires a thorough acquaintance with the basic subcomponents out of which it is constructed.

By far the easiest pollution strategy for the public to comprehend is the one based on standards. Some branch of the government acting on behalf of society's interests establishes upper limits on emissions for each polluting firm. It might appear at first glance that with standards cleanup costs are borne by the polluter, but in fact they will eventually be passed on to the consumer in the form of increased prices, reduced employment, and so on. The standards approach is popular in large part because it represents a direct assault on the problem that fits comfortably with legal, moral, and historical traditions. If the problem is that some identifiable group is overpolluting, the obvious remedy is to force it to clean up to a level more in keeping with society's needs as a whole. What could be simpler than decreeing that pollution be cut back to some level that approximates the public's interest?

An alternative strategy, one frequently favored by economists, is the effluent fee system. Effluent fees are prices that attach to effluents. A polluter pays for his effluent an amount proportional to his discharge volume. In controlling the fee, a regulatory agency indirectly controls effluent discharge by manipulating the incentive to engage in cleanup. A higher fee encourages polluters to clean up more whereas a lower fee elicits more pollution.

To many people a fee system is an unfamiliar and peculiar method of controlling pollution. To make a fair evaluation, it is important to understand exactly how it works. When a fee is imposed on emissions, it indirectly controls pollution in the following manner. A polluter, in order to maximize profits or minimize costs, will fix emissions at that level at which the incremental cost of cleaning up an extra unit of pollution equals the fee. If the fee exceeds the marginal cleanup cost, money could be saved by making the extra investment needed to cut back pollution slightly, and vice versa when the marginal cleanup cost is greater than the fee.

Effluent fees and standards differ in how the burden of cleanup costs is shared. A full analysis of the distributional implications of either system would constitute an excessively lengthy aside. However, it is worth noting one point: an effluent fee system generates government revenues, which can be used for public expenditures or to reduce the burden of taxes collected in other ways. How they are used will in large part determine the distributional impact of the control strategy. Economists regard these distributional issues as impossible to decide on purely economic grounds, but this difficulty does not mean they are unimportant. In our experience, many practitioners react negatively to the potentially large payments that must be made under a fee system.

In an uncertain world in which cleanup costs are not precisely known to the regulators at the time a decision must be made, the comparative advantages of standards or fees derive from the following basic observation: *standards fix pollution levels but leave cleanup costs uncertain;* in contrast, *fees fix (incremental) cleanup costs but leave pollution levels uncertain.* Which of these features is more desirable depends on the underlying economic situation. As we shall see, sometimes one feature is more important, sometimes the other is. We propose to examine a few extreme cases to illustrate the general principles that are involved.

The fact that effluents are fixed under a standards system tends to make that approach relatively more desirable as the damages function is more highly curved. When marginal damages rise rapidly around some threshold level, it would probably be foolish to use effluent fees to control pollution because the pollution level remains uncertain. If the marginal social benefit of clean air is low in some range but increases precipitously as pollution starts to become dangerous or uncomfortable, then for a wide range of costs the effluents should be at or near the threshold level. And under standards they will be. But if effluent fees are used, pollution levels will vary with costs. Should cleanup costs turn out to be higher than anticipated, profit-maximizing polluters could elevate pollution levels into the danger zone. If costs are lower than expected, polluters may be motivated to clean up well beyond the threshold level, to an extent that is not socially justified because marginal damages at that point are insignificant. In a world of cost uncertainty in which regulators must work with fixed fees or standards, a threshold effect in pollution damages makes a strong case for standards.

The opposite kind of conclusion holds with respect to the curvature of the cost function. The straighter the cleanup cost function, the stronger the case for standards. If incremental costs increase only slightly with cleanup levels, it is very difficult to control pollution by fees. Suppose a fee is named. A polluter will set emissions at the level at which marginal cleanup costs equal the fee. Now, when marginal costs vary little as pollution changes, it means that pollution varies greatly as marginal costs change. Because polluters set marginal cleanup costs equal to the effluent fee, even slight fee changes will be translated into large swings in the pollution level. If the fee is set correctly in the first place, everything will be fine. But, as we have tried to emphasize, there is a large amount of uncertainty in any real-world regulatory environment. With relatively straight cleanup costs, the slightest miscalculation of the fee will result in either much more or much less than the desired pollution level. In such a situation, standards tend to be better because a high premium is placed on the rigid output controllability that only they can provide under uncertainty.

Just as a more highly curved damages function favors standards, so a relatively straight damages function is more conducive to fees. If the damages function is close to being linear, it would be foolish to name standards. If the marginal social damage is approximately constant in the relevant range, a superior policy is to confront the effluent sources with an effluent charge equal to the marginal damages. Then the polluters will automatically bring themselves close to a social optimum by picking the emission level that equates marginal cleanup costs to the fee.[8] With a straight damages function it is much better to have the polluters find their own desired emission level on the basis of a fee than to have the regulators determine it for them by setting a rigid emission standard at a time when costs are uncertain. In this case, the fee system is more attractive because it gives the ability to fix marginal costs in an uncertain world.

Perhaps the main reason that some economists traditionally favor fees over standards is that fees automatically induce an efficient distribution of cleanup effort among different sources. The economist tends to view standards as piecemeal regulation that offers no guarantee that the overall level of pollution will be attained at least cost.

Here is an example. Suppose it has been decreed on the basis of crude cost calculations that mobile sources should cut back sulfur dioxide emissions by 30 percent and stationary sources by 50 percent. Because the regulators don't really know what the actual costs of pollution abatement will turn out to be (and maybe even if they do know), there is no guarantee that the incremental costs of a small further reduction will be the same for both sources. It would be better to require the source with the smaller incremental cleanup costs to pollute a little less and to permit the other source to pollute a little more. The same overall sulfur dioxide level would be attained but at less total cost.

It is important to understand that imposing a uniform fee on all emitters of a specific pollutant automatically guarantees that the overall cleanup level will be obtained at least total cost. Each polluter, in order to maximize profits or minimize costs, will set emissions at that level at which the marginal cost of cleaning up an extra unit equals the fee. Because the marginal cleanup costs of each polluter are equal to the same fee, they are equal to each other. This is the hallmark of a least-cost allocation of cleanup activity. Only when marginal cleanup costs differ would it be possible to obtain the same overall pollution level at less cost. As in the previous example, this effect would be accomplished by allowing less pollution from the source with low marginal cleanup costs and more from the high-cost polluter.

The automatic efficiency of a fee system is a definite point in its favor because it results in cost savings. If we lived in an infinitely flexible control environment where the regulators could continuously and costlessly adjust the fee, the efficiency argument might be overwhelming. But, in practice, adjustments are usually very costly. The consequences of any regulatory action are going to be with us for a while. And then the uncertainty about pollution levels that is inherent in a fee system can become troublesome for the reasons just discussed. Standards may be preferable to fees in a multiple-source setting even though standards are inefficient. It all depends. If it is important to hold overall pollution to some prescribed level, that need may take precedence over having a cost-minimizing way of achieving an uncertain level of pollution.

The cost-saving or efficiency argument for fees becomes more significant as the variety of polluters increases. A fee system enables the regulators automatically to screen out the low-cost polluters by encouraging them to clean up more relative to the high-cost polluters. This screening effect is more significant if there are many different types of polluters because the possible cost savings are greater. If there are three distinctly different types of sulfur dioxide emitters that have independent cleanup technologies instead of one large pollution source that yields the same aggregate effect, the case for fees is strengthened, other things being equal.

Note that the desirable cost-screening effect of fees doesn't work unless the different pollution sources really are different. A fee system for automobiles is not likely to permit much cost screening because cleanup costs are not likely to differ much from one auto to another. If costs of several polluters are highly correlated, as with automobiles, it is best to lump the units together and view them as one *type* of pollution source. The more different types of polluters there are, the greater the potential cost savings of a fee system due to the screening effect.

It is perhaps useful to summarize at this stage. The comparative advantage of fees and standards for controlling pollution depends on the shapes of the damages and cost functions, on the magnitude of the uncertainty, and on the number of effluent sources with relatively independent cleanup costs. Standards are favored as damages are curved or costs are straight. Fees are favored as damages are straight or there is a larger number of independent polluters.

The purpose of this section has been to give the reader a feeling for the way fees and standards work and when each one is likely to work better than the other. In the next section we are going to propose a mixed standards-fee system that outperforms either pure system.

The Pressure-Valve Approach to Regulation: Standards with Effluent Penalties

In the presence of limited information, standards are often preferable to effluent charges because they prevent the levels of effluents from running up when costs turn out to be higher than the initial estimates. This feature of standards is particularly important when the incremental damages increase rapidly with the level of effluents. On the other hand, standards are rigid and therefore unresponsive to situations in which cleanup costs turn out to be high. Under these conditions, of course, it is possible to relax the standards. Relaxation, in the form of a delay, has occurred in the case of automobile emissions. And it is done elsewhere when the need arises. But the conditions under which the standard is relaxed and the way it is relaxed are of extreme importance in determining the effectiveness of the control program.

The problem, in a nutshell, is to devise a mechanism that responds to high cleanup costs for individual sources and at the same time does not create an incentive for noncompliance with the standard. What is needed is a penalty, specified in advance and paid by the source in case its emissions exceed the standard set for that source. Moreover, to maintain the incentive to clean up, the penalty should increase with the amount by which emissions exceed the standard. A practical way to achieve this goal is to establish a penalty *per unit* of emissions in excess of the standard.[9] That penalty will act like a high effluent charge for the sources that turn out to have high costs. Those sources that have costs close to prior expectations will find it desirable to meet the standards. For them, and they will be in the majority, the system will function as if there were simple standards. For the very high cost polluters, the system will function as if there were fees.

The proposal is to add effluent penalties to the standards that take effect only when the standards have been exceeded. It is a system in which a set of standards is supplemented by *pressure valves,* which release only when the costs for an individual source exceed the estimated costs by a significant amount. Each individual polluter decides for himself whether his cleanup costs are high enough to justify paying the fee for exceeding the standard. Moreover, when the escape valve releases, it is not a complete release. The standard for a high-cost polluter is *not* reset at no cost to the source. Rather, the polluter pays a penalty, which can be avoided at a future date, in the event of a reduction in the costs of cleanup. Therefore the incentive to maintain the effort to reduce the cleanup costs is retained.

This approach has several attractive features. Most sources will have costs in the neighborhood of the average of prior expectations. They will therefore meet the standards. Only those sources whose costs are significantly higher than estimated will choose to exceed the standards, and they will pay the penalties. High-cost sources will clean up less than those with lower costs. This is one of the more important attractive properties of the effluent charge approach, but it is one that can be overridden by other considerations. The protection of standards against high pollution levels is substantially maintained. So long as the penalty is set above expected marginal damages, the escape valve will operate only when costs and benefits differ by a sizeable amount.

The penalty that the individual polluter faces more closely approximates the social costs of his contribution to pollution than under pure standards or effluent charges. The penalty is neither prohibitive, as in effect would be the case under rigid standards, nor is it too lenient, as when it is equal to the incremental damages at the anticipated outcome. In the latter case, the system would function like an effluent charge system and would therefore have the problem of suboptimally high levels of pollution that high costs would cause.

The setting of the optimal penalty is a matter of some complexity.[10] For practical purposes, however, there is only one important principle. The penalty should be related to the damages or social costs of pollution. As a rough approximation, the penalty per unit of effluent should be somewhat higher than the marginal damages at the levels of effluents that would obtain if all the standards were met.[11] It is important to remember that the incremental damages increase with the level of effluents. Therefore, as a first approximation, one could do worse than setting the penalty equal to the marginal damages at the level of effluents prior to the imposition of regulation. Or to put it in the form of a decision rule, assess the marginal damages of pollution at existing (precontrol) levels of effluents and set the unit effluent penalty equal to that number. If that is done, then the escape valve will function only for those sources whose costs are so high as to exceed the benefits of cleanup at relatively high precontrol levels of pollution. If marginal damages increase dramatically with effluents, then that penalty will be high, as it should be. On the other hand, if marginal damages do not increase rapidly, then the penalty will be lower, and that relationship also is desirable. The setting of the escape valve can and should be tied to the curvature of the damage functions. Setting the penalty at the precontrol marginal damages is one relatively simple way of accomplishing this end, although there are others.

Analysts have worried that the effluent charge system, even when it is desirable (for the reasons discussed earlier), possesses some embarrassing features on the financial side. It generates a lot of revenue that then has to be disbursed. And effluent sources pay double: once for cleanup and once for the effluent fees on the emissions after cleaning up. That system may impose a rather heavy financial burden on some sources. The standards and penalties approach does not have this pair of problems. Most sources do not pay anything to the government. They meet the standards. Some sources pay some penalties. But the penalties are only on the emissions in excess of the standards. The payments are therefore smaller by orders of magnitude than the fees that would be paid under a simple effluent charge system. Indeed, another way of describing this control strategy is as an effluent charge that doesn't take effect until certain targets or standards (one for each source) have been exceeded.

In most of the pollution problems with shich we are familiar, damages increase at an increasing rate with the levels of pollution. These are the circumstances under which the use of standards is preferred, when the choice is between standards and simple effluent charges. We are suggesting that one can improve upon standards by instituting an automatic escape valve that applies to individual sources and prevents the worst dislocations in the event of unexpectedly difficult cleanup problems. The escape valve takes the form of a high (but not prohibitively high) penalty that is proportional to the amount by which effluents exceed the standard.

The basic argument for this modification is that it makes the penalties for the individual polluter more closely approximate the actual damages he is causing than is the case under either rigid standards or effluent fees. The modified system is a form of flexible standard approach, but it is preferable to systems in which standards are simply relaxed without penalty (then no one would comply). The costs of exceeding the standards are nontrivial and nonzero. These costs are also specified in advance. The system therefore maintains the incentives to attempt to meet the standard and to reduce levels of effluents. It also removes the incentive to try to appear to have high cleanup costs and thereby impress the regulators to relax standards. In addition, the penalty-modified standards will not result in the collection of enormous volumes of revenue from the private sector. And implementation of the system does not require a major change in the direction of policy. All that is required is a redefinition of how to assess penalties for noncompliance. It is a practical and, we think, useful addition to the current system of control via standards.

There is another rather important set of issues that have emerged in discussions with regulators and interested observers. They concern the incentives for research and development in pollution control technology that the regulations create. It is clear that the regulations not only affect cleanup activity in the private sector, but they also determine the way in which technological development proceeds in this relatively new industry. The effects of regulation on the patterns of technological development are potentially among the most important long-run effects that regulations can have.

This subject can be rather difficult and complicated, but there are a few observations that can be made in support of the penalty-supplemented standards that are being proposed here. Both standards and effluent charges have potentially distortionary effects on the pattern of research and development. From a social standpoint, the R and D problem is one of investing in technologies that are good "gambles." Good gambles are technologies that have a significant chance of reducing the costs of effluent reduction and at the same time do not run significant risks of being ineffective in reducing emissions. Society wants resources and effort expended in attractive but nonspeculative ventures.

We can think of an R and D investment program as having a cost (the initial investment) and a distribution of outcomes defined in terms of the levels of effluent reduction that can be achieved at the conclusion of the program. It is useful to think of that distribution of outcomes as having a mean and a variance. What we want is a program with relatively low costs that has a high mean and a low variance. The low variance is desirable because it means that the chances of a disaster (no significant reductions in pollution) are reduced.

With these objectives in mind, what can be said about effluent charges and standards? Consider effluent charges first. Because the penalties facing the regulated sector are proportional to emissions under effluent charges, that sector will respond by trading off the costs of the R and D program on the one hand and the mean of the distribution of effluent reductions on the other. The variance is likely to be ignored as a factor in the choice of an R and D program, but the variance is irrelevant only when the damages function is linear. Therefore, the effluent charge approach in the context of the R and D problem has the potentially fatal flaw that it fails to provide the regulated sector with an incentive to respond negatively to the variance of the outcomes.[12] It therefore fails to provide protection against the risk of potentially high pollution levels. This problem is the analogue of the tendency of effluent charges to generate excessive levels of effluents when costs turn out to be high. Both problems result from the fact that a curved damage function is approximated by a straight line.

Standards pose a different problem. Under standards, provided they are not expected to be relaxed, the private sector will respond by selecting a program that minimizes the probability of failing to meet the standards. That may be a very costly program. And thus, in an important sense, standards may cause the R and D program to be excessively costly and conservative. A substantial reduction in costs with a small increase in the probability of noncompliance will be rejected even if it might appear to be a rather good decision from a social point of view. Costs figure in the R and D program under standards only if there are programs that assure the meeting of the standards. The private sector will then select the least-cost program that meets the standard.

As a rough approximation, standards respond first to the mean and variance, especially the variance, and then to costs if the standards can be met with certainty. This pattern also produces distortions. Costs and benefits may not be properly traded off.

Because the penalty-modified standards more closely approximate the damages, these sorts of problems will not arise in as severe a form. The private sector will hedge against uncertainty but not to the exclusion of cost considerations. Thus the case for the mixed approach is based in part on the need to structure the incentives for research and development in an appropriate way. The automobile air-pollution case is an example of the desirability of having a somewhat flexible standard in a situation in which the development of the control technology is a central feature of the problem. It is a complicated case, which Mills and White treat in detail elsewhere in this volume.[13]

The Number of Effluent Sources

The preceding analysis has assumed that firms are the source of pollution and that their number is fixed or varies only slowly over time. It further assumes that the number of sources is not particularly sensitive to the regulations, whatever form they take. There are, however, important problems in which this assumption may not be true. The best-known and perhaps most important is the automobile case. When the number of sources is fixed, the issue is confined to determining the optimal effluent level per source. When the number of sources varies there are two issues. One concerns the effluents per source; the other is the number of sources. In the automobile case, effective pollution policy involves controlling both the effluents per vehicle and the number of vehicles.

In the short space available to us, we cannot deal completely with this problem. However, some comments are in order. First, many of the considerations discussed previously are applicable to this situation and point to the merits of a mixed approach to controlling effluents per vehicle. One difference is that a strong case can be made that the control program directed at the effluents per vehicle should be supplemented with a tax on the vehicles themselves. Without such a tax, the pollution cost of an additional vehicle is not borne by the purchaser; therefore, unless the demand is inelastic, there will be too many vehicles on the road.

Even in the case of stationary sources, there is a long-run decision as to their number and locations. For these decisions to be made correctly, something more than the correct marginal incentives is required. The absolute magnitude of the penalties paid by an effluent source should equal the estimated incremental social cost of that source. Otherwise the long-run entry and exit decisions may not be the correct ones. This possibility suggests that the mixed system proposed earlier should have appended to it an effluent charge equal to or slightly below the marginal damages at the pollution levels implied by the standards. This charge has two properties. It is paid on all units of effluent, so that not only the marginal social cost but also the total cost of the source is internalized. And the effluent fee is below the supplementary penalty that takes effect only when the standard has been exceeded.

Thus, the appropriate regulatory response to cleanup cost uncertainty would include an effluent charge of the conventional kind and a standard accompanied by a penalty for emissions in excess of the standard. This system would act like a nonrigid standard system. It would also provide an incentive to clean up beyond the standard for those sources that turn out to have low abatement costs. And it would approximate the appropriate long-run incentives in the private sector for the correct entry and exit of effluent sources.

Notes

This paper is based on some recent work by the authors and others on the subject of controlling via quantities and prices. Weitzman was the first to study the advantages of quantity controls as a response to uncertainty. Spence and Roberts used the device of transferable licenses in the pollution context to examine the merits of effluent standards and of mixtures of fees and standards. The purpose of transferable licenses is to maintain efficiency in the distribution of effluent reductions among sources and to employ an overall quantity control. The authors would like to thank the members of an early workshop on the Clean Air Act for their encouragement and their aid in helping us to understand various facets of the problems. We want particularly to thank Douglas Allen for his help. Robert Dorfman, Peter Diamond, and Robert Solow pointed to some problems with which we have tried to deal. There remain some issues of a theoretical kind that are best confronted in the more technical papers listed in the references.

1. The reason is that the sources of pollution clean up to the point at which the marginal costs of cleanup equals the effluent fee. But then the marginal costs of cleanup are the same for all sources, the implication being that cleanup cannot be shifted among sources in a way that reduces costs.

2. The benefits, of course, are extremely difficult to measure and to agree on. And no regulatory approach is excused from undertaking this task.

3. It is not easy to predict how the regulated sector will respond to a situation in which regulations are anticipated to change in response to costs. It seems reasonable to surmise that the regulated sector's efforts might be sluggish under these conditions.

4. The decision about how far to go in collecting information is one of the more important policy decisions that arises. It is rarely made with a view to what difference further information will actually make.

5. Moreover, monitoring technology is capable of changing over time. It appears, for example, that the monitoring technology for particulates from stationary sources is insufficiently advanced to permit the use of effluent charges in that context. On the other hand, it is also true that this situation may change if the government devotes more resources to the advancement of monitoring technology. At present, this essentially public R and D investment seems to be deficient.

6. See, for example, the costs derived by Donald N. DeWees in Chapter 7 of this volume.

7. In the interest of focusing sharply on the essential economic differences between fees and standards, we are abstracting away a host of "noneconomic" factors. This treatment does not mean that we feel they are unimportant. On the contrary, there may be significant legal, administrative, or historical reasons for favoring standards or fees in one situation or another. Even such factors as the capability of the monitoring system play no small role. At the present time we do not seem to have an in-place technology capable of continually monitoring stationary particulate sources to the degree of accuracy required for making an effluent fee system work effectively. The desirability of having the government encourage research in the monitoring area is something that cannot be stressed enough. Certainly the regulated sector has little direct incentive to develop or improve the monitoring system.

8. Of course, effluent sources may first have to expend resources to learn about the costs and control technologies.

9. There are other possibilities. The penalty per *unit* of emissions in excess of the standard could increase with the amount of the excess.

10. For a technical discussion of related issues, see Martin L. Weitzman, "Prices vs. Quantities," *Review of Economic Studies,* October 1974; and Marc J. Roberts and Michael Spence, "Effluent Charges and Licenses Under Uncertainty," *Journal of Public Economics,* 5 (1976):193–208.

11. There is currently a movement to employ penalties as part of the enforcement programs in some states. Connecticut is one of them. These penalties are set to equal the cleanup costs avoided by those not in compliance. It is important to stress that this is a technique *for enforcing standards*. It should not be confused with the proposal put forward here and elsewhere. The penalties in the escape-valve approach are related to damages and may be translated into maximum tolerable abatement costs, as discussed earlier. But maximum tolerable costs and actual costs are very different.

12. It may be that the regulated sector behaves as if it were risk averse. If it does, then it will respond negatively to the variance even though the penalties are linear in effluents. This behavior will mitigate, to some extent, the tendency of effluent fees to cause risk to be ignored in R and D programs. A full analysis of this complex set of problems is unfortunately beyond the scope of this paper.

13. See Edwin Mills and Lawrence J. White, Chapter 8 of this volume.

Comment

Edwin H. Clark II

It is an opportune time for the Spence–Weitzman paper on regulatory strategies to be written. There is increasing dissatisfaction with present regulatory practices generally and with environmental regulations specifically. This dissatisfaction is shared, at least partially, by the environmentalists, the polluters, and the regulators. The environmentalists have retreated from their earlier opposition to economic incentives (although the industrialists have adopted the arguments the environmentalists formally used); a new president is coming to town who is said to look kindly upon economic incentives, and both the Clean Air Act and the Water Pollution Control Act are up for substantial amendment in the coming session of Congress.

The Spence–Weitzman proposal for a reasonable compromise between (or amalgamation of) the standards approach and the effluent charge approach thus finds a less hostile world than it would have found any time during the last six years. The paper has many virtues: it is reasonable; it seems to be practical; it is well written; and it presents articulate arguments. But for all its virtues, it is a disappointment to this reviewer.

We need a well-conceived program that skillfully deals with the many difficult issues associated with any regulatroy scheme, but we find a proposal that vaguely ignores too many of these issues and presents an incomplete scheme demonstrating somewhat less insight and analysis than other schemes that have already been proposed (and in some cases adopted).

I have some trouble in commenting on their paper for it is not clear exactly whether the authors' purpose is to translate the esoteric analyses of academia into a practical regulatory proposal or conversely to justify a possibly practical regulatory compromise to their colleagues in academia. There are many more competent than this reviewer to offer criticism if they are attempting the latter. Assuming, then, that theirs is the former intent, I would argue that the paper has several weaknesses.

First of all, I think the paper would have benefited from a more complete conceptual framework for defining the type and extent of regulations. Since it has not, I will digress briefly to do so here in order to give these comments some structure.

Economists are accustomed to thinking in terms of schemes that equate the marginal benefits of an action with its marginal costs. These are the traditional benefit-cost concepts, and with them social welfare (vaguely defined) is maximized and economists everywhere begin salivating noisily. But with most of our social programs it is very hard to identify marginal benefit curves sufficiently well to support a regulatory scheme based on them.

For this reason, many economists have retreated with some dignity to a cost-effectiveness concept. We would allow the political process to define a goal and then would devise programs for achieving that goal at the lowest cost. Such an approach dampened expressions of ecstasy but was nevertheless acceptable in semipolite company and provided economists with employment.

During the last decade however, Congress seems to have been designing regulatory programs in a third conceptual framework, which can be called a "cost-impact" approach. In simplest terms, the concept seems to be that you squeeze until it hurts or until it really hurts or until it hurts so much that they can't stand it. The emphasis is more on the capacity of the regulated group to absorb the costs of regulation and less on whether it is the least costly way of achieving a given goal or whether the costs are necessarily consistent with the benefits. This cost-impact concept is expressed most clearly in the Water Act, which contains such standards as "best practicable technology" or "best available technology, economically achievable." The modifiers "practicable" or "economically achievable," serve to give some indication of how hard Congress wants to squeeze.

There are a number of reasons why this shift has taken place. The first is that the previous system tended not to work very well. A second, which is usually not recognized by economists, is that Congress often incorporates a number of divergent values into its legislation, several of which are associated with vague notions of fairness or equity. Thus Congress concentrates on the economic capacity of the polluter to make the abatement expenditures because it seems fair that the wealthy and profitable firms should absorb a larger portion of the abatement costs. Congress also established national standards in order to avoid the problem of firms moving away from particular areas (with a resulting adverse impact on the economy of that area) in order to avoid environmental regulations. Such concepts are not appealing to economists who focus sternly on the glittering promise of "economic efficiency," but it is not clear whether the problem is that the economists' criteria are too narrow to include such "noneconomic" considerations or whether economists aren't very good at measuring real social costs and benefits. Whatever the problem, ignoring these trends and values will continue to result in the economist being a rather ineffectual contributor to environmental (and much other) regulatory policy making.

With this background, we can go on to observe some problems with the Spence–Weitzman proposal. From a practical standpoint, a major problem is that it is woefully incomplete. It is little more than a general proposal for the implementation of an effluent or emissions charge scheme related to some necessarily vague definition of benefits. The major difference between it and most previous general recommendations is that the Spence–Weitzman proposal would have the charge start at the effluent standard rather than at zero discharge. But this provision is not sufficient to stimulate the nonbelievers to embrace this proposal if they have scorned all the others. In terms of completeness, the Spence–Weitzman proposal compares very badly to the Mills–White proposal outlined in Chapter 8 of this volume. Mills and White have come up with a very specific proposal and have directly addressed many of the questions raised by opponents to effluent charges. The Spence–Weitzman proposal would receive more consideration were it to do the same.

More consideration, of course, would not necessarily mean more acceptance. However specific, the scheme would still have a number of problems. An economic incentive scheme can do any one of a number of things depending on how it is designed. Most effluent-charge schemes are perceived in terms of driving society quickly, effortlessly, and ineluctably toward the nirvana of economic efficiency, where the marginal costs of abatement are exactly equal to the marginal damages caused by the pollution. Economic incentives in this guise are directed at achieving complete allocative efficiency. We can call this form of charge the "nirvana effluent charge."

Economic incentives can play other roles, as well. These roles, though they may be considered inferior by the economist, are perhaps more acceptable to the nonbeliever. If, for instance, one is willing to allow the political process to establish the goals of a program, one can still propose effluent charges as a scheme for achieving these goals efficiently. Here the effluent charge serves to stimulate the most effective response to achieving the goal. We can call this form the "cost-effectiveness" effluent charge.

Finally, economic charges can be imposed in such a way as to provide no allocative efficiency at all but still serve a useful purpose in providing a strong incentive to comply with laws or regulations—an incentive that may be seriously lacking under a standards approach. In this third form, the economic charge is called a "compliance incentive."

These are the three general roles that economic charges can play. The nirvana effluent charge will accomplish complete allocative efficiency. The cost-effectiveness charge will generate a least cost of compliance. And the compliance incentive will stimulate violators to obey the law. The nirvana effluent charge also provides a cost-effective solution and stimulates compliance and thus accomplishes all three purposes. Similarly, the cost-effectiveness effluent charge also provides a compliance incentive. But the compliance-incentive charge accomplishes only the one goal. It is not clear where the Spence–Weitzman proposal falls in this scheme. It looks like a compliance incentive but tries to act like a nirvana charge. Let us assume that it is supposed to be a nirvana charge and see what the problems are at that level and then work through the list.

A number of problems have been raised with respect to designing a nirvana effluent-charge system that sets marginal abatement costs equal to marginal damage costs. The most fundamental is our inability to estimate damage costs accurately, particularly as they are related to a single source. Some benefits are particularly hard to estimate (for example, the cost associated with the destruction of a historical monument or the psychic costs associated with illness and death). Others may vary substantially even within small geographical areas (for example, the value of a view depends on whether you live on an upper floor of a high rise, in the house behind the high rise, or somewhere down in the valley). Some pollutants travel particularly long distances and may be transformed in the process; Norway and Sweden have claimed that as much as 20 percent of their acid rain is associated with pollution occurring in the United States, and much of the sulfate load in New York's air is thought to come from widely dispersed sources. All of these factors make it virtually impossible to determine accurately the marginal cost of damages caused by pollution from a particular source. The Spence–Weitzman proposal does not address any of these problems.

A second problem with the attempt to achieve general allocative efficiency is that, as Spence and Weitzman argue, the marginal damage costs curves should be continuously increasing. They express the belief that this is the case, but it frequently may not be. Figure 1, for instance, shows the relationship between the distance one can see and the concentration of particulates in the air. Views diminish rapidly at low concentration and very little at higher concentrations—that is, the marginal damage cost is downward sloping, just the opposite of what Spence and Weitzman presume.

More generally, the concept of thresholds implies that the marginal cost curve may not be at all well behaved. Our pollution control laws are based on a concept of thresholds, a concept that implies a marginal pollution cost

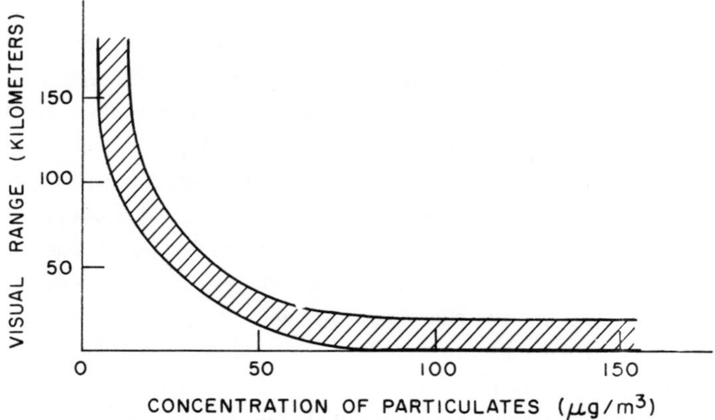

Figure 1. Relationship between particulates concentration and visibility. (Source: Public Health Service, U.S. Department of Health, Education, and Welfare, *Air Quality Criteria for Particulate Matter* (Washington, D.C.: U.S. Government Printing Office, 1969, p. 57).

curve that is zero up to the "threshold" pollution concentration, at which point it becomes very high (the laws and regulations suggest it is close to infinity) and then lower (at least in the law) until the next threshold is reached. Although there may be good reason to reject the concept of thresholds in air pollution,[1] in water pollution it probably does have some validity: water is reasonably safe to drink up to a certain point, beyond which it is safe to swim in until you reach a higher level of pollution, beyond which it may still support aquatic life until a still higher pollution level is reached. The implied marginal-damage-cost curve apparently looks like figure 2. Such misbehaved marginal-damage-cost curves create very messy problems for the designer of an effluent charge system, and this problem is not solved by the simple assumption that the curves are well behaved.

These and other problems have convinced many advocates that the nirvana effluent charge is not likely to be adopted, at least in its purest form. A fallback is to propose the cost-effectiveness effluent charge. In this scheme the environmental goal is first translated into a required amount of pollution reduction, and then the marginal cost of attaining this amount of abatement from all sources (an industry supply curve if you wish) is estimated, and the effluent charge is set equal to this marginal cost. In this form the effluent charge has several advantages: (1) all the problems of estimating benefits

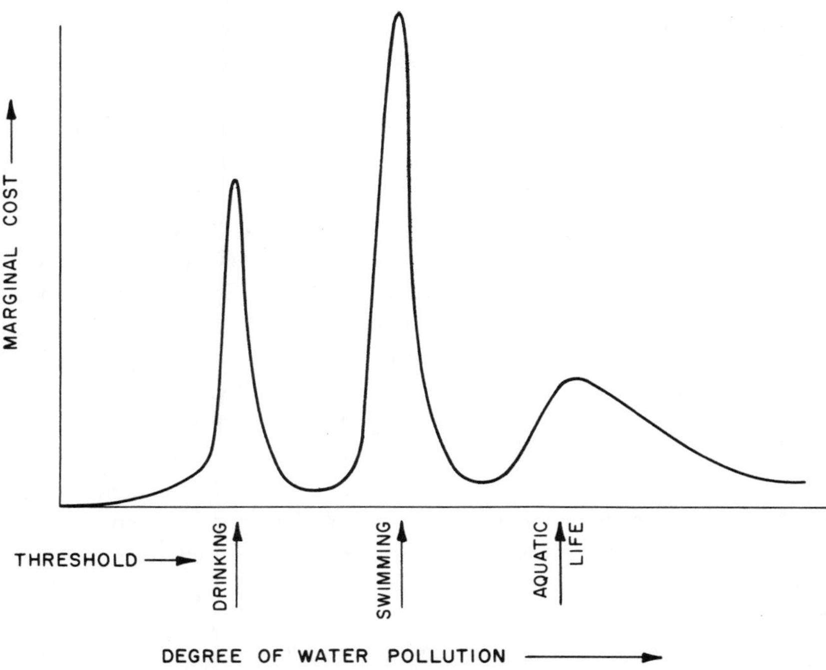

Figure 2. Representation for marginal pollution cost curve for water pollution.

described above are eliminated; (2) at the same time, there is a clear public signal as to how much we have to pay (at least on the margin) to achieve our goal; (3) such a program will achieve a given pollution reduction at least cost; and (4) if the initial guess is wrong about how much abatement has to take place or how high the marginal abatement cost is, the effluent charge can be changed (how easily is debatable) to stimulate a more desirable level of abatement. Most effluent charge proposals fall into this mode. Under such a scheme the effluent charge can vary from air shed to air shed to reflect different air quality goals (such variations may make the charge look like the ''pure'' effluent charge described above), different pollutant loadings, and different ''assimilative'' capacities.

One argument raised against the cost-effectiveness effluent charge is that it requires more monitoring than a standards approach. This argument is dismissed by the economists on the basis that it requires no more monitoring than should accompany a standard approach. Although I believe that the monitoring problem can be solved (although with somewhat more difficulty than the economists' argument would imply), it is in fact easier to prove that someone is out of compliance than it is to prove how much pollution he has discharged, given the problems of sampling and analysis errors. Continuous monitoring is always expensive and often infeasible. A standards system can get along on much less monitoring because the expected cost of noncompliance to the discharge can be just as high with infrequent monitoring (if the penalty associated with noncompliance is high enough) as it would be with continuous monitoring combined with an effluent charge.

But moving away from the various general arguments raised against the cost-effectiveness effluent charge, we can see that the Spence–Weitzman proposal linking effluent charges and standards loses some of the advantages that a straight effluent-charge scheme would have. The most important of these weaknesses is that it promotes only one-sided efficiency. Dischargers that find it more expensive than the average polluter to abate pollution will discharge more than the standard allows, but those for whom it is less expensive than the average have no incentive to discharge less than the standard allows. Thus, at any given charge there will be more pollution than there would be with an effluent charge starting at zero discharge rather than at the standard; or, put another way, to achieve a given degree of pollution abatement the Spence–Weitzman proposal will require a higher charge than an effluent charge starting at zero emissions and, therefore, will be clearly less efficient.

In the first part of their paper, Spence and Weitzman present some arguments about the shape of abatement cost curves that suggests that this inefficiency may not be a serious problem. They make an important point here. If the marginal abatement cost curve is very steep (and it tends to be as one approaches zero emissions), then a small difference in standards can make a large difference in costs, but a small difference in the level of the effluent charge will not result in a large difference in the amount (in absolute terms) of emissions. Thus there is a much greater chance of making a mistake in estimating the costs associated with standards than there is of making a mistake in estimating emissions when setting charges. This is one of the more powerful reasons in favor of charges, particularly as we tighten up on allowable emissions.

Another possible answer to the problem of potential inefficiency is that, if it looks like a serious problem (that is, if the marginal abatement curves are relatively flat), we can offer bribes to dischargers cleaning up more than the standard calls for. Such a bribe scheme, although theoretically as efficient as a straight charge scheme, is likely to be more difficult to legislate and implement.

A problem that does remain with the cost-effectiveness effluent charge is that, if the same charge applies to all polluters in an area, it ignores the differences in the economic capability of different polluters to pay the costs of pollution abatement. This is the "equity" concern mentioned at the beginning of the paper. Under the present system, the standards can be set so as to reflect the financial health of an industrial firm. Under an effluent charge scheme this flexibility would not exist. There are two responses to this problem. The first is that the present schemes also create substantial inequities, particularly among different firms within an industry, and these inequities may be as serious or more serious than any interindustry inequities under a charge scheme. I know of no evidence that either supports or refutes this assertion. However, under the present scheme many potential inequities seem to be eliminated, or at least significantly reduced, as the general abatement requirements are imposed on individual facilities, a process that does take into account the particular physical and financial characteristics of the individual polluter.

A second possible response to this inequity problem is that the level of the charge, or the emission level at which the charge begins to be incurred, could be adjusted to reflect inter- or even intra-industry distinctions. This response would suddenly make the cost-effectiveness effluent charge look much like the economic compliance incentive, the third type of economic incentive scheme listed above.

An economic compliance incentive incorporates only one of the several benefits claimed for economic incentives. It does not serve to allocate resources more efficiently (except, of course, to the extent that the regulations it supports are efficient) but does remove the profit associated with noncompliance. A major problem with the present regulatory scheme is that it is more profitable for firms to delay compliance than to comply. If they succeed in delaying or avoiding compliance, they avoid costs that do not contribute to revenues and gain an economic advantage over their competitors who have complied and are experiencing these costs. Most firms, in fact, seem to be complying in spite of these perverse incentives, but enforcing the present regulatory system with respect to the recalcitrant firm is cumbersome and usually results in little more than the court establishing a new, delayed compliance date. Except for the litigation costs, there is little or no economic penalty imposed on the firm and thus they gain the benefits of the delay even though the regulatory agency has followed the full enforcement procedure. Often, of course, these firms escape the enforcement procedure entirely.

The purpose of the economic compliance incentive is to ensure that, regardless of what happens, the firm will not reap economic benefits from noncompliance. Connecticut has adopted such a system, and the Connecticut plan has become the prototype for all such schemes in this country. Very simply, under the Connecticut plan, if a firm misses a compliance deadline, the regulatory agency can impose an assessment on it that is exactly equal to the amount of money the firm is saving by not being in compliance. This assessment takes into account savings on delayed capital investment, avoided operations and maintenance costs, the rate of return on revenue-producing investments for the industry and the firm's particular tax status.[2]

The advantages of the Connecticut plan are: (1) it accomplishes its limited goals of eliminating the perverse incentives associated with a normal regulatory scheme; (2) it provides an administrative enforcement tool rather than relying upon more cumbersome judicial enforcement; (3) it is clearly acceptable to environmental advocates because it does not substitute for but rather reinforces the existing systems and thus avoids the divisive issue of economic incentive as opposed to regulations and standards; (4) it does not violate the vague concept of equity that seems to be a concern to the legislature; and (5) if presented as a scheme for eliminating unfair competitive advantage rather than as a penalty for noncompliance, it neutralizes most industrial opposition to economic incentives.

The acceptability of the Connecticut plan is one of its strongest virtues. The Senate and House were sufficiently impressed with and accepting of this concept that they both proposed amendments to the Clean Air Act that incorporated a delayed compliance scheme similar to the Connecticut plan. An important factor in this acceptance is that industrial opposition to effluent charges can be largely neutralized with a Connecticut-plan approach.

Recognizing the political viability of the economic compliance incentive, economists (and others who support economic incentives) are left with the choice of supporting a scheme that has a high probability of being adopted but lacks many of the advantages of an effluent change or arguing for a scheme that has more of these advantages but a smaller chance of being adopted. Before making this judgment, however, we should recognize that there are some serious problems with the Connecticut-plan approach beyond the fact that it does not incorporate the various advantages of the other types of economic incentives.

The major problem has to do with the application of the Connecticut plan after all abatement facilities have been built and the problem becomes one of ensuring that they are operated properly. The Connecticut plan is direct, clear, and relatively simple to design and implement during the construction phase. Not so during the operating phase, for it is much more difficult to estimate the cost savings associated with improper operation. In fact, improper operation can cost as much or more than proper operation. One answer to this problem would be to create a legal presumption that if the firm is out of compliance it will be presumed that nothing has been spent on operation and maintenance and to set the compliance incentive equal to the total estimated O and M costs. Such a presumption, of course, would have to survive numerous legislative and judicial challenges.

A variation on this proposal would be to create a legal presumption relating the extent of noncompliance to a presumption about how much has been paid from O and M. For instance, if the firm were emitting 10 percent more than the allowed pollution, there could be a presumption that it was saving 10 percent of its total O and M costs; 20 percent more would represent a 20 percent savings; and so forth. If we choose this option, the compliance incentive begins to look like an effluent charge, except that the amount of the charge would relate to the firm's abatement costs and then would probably be different for every firm. But this problem may create an opportunity. If the only reasonable answer to the O and M problem looks like an effluent charge, why not adopt an effluent charge scheme directly to deal with the O and M problem?

The challenge is to design a scheme that will convert the Connecticut-plan approach into an effluent charge and thereby gain some of the other advantages that effluent charges can provide. At least initially, such a scheme would probably have the characteristic of the Spence–Weitzman proposal that the charge begins at some level above zero emissions. Over time, the amount of "free" emissions could be reduced and the level of the charge increased if there were a desire to stimulate increased abatement. During this process the differences among the charges faced by different firms could be eliminated to the extent desirable. The level of the charge could also be adjusted, at least roughly, to reflect the relative value society places on maintaining different levels of environmental quality in different locations.

The advantages of this proposal—beginning with a compliance incentive and then shifting into an effluent or emissions charge—are: (1) it reinforces and extends the current program so that there is not the disruption involved in scrapping one program halfway through and suffering the delay and frustrations of beginning a new program; (2) it explicitly solves the problem of forcing industries to pay a charge while at the same time they are investing to comply with the law; and (3) it allows the regulatory agencies time to develop the procedures and techniques necessary for implementing an effluent charge scheme without removing the incentive to do so (as the existing programs do) or interfering with the progress being made in pollutional abatement (as would result from an immediate change in strategies). The most important advantage is that such a proposal would seem to have a higher chance of acceptance by the various parties involved and still incorporate, at least in the long run, the advantages of an effluent or emissions charge. However, a significant amount of detailed analysis would still have to be carried out in order to turn this general proposal into a specific adoptable scheme. In this respect, it has the same failings as the Spence–Weitzman proposal. In other respects, however, I think it is superior.

Notes

1. For any individual, both the health and welfare effects of air pollution most likely increase continually with increasing air pollution. And even if individuals experienced thresholds, the fact that different individuals are susceptible to air pollution damages at different ambient concentrations would most likely cause society's aggregate damage curves to increase continually.

2. The Connecticut plan is discussed in detail in a series of reports entitled *Economic Law Enforcement,* prepared by the Connecticut Enforcement Project, Connecticut Board of Environmental Protection, September 1975. These reports were prepared for the Region I office (Boston) of EPA and have the EPA document number EPA-901/1-76-003. Volume 1 of this report provides an overview of the essential features of the Connecticut plan. A briefer overview is provided in a paper prepared by David W. Tundermann for the Environmental Study Conference entitled *Economic Enforcement Tools for Pollution Control: The Connecticut Plan''* (copies available from the Council on Environmental Quality, 722 Jackson Place, N.W., Washington, D.C. 20006).

Comment

William Drayton, Jr.

For well over a decade, economists—with almost unprecedented unanimity—have argued that by far the best way to reduce pollution is to charge for it, that is, to impose effluent fees. By charging, say, $1.00 per pound of particulate emitted, they argue, government would cause sources that can cut particulate emissions for less than $1/lb. to do so. The net result: any level of environmental cleanup would be achieved at the least possible cost. By contrast, government now typically requires all sources to respect specified emissions ceilings regardless of costs.[1]

Many books, over one hundred articles, and inumerable speeches later, however, no government in the United States has enacted a general effluent tax. Legislators, budget staffs, and even environmental administrators, although quite familiar with the idea, have not bought it. In fact, they typically dismiss it as another annoying piece of academic impracticality.

For over ten years these two groups have been locked in an uncomprehending, increasingly ideologized debate. The environmental movement has suffered because this long-frozen confrontation has consumed so much of its attention and intellectual resources.

Practical Alternatives

The MIT meeting and the Spence–Weitzman paper reflect the beginning of this stalemate's long-overdue breakup. Just as some environmental administrators are recognizing the inadequacies of the current simple-standards-plus-enforcement approach, some economists are beginning to understand the problem with simple fee systems. As these two groups work toward one another, the environmental movement can look forward to the creation of a number of hybrid improvements in the current regulatory regime that will make it more economic, equitable, and effective—and still practicable.[2]

The most fundamental problem environmental regulators face is that of compliance. It is much easier to define standards than it is to ensure that they are respected. When faced with noncompliance, regulatros now have a very limited range of possible responses: they can bluff or they can try to go to court. Going to court is not much of an option: it entails extensive delay and considerable expense, and even when the government prevails, it is most unlikely to obtain an adequate remedy. Courts generally refuse to close businesses down or to impose large—let alone criminal—penalties for environmental infractions, especially when they have been given no practical standards to guide their remedy setting.

EPA and the state of Connecticut have developed a hybrid form of economics-based regulation that effectively solves this problem. Those who violate emissions standards or abatement orders are subject to instant, agency-imposed civil assessments equal to what they have saved by not complying (including what they can be assumed to have earned on the amount saved at their industry's cost of capital rate).[3] The correct assessment can be calculated in ten or fifteen minutes by state inspectors and engineers. It is legally and philosophically defensible to delegate the power to impose such assessments to the agency, even though it is both prosecutor and judge in such cases, because no assessments can be larger than the amount that the regulatee has saved by not complying with the law. The formula makes the delegation ministerial.

During the two years this new approach has been in operation in Connecticut, it has cut both noncompliance and administrative costs significantly. Several other states have already moved to adopt all or part of the approach, and, had the Clean Air Act Amendments not been filibustered to death in the last days of the 94th Congress for unrelated reasons, federal air compliance would now be relying on this new approach. The speed with which this quite radical innovation is being adopted suggests that the legislators and environmental administrators that have long stonewalled generalized effluent fees may not be as simple-mindedly unimaginative as many effluent fee economists suspect.[4]

The Effluent Fee Chimera

The Spence–Weitzman proposal resembles the Connecticut approach in several key regards. It adds a universal, principled charge to an underlying set of standards. If this charge could be easily calculated and applied, the approach would have many of the advantages now being demonstrated in Connecticut.

Unfortunately, the paper proposes that these charges be defined in terms of the "expected damages function." This is the oldest, most classic, most unworkable (even theoretically), and longest-and-most-firmly-rejected version of the effluent fee. Even assuming that one could measure all the costs of a polluter's emissions (crop damage, dead birds, auto finish deterioration, aesthetic losses, and so on), society has never been able to agree on a measure of the most critical of costs—human disease, suffering, and death. (By contrast, the costs of compliance used to determine the size of remedies in the Connecticut formula are easily determined.)

Most effluent fee advocates, recognizing that the damage function is a chimera, simply suggest that pollution charges be set at whatever rate will achieve the level of cleanup decided upon by the political process. At least the cleanup will then be cost effective.

Setting the right incentive is, however, a bit more difficult than most of these advocates suggest. (The Spence–Weitzman paper demonstrates just how serious the consequences of a slight error here can be.)

Results are unpredictable. An administrator would have to have an almost infinite amount of information to know how much any particular charge (or, to take account of regional variations and the number of different pollutants, any combination of charges) would affect water or ambient air quality. How can the government obtain or process the company and industry data needed to predict the responses, let alone the environmental effects thereof, of hundreds of thousands of sources? (A standards approach also requires extensive data if it is to be cost effective.) Such uncertain environmental results hardly encourage environmentalists to support the approach.

Iteration won't work. The public hates taxes. Legislators fear them and will not delegate so explosive a power. It is consequently unrealistic to suppose that incorrectly set effluent fees can iteratively be adjusted until the right response is elicited. (This is one of the standards system's great advantages; administrators can adjust their regulations more or less as needed, and they can make them fit the needs of particular groups of regulatees.)

Revenues are unpredictable. Tax proposals not supported by administration budget officers are unlikely to go far, and budget officers think of taxes as sources of revenue. They need to know what these revenues will be, and they strongly prefer revenue streams that will grow with time. An effluent tax's revenue will be as uncertain as its impact on emissions is unpredictable. Moreover, its revenue stream is likely to decline, at least during its first years.

Companies pay twice. Does it make sense or is it fair to add the cost of effluent fees to the burden of cleaning up?

These problems of rate setting and enactment are relatively familiar and, though difficult, conceivably could be overcome, given the possible savings to the economy of using economic charges.

However, effluent fees entail further, much more serious administrative, enforcement, and political difficulties. Unless a fee proposal overcomes these critical barriers, no environmental administrator should consider it.

Connecticut has over 12,000 registered sources of air pollution. The state environmental agency has the staff to make several thousand inspections each year (and it has to impose enforcement sanctions on less than five regulatees a year to obtain one of the best compliance rates in the country). If it had to administer an effluent fee, the agency would have to deal with all 12,000 sources each year. The staff requirements, especially taking into account the great difficulty of measuring emissions (discussed below), would probably be crippling.

Much more serious, effluent fees are virtually unenforceable. EPA does not have the staff or technical capacity to monitor emissions. Neither EPA nor its regulatees know or can prove how much of what pollutants are emitted. Continuous monitors exist for only a token few percent of all sources. (Furthermore, emissions factors (or effluent guidelines) expressed in terms of product volume—for example, the number of square inches an electroplater plates or the average number of ducks owned by a duck farmer (two examples from EPA's effluent guidelines)—are likely to be unenforceable for different but closely analogous reasons: How can an EPA inspector prove that an electroplater's estimate of the square inches of forks, cleats, atomic submarine parts, and so on, that he has treated in the last year is too high? Or that a duck farmer usually has fewer ducks in residence than he now does?)

How is the state agency charged with collecting such a tax going to do the job if it has no credible method of measuring emissions? Once an emissions tax is in place, the agency has no choice but to try to collect from every source, specifically including the 95 plus percent of all sources for whom continuously recording, reliable monitors do not exist. Imagine the conversation between the state environmental inspector attempting to assess this tax and a plant manager (neither of whom knows what the plant's emissions have been). Imagine the chronic but unequal underestimating that would result. Imagine the legal fights. Imagine the impact on the state agency's staff's morale.

Even more serious, imagine the impace such an ill-fated effort would have on voluntary compliance, by far the most critical factor for any regulatory program's success. Widespread confidence that the rules are fair, that they are applied to everyone equally, and that noncompliance does not pay are all prerequisites if needed citizen cooperation is to be assured.[5] Forcing the agency to try to impose an unenforceable (and therefore probably unadministrable) tax on everyone would expose the agency's impotence for all to see. Almost every source would quickly learn how easy it is to cheat and how probable it is that their competitors were guessing low or cheating. There would be almost no hope of bringing the inevitable small number of hard-core recalcitrants to heel, and failure to do so, a critical test as far as the staff is concerned and not long a secret to the relevant public, typically causes the whole fabric of voluntary compliance to begin unraveling.

The consequences of such unraveling could not be more serious:

Administrative costs. Declining voluntary compliance means: (1) an increasing number of citizen complaints requiring attention, that is, a reduction in the proportion of inspector and follow-up resources that can be targeted according to the agency's priorities; (2) more enforcement problems to detect and handle; and (3) increased cost to the agency per case handled as recalcitrance spreads. Even small shifts in compliance rates can have disastrous implications for a regulatory agency: If 90 percent comply voluntarily instead of 95 percent, the agency's workload, taking per-case-recalcitrance into account, will probably triple. Delays will increase and credibility decline, providing further impetus to this vicious circle.

Environmental impact. Environmental quality and public health will suffer at least as much as the regulators.

Equity. It is unfair to those who have already or will, as a matter of principle, obey the law when their peers and competitors can safely ignore it.

Business planning. Business cannot plan capital expenditures sensibly when standards, tax rates, and/or enforcement levels are in flux.

Environmental regulation either works, because 90 plus percent of the regulatees comply voluntarily, or it does not work; there is absolutely no way that an administrative agency can police everyone, let alone everyone all the time. And, once lost, it is very difficult to reestablish voluntary compliance.

Imposing a probably unadministrable, unenforceable tax on a great many people also has disturbing political implications. People do not like taxes. They like them even less when they impose significant administrative burdens (as an effluent tax with its difficult measurement problems surely would) and when they force the taxpayer either to pay, even though he knows that others are not, or to cheat. An effluent tax's opponent could easily attack it as impractical, prone to abuse because it de facto standardless, an invitation to cheating and litigation, and so forth. Understandably, there are few commissioners of environmental protection (one of whose chief responsibilities is to build and maintain a strong constituency for environmental programs among the general public, in the state government, and in the legislature) who would welcome such legislation.

The classical standards-plus-enforcement approach is much less vulnerable to these critical administrative, enforcement, and political problems. Because it is generally accepted that the commissioner should exercise prosecutorial discretion in taking enforcement action, he or she can carefully pick and pursue a small number of cases for which the agency has the necessary data and for which the equities are clear. In contrast, selective application of a tax would be considered entirely inappropriate.[6]

Environmental regulators cannot police, whether or not sources are operating and maintaining their control equipment properly and regardless of whether the regulators adopt a standards or fee strategy. This a growing, critical problem for environmental law enforcement. However, until environmentalists solve the problem, they had better stick with a standards-plus-discretionary-enforcement approach. Shifting to a fee strategy now would force them to parade naked into every polluter's front office, with the dismal results outlined above.[7]

Current Priorities

Environmental economists, lawyers, and administrators are beginning to understand one another's perspectives just in time. Environmentalists must find new methods for ensuring that the objectives carefully developed over the last decade are reached. There are at least two critical problems that must be solved: (1) giving regulators adequate tools to enforce the law and (2) finding a practicable means of policing proper operation and maintenance.

The Connecticut approach provides a set of easily used, highly flexible tools that will allow administrators to enforce the law once a violation is detected and can be proven.

The most difficult problem remaining, then, is the operations and maintenance problem.[8] Over the last several years, environmental regulators have been focusing on the problem of getting sources of pollution to install often expensive control facilities. Now, especially in jurisdictions that have had vigorous environmental programs and good citizen support, most significant sources have control equipment in place. Consequently, in the future environmental regulation is going to have to focus much more than it has so far on ensuring that sources properly operate and maintain the equipment they have installed. If this objective cannot be achieved, much of the benefit of the last half-decade's effort to install control facilities may be lost.

It is also clear that the operating and maintenance problem is a tough one. It is going to be difficult to get sources to operate and maintain properly because it is much more in their economic interest not to operate and maintain than it ever was not to invest in the control equipment in the first place. Figure 1, which compares the capital and operating-and-maintenance costs of each of the major types of air pollution control equipment on an annualized basis, makes this clear. In every case, the costs to the regulatee of operating and maintaining its control equipment is much greater—usually *many* times greater—than the annualized capital costs of the equipment.

Though difficult, the problem is by no means impossible to solve. Even if direct emission monitors are not available, for example, it will often be possible to install simple parameter monitors. (It should be possible to determine whether or not a plant has been using its baghouse as much as it should by attaching a sealed electric power meter to the baghouse fan.) EPA's lawyers will have to work closely with its engineers and enforcement staff to develop procedures for defining and holding sources to such indirect standards.

At last the long, ideologized stalemate between simple standards and simple fee proponents seems to be breaking up. This is fortunate. It may allow environmentalists to focus more clearly on the difficult, unresolved issues of how best to pursue the objectives defined over the last decade. And if these erstwhile protagonists begin to understand one another's perspectives, they

TYPES OF CONTROL EQUIPMENT	ESTIMATED PERCENT OF ANNUALIZED CAPITAL COSTS	
	ECONOMIC BENEFITS OF CONTROL	OPERATING AND MAINTENANCE COSTS
BAGHOUSE	%N OXIDE 10 CONCRETE 40	220 – 350
ELECTROSTATIC PRECIPITATOR		380 – 1,500
HIGH-TEMP. AFTERBURNER (HC)		1,800 – 6,000
CATALYTIC AFTERBURNER (HC)		700 – 2,200
LOW-ENERGY SCRUBBER (PART.)		90 – 160
CARBON ADSORBTION	150 SOLVENT RECOVERY	80 – 160

Figure 1. Operating and maintenance costs as percentages of estimated annualized capital costs.

should be able to produce the new, practical, and probably hybrid tools that environmental regulation now needs.

Notes

1. The current standards-plus-enforcement approach to abatement can and often does take cost into account when standards are set in the first place. Old and new plants are held to different standards, and individual industries typically face different emission/effluent limits. Adjusting standards in this fashion requires roughly the same information as is required to predict the impact of any level of effluent charge.

2. Why have these groups sailed past one another for so long? The problem is a recurrent one and reflects basic institutional arrangements in our society. Those who specialize in developing ideas—chiefly those working in our universities—specialize by discipline, have no career incentive to participate in problem solving (which is necessarily a supradisciplinary process) or to become intimately familiar with institutional constraints, and typically have little contact with and are of a different class than most managers. Managers, by contrast, have little incentive to question the basic rules by which they play. Time is too valuable.

Especially as change accelerates, it will become increasingly important to find new methods of bridging this division between conceptual capacity and problem-solving responsibility.

3. This additional charge, typically a 12 percent compounding rate, makes compliance as profitable as normal commercial investments. (The penalties saved equal the revenues from a profitable investment.)

4. The "Connecticut approach" is explained in *Economic Law Enforcement* (6 vols.), U.S. EPA publication EPA-901/9-76-003 (September 1976). (A brief summary of this work is available from the author, Littauer Center, Harvard University, Cambridge, Massachusetts 02138. The author directed the consultant team that helped EPA and the Connecticut Department of Environmental Protection develop this new approach.)

5. The Connecticut Enforcement Project studied a number of different enforcement systems, both in Connecticut and elsewhere, and found, not surprisingly, that enforcement credibility and voluntary compliance correlated very closely. Connecticut Enforcement Project, *Economic Law Enforcement* (vol. 1), U.S. EPA Publication EPA-901/9-76-003, p. 20 (1975).

6. As mentioned, Connecticut has imposed sanctions on less than five air compliance cases a year; it would have to tax over 12,000.

7. The only effluent fees that have actually been enacted in the United States over the last decades have all resolved the administration-enforcement problem. New York City's tar and nicotine tax piggybacked on top of (was collected at the same time and through the same mechanisms as) the city's flat-rate cigarette tax. See William Drayton, "The Tar and Nicotine Tax: Pursuing Public Health Through Tax Incentives," *Yale Law Journal* 81 (1972): 1487. New York's leaded-gasoline and recycling incentive taxes similarly used familiar fuel and excise tax vehicles.

8. Although discussed here strictly in an environmental context, how to police continuous processes (versus easily reviewed capital purchases or particular actions) is a generic problem that plagues a great many other regulatory schemes. Government's inability to deal with this problem explains why it tends to focus on capital investments and other easily policed actions.

6 Market Approaches to the Measurement of the Benefits of Air Pollution Abatement

Daniel L. Rubinfeld

Introduction and Summary

Recent policy decisions concerning the nature of standards and other regulatory policies to control air pollution have made little or no use of explicit cost-benefit analyses. This reality may not be surprising given the political setting in which these policy decisions were made.[1] However, there are economic as well as political factors that might argue against the explicit use of cost-benefit analyses. One argument, in its most simplistic form, states that it is impossible to obtain quantitative estimates of the benefits of air pollution control, first because individuals (as well as researchers) are simply unaware of all of the potential benefits and, second, because it is likely to be difficult to quantify those benefits about which individuals are aware. Despite the admittedly serious difficulties, economists are generally reluctant to give up the use of such a potentially powerful tool as cost-benefit analysis. Clearly some of the benefits from air pollution control are likely to be extremely difficult, if not impossible, to quantify. But even a crude quantitative estimate of these benefits, with the appropriate caveats, would be more useful for public policy than no estimate at all.

This paper seeks to shed some light on this issue by assessing the methods of measuring pollution control benefits that rely most directly on information obtained from the market process. This is clearly a difficult problem because there is no explicit market for the amenities associated with clean air, and thus there are no directly observable market prices that can be interpreted as the rate at which individuals are willing to trade clean air for other commodities. However, two approaches have received serious attention from economists and have proved successful in providing quantitative benefit estimates. The first approach involves *property-value studies,* in which the willingness to pay for clean air is inferred from the housing market (usually in an urban area), on the presumption that individual households will pay more for a housing unit located in an area with good air quality than for an otherwise identical unit located in an area with poor air quality.[2] A second approach involves *wage-rate studies,* in which wage differentials among urban areas (adjusted for labor quality and other factors) are used to impute the willingness to pay for clear air, on the presumption that individual households will accept a lower wage to work in an urban area in which the air quality level is good.[3]

There are, of course, other approaches that have been and are being used to obtain quantitative benefit estimates, the primary of which focuses on the health benefits of pollution control.[4] The health, market, and other approaches all involve substantial difficulties, both in terms of the conceptual framework and in terms of the specific technical and statistical problems involved in the analysis. However, the two market-oriented approaches (wage-rate and property-value studies) are perhaps the most controversial because there appears to be a wide range of opinion about whether one can place any reliance on the benefit estimates that they yield. For example, Lave argues, "Certainly the former [property-value] approach is simpler and closer to a classical economic approach. It focuses our attention on the things that economists have the most experience with, namely estimating utility functions, demand functions, and splitting the pollution component of a price from other factors. However, I believe that this approach is not likely to lead to fruitful results."[5]

On the other hand, Waddell concludes, "Because of its general ease of measurement and inclusiveness, the property value, or site value differential technique, is one of the most promising approaches to the estimation of the economic losses due to air pollution. The advantage of this method is that the investigator does not have to discover and evaluate the pollution sufferers' adjustment possibilities, nor does he have to worry about how to make individual properties commensurate so that he can aggregate them."[6] This wide variety of opinion arises not only because of the difficult set of empirical problems but also as a result of the difficult theoretical problems that are implicit in these approaches.

Perhaps the most important issue surrounds the so-called double-counting issue—whether and to what extent wage- and property-value benefits overlap the benefits obtained from health and related studies. To be specific, consider the property-value approach. Most studies of this sort have specifically focused on the residential-housing market. As a result, they do not provide estimates of the air-quality improvement benefits associated with households' work places or shopping areas. In addition, they do not measure benefits that might accrue to businesses in the form of lower costs of production. In order for the housing market to capture some of the effects of air pollution, detailed knowledge of the specifics of air pollution is not necessary. However, it is important that some households are willing to pay more for houses in less polluted areas than for equivalent quality houses in highly polluted areas. Thus, it seems likely that households will be aware to some extent of the extra cleaning and maintenance costs associated with poor air quality, as well as the soiling, odor, eye irritation, and lack of visibility that it causes. What is not clear is the extent to which these measured benefits overlap the benefits obtained from health studies. Under the assumption that households don't perceive well the health effects of air pollution and have no adverse expectation about these effects, the property-value benefit estimates ought to be added to the benefit estimates obtained through epidemiological studies. However, to the extent that households respond to known adverse health effects or simply expect health effects to be serious in highly polluted areas without specific knowledge, the property-value and health benefits are likely to substantially overlap.

The automobile emissions study of the National Academy of Sciences was concerned with these issues and argued, "Some double counting may be involved, however, in the simple addition of the two estimates. A point estimate nearer the lower middle of the range . . . would thus seem to be a reasonable guess of total benefits due to the federal emissions program."[7]

Waddell, in his survey of air-pollution damage studies, took the view that double counting is not a serious problem and argued, "While the evidence is far from clear, it is reasoned that as interpreted in this study, the estimates determined via the site differential and technical coefficients methods should be considered additive, with only minor adjustment for obvious areas of overlap."[8]

Waddell is correct in stating that the evidence is far from clear. Some of the acute symptoms associated with poor air quality are likely to affect households' location decisions, but many other known adverse health effects seem unlikely to show up in property-value studies. This expectation seems especially reasonable in the case of carbon monoxide because CO is odorless and tasteless, but it is less clear in the case of particulates and sulfur oxides, for example. In this author's view, however, Waddell's point of view is likely to involve less error than an approach that argued that health and property-value benefit estimates entirely duplicated one another. A good portion of the benefits that show up in property-value studies are aesthetic benefits that are unrelated to households' direct fears about health effects. As a result, the addition of a *correctly* estimated property-value benefit measure to the benefit measures obtained from health and other cost studies may not seriously overestimate benefits. Of course, this judgment may change as information improves and households respond more directly to the health hazards associated with air pollution. Whatever one's opinion on the double-counting issue, it should be clear that quantitative estimates of both health and market-related benefits should provide useful inputs into cost-benefit analyses of air pollution control and therefore should be seriously evaluated by policy makers involved in the regulatory process.

This paper summarizes some recent evidence concerning the nature of air-pollution-control benefits that can be obtained from property-value data and assesses the theoretical and empirical difficulties that are inherent in such studies. Despite the rather substantial list of complexities involved, these recent results are quite promising from a policy viewpoint. Not only are quantitative estimates of the benefits of reductions in air pollution concentrations obtained, but more important, quantitative estimates of the sensitivity of the results to some of the underlying assumptions are also available. As a consequence, policy makers should have sufficient perspective to make proper use of the numbers that are available. Even a crude guess about the facts, with the appropriate caveats, should be more useful than a guess based purely on whim. This guarded optimism must be qualified in one important respect, however. The results of the study are discouraging in that they suggest that property-value studies are likely to fail at providing policy makers with a list of benefits associated with the reduction of concentration levels of individual pollutants. Further information is likely to be necessary if reasonable guesses are to be made along these lines.

The organization of the paper is as follows. The first section discusses in some detail the conceptual issues involved in the market studies. An examination of the extent to which property-value studies can be utilized to estimate individual households' willingness to pay for both small and large improvements in air quality in one urban area and throughout the United States leads into an analysis of the conditions under which wage-rate differentials can be used to estimate benefits and the relationship between the wage-rate and property-value approaches. The second major section considers the empirical results of property-value studies.[9] Rather than reviewing a long list of studies that are not comparable, the section provides a focus by concentrating on the recent analysis of a data set for the Boston metropolitan area. This focus allows estimates to be obtained for the magnitudes of some of the biases that arise in the property-value studies. The objective is to obtain some reasonable statistical bounds for the property-value-related benefits of an improved environment and to see to what extent these benefits can be attributed to improvements in specific pollutant levels. The final section summarizes the major conclusions.

Conceptual Issues

Economists argue that environmental decision making must involve a comparison of benefits as well as costs. Typically, the set of available policy choices may involve various levels of technology and various corresponding degrees of pollution control. As a consequence, the analysis of such policy choices involves a calculation of the benefits of pollution reduction for various degrees of reduction. A useful first step in the analysis of the benefit side of a cost-benefit analysis might involve the determination of the benefit to society of a small amount of pollution abatement, that is, the marginal benefit. However, this step is only a first step, because most reduction programs involve substantial abatement. The marginal benefit of pollution reduction may change when reductions occur, so the estimation of the total benefits associated with a large, or *nonmarginal,* reduction in air pollution levels is a more important and yet a more complex problem.[10]

This section addresses the question of whether market approaches can be used to derive estimates of the benefits of improved urban air quality. The first issue discussed is whether property-value studies can be utilized to measure the willingness to pay for nonmarginal air quality improvements. Most existing studies yield estimates that provide correct measures of the marginal benefits of environmental improvements but provide biased estimates when the improvements are nonmarginal. Following this discussion, the methodological problems associated with extrapolating benefits calculated for one urban area to benefits for the entire United States are briefly outlined. Finally, the wage-rate approach to benefit measurement is analyzed and special attention given to the methodological problems involved and to the relationship between wage and property-value studies.

The Willingness to Pay for Clean Air

Implicit in the property-value approach is the assumption that market prices (or imputed market prices) provide a meaningful basis for measuring the benefits of an environmental improvement.[11] To correctly utilize this price information, the concept of willingness to pay for an air quality improvement must be explicitly defined. Consider a household at a particular location in an urban area in which the level of air pollution is represented by the index A. Hold all prices and the household's money income constant and gradually decrease the level of air pollution. The household's *total* willingness to pay for the air quality improvement is defined as the maximum amount of income that can be taken after the air quality improvement without leaving the household worse off.[12] The household's marginal willingness to pay for small improvements in the environment can be found by calculating the derivative of total willingness to pay with respect to the level of air pollution. This relationship is called the household's *willingness-to-pay function* because it emphasizes the link between the physical level of air pollution and the marginal benefit associated with small improvements at each level. It is also known as the *marginal damage function* because it measures the household's valuation of marginal reductions in pollution damages.[13] In economic terms, the willingness-to-pay function is equivalent to a compensated demand curve for reductions in air pollution because money income has been allowed to adjust so that real income (utility) remains constant.

Calculation of Total Benefits If one can statistically estimate the willingness-to-pay function for each household, it is relatively straightforward to calculate the aggregate benefits associated with both marginal and nonmarginal improvements in air quality throughout an urban area. If $W_i(A_i)$ represents the marginal willingness to pay for a unit reduction in air pollution by the household residing at location i (where air pollution is at level A_i), then the dollar value of a marginal reduction in air quality is simply $W_i(A_i)$. To obtain the benefits associated with a nonmarginal change, the values of these marginal willingnesses to pay must be summed from the original to the new level of air pollution, that is, the W_i function must be integrated from the original to the new level of air pollution.[14]

If the willingness-to-pay function is independent of the level of air pollution, then the total benefits will simply equal the product of the marginal willingness to pay and the total reduction in air pollution. This equality is illustrated as Case 1 in figure 6.1.[15] However, if the marginal willingness to pay decreases as the level of air pollution is reduced, the total benefits will be lower than in Case 1,[16] so that the incorrect assumption of a constant marginal willingness to pay will cause total benefits to be overestimated. Finally, if (as in Case 3) marginal willingness to pay increases as the level of air pollution is reduced, the assumption of a constant marginal willingness to pay will lead to an underestimate of total benefits. The analysis suggests that, for small air quality improvements, the shape of the willingness-to-pay function is not likely to be relevant, but when the improvement is large, the relationship between a reduction in pollution and willingness to pay may be important. Thus, one of the objectives of any market analysis of air pollution (as with nonmarket studies) should be to ascertain the shape of the air-pollution willingness-to-pay or damage function.

One important methodological question is whether, and if so under what conditions, it is possible to estimate the willingness-to-pay function for urban households. The problem is a difficult one because marginal willing-

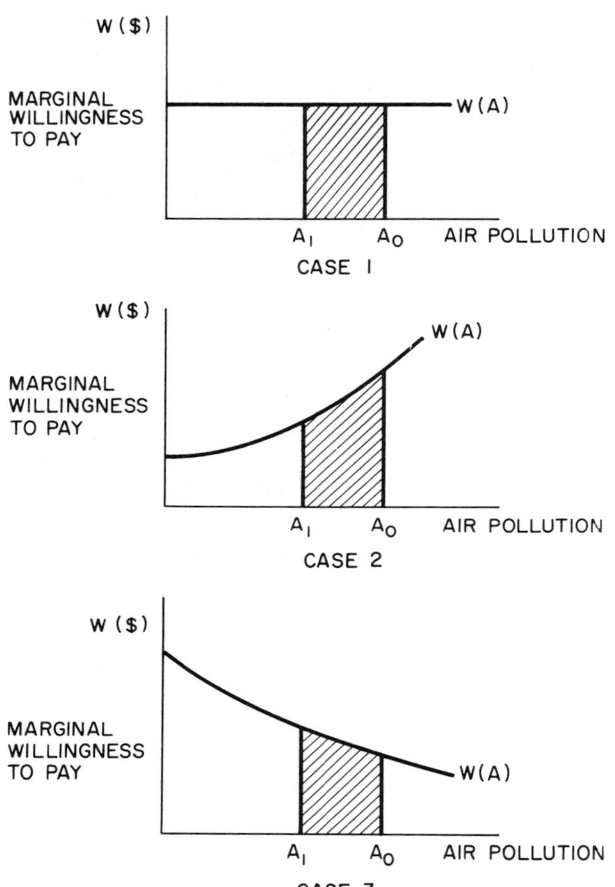

Figure 6.1. Willingness to pay for reductions in air pollution. (A_0 = original pollution level; A_1 = improved pollution level)

ness to pay will depend not only on the level of air pollution (a movement along the "demand" curve) but also on taste variables such as income (a shift of the "demand" curve). For example, if air quality is a normal good, then the marginal willingness to pay for clean air will increase as household income increases. Unfortunately, it isn't possible to carry on a controlled experiment, that is, to study identical households simultaneously consuming (purchasing or renting) housing at locations with different levels of air quality at the same point in time. However, if households with similar taste structures are known to consume different levels of air quality and other attributes associated with housing, it may still be possible to determine the willingness-to-pay function.[17] This section describes how such an estimation procedure might be accomplished and at the same time stresses the rather strong set of assumptions that are necessary for such a procedure to be appropriate.[18] A later section includes empirical estimates of the magnitude of the difficulties that arise when some of the underlying assumptions are not satisfied.

Estimating the Willingness-to-Pay Function Assume that the housing (owner plus rental) market is in short-run equilibrium so that housing values (and rental prices) equate the demand and supply of housing services.[19] To simplify, assume that in the short run the supply of air pollution and the housing stock at every location in the urban area is fixed, so that the movement of households among locations serves as the mechanism that equates housing demand to housing supply. Individual households are assumed to maximize their utility by selecting their residential location and, in the process, trading off variations in housing and neighborhood attributes with accessibility to work place.[20] Rather than imagining housing as a commodity with a known, measurable output, it is more useful to conceive of housing as composed of a bundle of attributes, which include not only structural characteristics but also neighborhood and accessibility characteristics and, of course, one or more indices of the level of air pollution. The relationship between equilibrium housing values and the level of each housing characteristic (alternatively called a *housing-value, property-value,* or *hedonic-value* equation) might be represented as:

$$V_i = f(S_i, N_i, A_i)$$

where V_i is the residential property value[21] (that is, value of land and improvements) at location i, S_i, the structural characteristics of location i, N_i the neighborhood and accessibility characteristics of location i, and A_i the air pollution level (or levels) at location i.

Because individuals are assumed to take into account air quality variations as they are making their housing choices, we expect that, everything else equal, the change in property values with respect to a change in air pollution ($\partial V_i/\partial A_i$) will be negative; in other words, households living in low-air-pollution areas (other things equal) will pay more for their houses than households living in high-pollution areas. A necessary condition for equilibrium to be attained is that, at each location, the household's marginal willingness to pay for a unit reduction in air pollution must be identically equal to the marginal cost incurred in purchasing a different house if all attributes are the same except that air pollution is reduced by a unit.[22] In mathematical terms,

$$W_i(A_i(= - \partial V_i/\partial A_i = V_{\text{Ai}}$$

If the housing-value function (f) were linear in each of the housing attributes and households were assumed to be identical, then the marginal willingness to pay for air pollution reductions would be identical for each household and at every location. Thus $W_i(A_i)$ would be the (horizontal) willingness-to-pay function from which we could accurately measure the benefits of both marginal and nonmarginal improvements in air quality. However, there is reason to believe that the housing-value function may be nonlinear. Nonlinearities arise because the housing market may not be in long-run equilibrium. Unlike the attributes of less durable commodities, housing attributes cannot be untied and repackaged to produce an arbitrary set of attributes at all locations. As a consequence, the marginal valuation placed by consumers on a given housing attribute (including air quality) may depend upon the particular combination of housing attributes at that location. For example, good air quality may be more highly valued at locations in which there is a view of mountains than at locations in which the view is blocked by a series of high-rise apartment houses.

In the general case, the slope of the housing-value equation does not describe a demand or willingness-to-pay function; rather it represents the locus of equilibrium marginal willingnesses to pay for all households. In order to identify the willingness-to-pay function for each household, it is necessary to identify household groups on the basis of tastes and incomes.[23] Only by examining variations in $W_i(A_i)$ within each household group is there hope of identifying the willingness-to-pay function of that group. For example, assume that all households have identical tastes but different incomes. By accounting for the fact that the willingness-to-pay function shifts as income changes it may be possible to estimate the relationship between $W_i(A_i)$ and A_i. Specifically, consider the functional relationship

$$W_i(A_i) = g(A_i, Y_i, Q_i)$$

where Y_i equals income and Q_i a vector of housing attributes that are complementary or substitutable with air pollution. Given a level of income for the household located at i, the function g associates a marginal willingness to pay with each level of air pollution A_i. If g is properly specified, it will in fact represent the desired willingness-to-pay function. However, if taste differences are not properly taken into account, g at best will represent an average willingness-to-pay function over several household groups.[24]

To the extent that any of the assumptions in the willingness-to-pay approach are violated, empirical estimates of the benefits of improved air quality may be biased. If, for example, the housing market were not close to short-run equilibrium (due to factors such as high moving costs), the change in house values with respect to a change in air pollution might not reflect the benefits of better air quality. In addition, to the extent that households do not have perfect information about air quality (and other data), low values of marginal willingness to pay may reflect imperfect knowledge rather than a lack of concern for air quality. Finally, the theory presumes that the levels of air pollution are correctly measured. To the extent that the level of air pollution varies over time and with climatic conditions, the use of any single index of air quality may lead to biased results.[25]

Assuming that the air-pollution damage function has been correctly estimated, the final conceptual step is to use the damage function to calculate the dollar benefits associated with a particular scheme for improving air quality. The per-household willingness to pay for nonmarginal reductions in air pollution (in principle) can be calculated for households at each location in the urban area by finding the area under the willingness-to-pay curve from the old to the new level of pollution, that is, by integrating the function g and evaluating the integral at the appropriate pollution levels. Because the conceptual model has been written with property value as the dependent variable (rather than rental price), the integrated willingness to pay is an estimate of the capitalized (as opposed to annual) value of the air quality improvement to each household.

Implicit in the willingness-to-pay approach is the assumption that households living in an improved area do not move when the improvement occurs. This is a reasonable assumption, given the difficulties involved with the prediction of household moving behavior, but it does suggest a possible bias in the benefit calculations. Assume, for example, that households living in highly polluted areas before the improvement in air quality do not place great value on clean air. After the improvement, families that place greater value on clean air may move into the area. As a result, the estimated willingness to pay of the original occupants of the area will be lower than the willingness to pay of the new residents. Although the general equilibrium nature of the economic system makes generalization of this example difficult, it seems reasonable to expect that benefit estimates that fail to account for such household adjustment will underestimate benefits.

Obtaining an Aggregate U.S. Benefit Measure

One difficulty associated with the property-value approach is that in order to obtain an estimate of the benefits of a national air-pollution-reduction program, data and cost limitations force one to extrapolate the experience in one or several urban areas to all urban areas in the United States. For example, in the National Academy of Sciences study of the benefits of automobile emissions controls, a national estimate of benefits was obtained by multiplying the estimated average benefits per household in Boston by the estimated total number of households living in U.S. urban areas and thus likely to be affected by air pollution.[26] Unfortunately, however, there are strong reasons to believe that such linear extrapolation may result in serious biases. Urban areas throughout the United States clearly differ in the levels of various pollutants as well as in the distribution of those pollutants over space. In addition, the degree of reduction of pollutants in each urban area is likely to vary substantially depending upon the specific nature of the legislated air pollution controls. Finally, there is reason to believe that individuals have to some extent sorted themselves out geographically on the basis of their specific tastes for climate, location, and possibly the level of air quality. All of these difficulties suggest that one ought to base national estimates of benefits upon studies of several urban areas that differ in terms of population characteristics and the levels of various pollutants.[27]

Predicting Changes in Residential Property Values

A number of empirical studies of the benefits of air quality improvements within an urban area have argued that the sum of the changes in property values arising from an air quality improvement provide a good estimate of the total benefits of the improvement.[28] The argument for this approach is stated clearly by Ridker as follows:

> If the land market were to work perfectly, the price of a plot of land would equal the sum of the present discounted streams of benefits and costs derivable from it. If some of its costs rise . . . or if some of its benefits fall . . . the property will be discounted in the market to reflect people's evaluation of these changes. Since air pollution is specific to locations and the supply of locations is fixed, there is less likelihood that the negative effects of pollution can be significantly shifted onto other markets. We should therefore

expect to find the majority of effects reflected in this market, and can measure them by observing associated changes in property values.[29]

The procedure utilized to predict the change in property values is simply to multiply V_{Ai}, the change in property values resulting from a change in air pollution obtained from the estimation of a hedonic-value equation, by the change in air pollution at every site in the urban area. As outlined in the previous section, this is the correct procedure for determining the total benefits associated with a marginal improvement in the environment, but it is likely to lead to biased results if the environmental change is nonmarginal. Thus, for marginal changes, the procedure utilized by most empirical studies of air pollution correctly estimates benefits.[30]

Two questions remained unanswered. First, do the empirical studies accurately predict changes in property values; second (and more relevant for this paper), would the correctly estimated predicted change in property values provide a useful measure of the willingness to pay for nonmarginal air quality improvements. Unfortunately, both questions must be answered in the negative. The problem with attempting to predict property values from a hedonic-value equation is that the V_{A_i} are calculated on the presumption that all housing attributes (including air quality) at other locations in the urban area are held constant. As a result, the hedonic approach cannot take into account the important general equilibrium impacts likely to occur when there is a substantial improvement in air quality throughout the urban area. This point was clarified by Polinsky and Rubinfeld, who argue:

> Consider an urban area with a fixed population and fixed boundary in which air quality improves with distance from the center. If air quality is then raised throughout the area to a uniform level, say to the level at the boundary, this would induce an excess demand for land in the inner part of the urban area bidding up property values there, but also an excess supply of land in the outer area, lowering property values there (even though air quality also improved). . . . Thus, it appears that the estimated coefficients of the (hedonic) equation cannot be used in this way to predict the pattern of property value changes throughout the urban area.[31]

Even if the aggregate change in property values could be predicted, this change will not in general correspond to a correct measure of total benefits. The difficulty arises because, as a rule, the improvement in air quality throughout the urban area generates not only changes in property values (surpluses to landlords) but also producer surpluses and consumer surpluses. Only in the special case in which consumer and producer surpluses are identically zero will the aggregate change in property values correctly measure benefits. In the general case, changes in property values may provide seriously misleading benefit estimates. For example, Polinsky and Shavell cite a hypothetical example in which air quality is improved everywhere but the aggregate change in property values remains constant.[32]

Polinsky and Shavell and Polinsky and Rubinfield point out that the validity of using cross-section hedonic equations to predict changes in property values depends crucially on one's assumption about the mobility of households among (as well as within) urban areas.[33] In the special case in which the labor market is exogenous so that wage rates are fixed and migration is costless between any two urban areas, property-value changes will accurately measure benefits. However, to the extent that interurban migration is costly or labor-market adjustments are taken into account, changes in residential property values may provide seriously biased estimates of benefits. The costless mobility assumption is crucial because it guarantees that all of the benefits of the environmental improvement will appear in property values, whereas consumer and producer surpluses will not change. The exogenous labor-market assumption is crucial because it guarantees that wage changes will not absorb a portion of the benefits of the environmental improvement.

Interurban Wage-Rate Differentials

Several authors have argued that, when adjusted to hold labor quality and certain other factors constant, wage differentials among urban areas may provide an estimate of the willingness to pay of households for an environmental improvement.[34] For example, the National Academy of Sciences report on Automobile Emissions argues, ''Returns to labor consist of real wages and non-monetary consumption goods. Assuming that information on wages and price differences between cities is available, and that there are no barriers to mobility, wages should be negatively correlated with amenities in order to equalize returns to labor between cities and ''clear'' the labor market. For example, one expects higher real wages to compensate for living in a city where there is a higher probability of personal injury. Hence, we can place at least a rough value on amenities by comparing real wages and amenities among cities.''[35]

The empirical difficulties with such cross-sectional studies (in which the units of observation are usually SMSAs) are great. Such studies rely explicitly on the assumption that migration among urban areas is costless and information is sufficiently perfect that the general system of urban areas is in equilibrium. This assumption seems to be more questionable than the assumption that a single urban housing market is in equilibrium. Such studies must also (at a highly aggregated level) hold constant differences among urban areas in climate, transportation, the provision of public services, and other non-air-quality amenities.

In addition to these empirical difficulties, there are some serious methodological problems inherent in the wage-rate studies. The most important weakness of the wage-differential argument is that the underlying model does not account explicitly for the factors that might restrict mobility. The authors of the National Academy report were aware of this difficulty: "With factor mobility, prices, or in this case wages, ultimately adjust so as to equate the utilities of similar households or workers across different urban areas. If there are certain limitations upon factor mobility, some urban productivity and amenity differentials could be captured as land rents. Similarly, certain kinds of "frictions," particularly transportation cost differences within and among urban areas, could cause any potential rents attributable to factor immobility to be shared between land and other sectors, such as transport."[36]

To clarify this issue consider a simple long-run model of an urban area that includes a single homogeneous labor market and a residential-housing market.[37] The wage that firms pay to their employees depends on the marginal product of the workers, which depends, other things equal, on the size of the employable population in the urban area. Now assume that the environment is improved throughout the urban area. One possible outcome is that property values will rise throughout the urban area. However, if migration among urban areas is costless, additional households will be attracted to the cleaner environment. Other things equal (that is, the land area of the urban area allocated to business use), the increased supply of labor will cause the marginal product of labor and thus the wage rate to fall. Immigrants will be willing to accept this lower wage, however, because they are receiving improved air quality in exchange. Thus, the environment improvement results in *both* a fall in the wage and an increase in property values. Only in the special case in which property values remain unchanged would the full extent of the benefits associated with the environmental improvement show up as a reduced wage in the improved urban area. Depending upon the labor-market conditions and the nature of demand for residential housing, it seems entirely possible that most of the adjustments to the air quality improvement will occur in the local land market and not through the labor market.[38]

To the extent that the wage rate falls, business property values must increase to keep firms in their equivalent competitive positions. This relationship suggests that, to the extent that migration among cities in the urban system is costless (that is, the urban areas are *open*), one can measure benefits by adding the benefits associated with the change in residential property values to the benefits associated with the wage differentials. Because wage-related benefits correspond to benefits associated with business property values, in this special case the sum of the change in residential and business property values within the impacted urban area will provide a correct benefit measure. Unfortunately, this result depends crucially upon the assumption that migration among the system of urban areas is costless and that the number of urban areas is "large." As mentioned previously, the strong mobility assumption guarantees that the environmental change will not alter producer and consumer surpluses.[39] The assumption of a large number of urban areas is important because, for nonmarginal environmental improvements in a finite system or in urban areas, capital market adjustments throughout the system of urban areas may cause a substantial portion of the benefits to be obtained by those residing outside the area in which the improvement occurred.[40]

The previous analysis suggests that benefit estimates based on wage differentials will understate the true willingness to pay for cleaner air. This conclusion seems quite plausible, but given the empirical difficulties, it does not necessarily follow from the theoretical model just outlined. Imagine an improvement in air quality in a situation in which clean air and housing are highly substitutable. As air quality in one urban area is improved, households (over the long run) tend to purchase houses with smaller lots (for example, they enjoy greater use of parks). The net outcome of the market process might be that property values fall throughout the residential urban area. Because each household consumes less housing, there is sufficient space in the urban area for a large immigration of those who value highly the cleaner environment. This influx results in a decline in the wage rate, which is greater than the decline that would occur if property values were unaffected by the air quality improvement. The result is that the correctly measured wage differential will overstate the true measure of benefits. Although admittedly somewhat artificial, this example is useful because it suggests that to be fully correct any wage-differential measurements should be made in combination with estimates of the predicted change in property values in the affected urban areas. Given the previous discussion, this is likely to be a very difficult task.

Empirical Issues

Relatively little is known about the shape of the air-pollution damage function. However, information concerning the aesthetic and other market-oriented damages from air pollution can be obtained from the housing-value approach described in the preceding section. The first part of this section presents empirical results of a recent analysis of housing data in the Boston metropolitan area conducted by the author and David Harrison.[41] These results not only suggest that air-pollution damage functions are nonlinear but also provide estimates of the magnitude of the bias involved when benefit estimation fails to take account of such nonlinearities.

The first two parts of the section are generally self-contained and serve to provide a general overview of the empirical results. However, such results must be placed in their proper perspective by giving serious consideration to the underlying empirical problems. This task is done in the remaining part, which concentrates on more technical issues relating to the air pollution data and to the nature of the housing market. Most of the empirical issues raised in this secion are not new, but little is known about the quantitative importance of the specific problems associated with the property-value approach. This recent information partially fills the gap.

The Data

The basic unit of observation for the Boston study (as in most other studies) is the census tract.[42] The use of census-tract data rather than individual housing data poses certain empirical problems for property-value studies, first because the use of median census-tract observations eliminates information about the variance of individual variables within census tracts (and thus may cause correlations between variables to differ at the micro and census-tract levels) and, second, because it is difficult to properly control for certain nonpollution variables, such as structural and neighborhood characteristics, as a result of data limitations.[43]

Measuring Air Pollution Most property-value studies rely exclusively on monitoring data to obtain quantitative measures of pollutant levels. Aside from the technical difficulties associated with attempts to monitor pollutant levels, there are some important problems that arise when such data are used. Property-value studies typically utilize annual averages of pollutant levels (often geometric means) as point estimates of pollutant levels. Conceptually this practice appears to be a serious source of misspecification because there is reason to believe that households are more sensitive to unusually bad air pollution episodes than they are to average levels. Thus, a measure of the variance in pollution levels might also have an impact on property values. As a practical matter, however, this issue may not be very important. In the Boston data set, the distribution of pollutant levels is approximately lognormal and variances are roughly constant over space.[44] To the extent that this is true, the inclusion of variance in the property-value studies would have no effect on the results. A more serious problem arises because of the relative scarcity of monitoring stations in each urban area.[45] As a result, researchers are forced to extrapolate to obtain estimates of air pollution levels in many of the census tracts in the urban area. Without a proper correction, the statistical significance of the results will be overstated.

The problem of relying directly on monitoring data for a few stations was circumvented in the Harrison and Rubinfeld Boston study by using information on 1970 air pollution concentrations obtained from a meteorological model of the Boston air shed. The Transportation and Air Shed Simulation Model (TASSIM) generates information on the mean air pollutant concentrations in 122 zones in the Boston SMSA from internally generated estimates of vehicle miles of travel by zone, emissions characteristics of the 1970 automobile fleet supplied by the U.S. Environmental Protection Agency, and estimates of emissions from nonautomobile area sources, including almost 400 stationary sources.[46] The air shed model was calibrated using Boston monitoring data, so that some difficulties inherent in using monitoring data still remain.

In addition to providing more degrees of freedom in the air pollution data, the meteorological model provides estimates of concentrations for a broad set of pollutants. Use of the meteorological data allows for a close examination of the degree of multicollinearity that is likely to arise when one attempts to distinguish among various pollutants in a property-value analysis. Table 6.1 lists the simple correlations among the various pollutant concentrations predicted by the Boston TASSIM model. It is not very surprising to find that NO_x and HC, two of the automobile emission pollutants, are highly correlated, as are PART and SO_2, two of the major stationary-source pollutants. What is somewhat surprising, however, is the high correlation between mobile-source and stationary-source pollutants (see note b of table 6.1 for a partial explanation). These simple correlations suggest that it is likely to be very difficult to distinguish between the impacts of individual pollutants on property values in a cross-section study of an urban area. The analysis of the Boston data set confirmed this expectation. When two or more individual pollutant variables are included in the hedonic housing equation, multicollinearity becomes a serious problem. One of the pollutant variables often takes on the wrong sign, and the values of the pollutant coefficients become very sensitive to the specific functional form of the hedonic equation. The pollutant that is least highly correlated with the others is carbon monoxide. However, the CO variable should not enter significantly in the housing-value equation as a pollutant measure because CO is odorless

Table 6.1 Correlations among Pollutants

	NO_x	HC	O_x	PART	SO_2	CO
NO_x	1.00					
HC	.93	1.00				
O_x	.97	.98	1.00			
PART	.96	.90	.92	1.00		
SO_2	.96	.85	.90	.97	1.00	
CO	.88	.93	.93	.79	.78	1.00

[a]NO_x = nitrogen oxide concentrations in parts per hundred million (pphm) (partially from automobile emissions and partially from stationary sources); HC = hydrocarbon concentrations (mg/cm^3) (partially from auto emissions and partially from stationary sources); O_x = crude index of oxidant levels = HC*NO_x (attempts to account for the fact that hydrocarbons and nitrogen oxides react to form ozone and other oxidants); PART = particulate concentrations (mg/cm^3) (primarily a stationary-source pollutant); SO_2 = sulphur dioxide concentrations (ppm) (primarily a stationary-source pollutant); and CO = carbon monoxide concentrations (ppm) (primarily due to automobile emissions).

[b]The correlation matrix tends to overstate the true correlations among pollutants somewhat because the TASSIM model generated data for 122 zones rather than 506 census tracts. Translating zonal data into census-tract data tends to overstate the correlations because relatively more census tracts are located in center city zones in which pollutant levels are most highly correlated.

and colorless and thus would not be recognized by households. In fact, when CO is entered in the equation, it typically shows up (insignificantly) with a positive sign. The CO variable probably represents a combined measure of accessibility and the noise levels (and other disamenities) associated with automobile traffic because CO provides a good measure of access to major highway arteries.

Even if the simple correlations among pollutants were not so high, it would be very difficult to distinguish between the individual effects of various pollutants because of measurement problems and because the pollutants often act in combination to cause damage to the environment. According to Waddell,

Research has shown that SO_x, NO_x, and HC all break down to the particulate state; thus, any individual particulate air quality measurement might also be representative of those pollutants that were originally emitted as gases. This possibility, then, complicates and raises serious questions of the validity of allocating costs by pollutant. . . . Also, these pollutants act synergistically to cause damage that perhaps would not occur when acting independently. So again, we have the problem of attaching weights to the different pollutants, which, by themselves are perhaps harmless, but which, in the presence of other pollutants become harmful.[47]

The Air-Pollution Damage Function Most property value studies implicitly assume that the air-pollution damage function is linear, that is, the marginal willingness to pay for an improvement in air quality is independent of the level of air quality. To test whether this assumption is reasonable, the methodological procedure outlined earlier was followed. As a first step, a nonlinear housing equation was estimated and used to calculate the marginal willingness to pay for an air quality improvement at each location. The data used were identical to the data used in the National Academy of Sciences Boston study, so that a direct comparison between the damage function estimated here and the damage function obtained in the academy report is possible.[48] The dependent variable is the median value of the owner-occupied homes in each census tract and the independent variables include two structural attribute variables, eight neighborhood variables, two accessibility variables, and one air pollution variable.[49] To make the results comparable to the National Academy study, the nitrogen oxide level NO_x was chosen as the pollution variable. However, NO_x is used solely to proxy air pollution because the high correlation among pollutants makes it extremely difficult to separate the independent impacts of pollutant variables.

Nonlinearities in the Air Pollution Variable Because the marginal willingness to pay for clean air may not be constant over urban space, nonlinear forms were estimated for the housing-value equation and special focus was given to nonlinearities in the pollution term. The best-fitting equation yielded a negative coefficient on the pollution variable suggesting that (on the margin) increases in the level of pollution result in decreased property values.[50] Because the equation was nonlinear, the calculated willingness to pay varied spatially. Specifically, when air pollution and the other variables take on their mean values, the average willingness to pay for a marginal improvement (one part per hundred million decrease in pollution) was $1613.[51] This figure was substantially lower (21 percent) than the $2052 (assumed constant over space) obtained in the National Academy study from a linear specification of the housing-value equation. Thus, allowing for the nonlinear interaction between pollutant levels and property values, the estimate of marginal benefits falls roughly 20 percent.

A test for the sensitivity of the estimated marginal willingness to pay suggested that with reasonable changes in the functional form of the pollution variable, the estimated marginal willingness to pay (averaged over all individuals) could easily be as high as $2040 or as low as $1186. In order to illustrate the importance for policy of differences in these functions, alternative marginal willingness-to-pay relationships were used to estimate the benefits associated with a program to improve Boston air quality. To distinguish this sensitivity experiment from others, all damage functions faced by households were assumed to be linear (of course, the average damage level was allowed to vary by location). To make the results comparable to the National Academy of Sciences study, the expected pollutant level was chosen to be the level of nitrogen oxides predicted by the TASSIM model for 1990, on the presumption that federal automobile emissions standards would lead to a 90 percent reduction in the 1970 levels of nitrogen oxides emitted from mobile sources.[52] However, the benefit measures used do *not* determine the dollar value of benefits for a particular emissions control strategy. Much greater care would have to be taken to separate out the independent influence of the automobile pollutants from the overall air pollution in Boston if a precise dollar value for the benefits of the federal automobile-emission-control program were needed.[53]

Under the assumption that hedonic-value equation was linear, total (not marginal) benefits were determined to be $1180 per household in the Boston SMSA (the average improvement in nitrogen oxides throughout the urban area was approximately .06). Using the best-fitting nonlinear functional form, reasonable bounds on the benefit estimate were calculated to range from a low of $783 per household to a high of $1053 per household. Thus, solely on the basis of the functional specification of the pollution variable, benefit estimates fall approximately 22 percent below those based on the linear specification and may fall as much as 34 percent. This sensitivity analysis is not complete, however, because it does not account for biases due to nonlinearities in household willingness-to-pay or *damage* function. This omission will be explained.

The Shape of the Damage Function On the assumption that changes in supply (levels of pollution) from location to location are sufficient to identify the willingness to pay for clean air, the willingness-to-pay or damage function was estimated by regressing $W_i(A_i)$, the marginal willingness to pay in each census tract, on the level of air pollution (proxied by the level of nitrogen oxide in each tract, NO_{x_i}), and median household income (INC_i) in logarithmic form.[54] The results are depicted graphically in figure 6.2.

The horizontal line indicates the linear air-pollution damage function obtaining from a linear specification of the housing-value equation. The three curves illustrate the willingness to pay as a function of air pollution for three income levels: ($8,500 per year), middle ($11,500), and high ($15,000). The positive slope for all curves implies that households perceive the damages from air pollution and thus the willingness to pay for marginal reductions to be greater at higher pollution levels. Moreover, these differences are substantial for the pollution levels in Boston in 1970, where nitrogen oxide

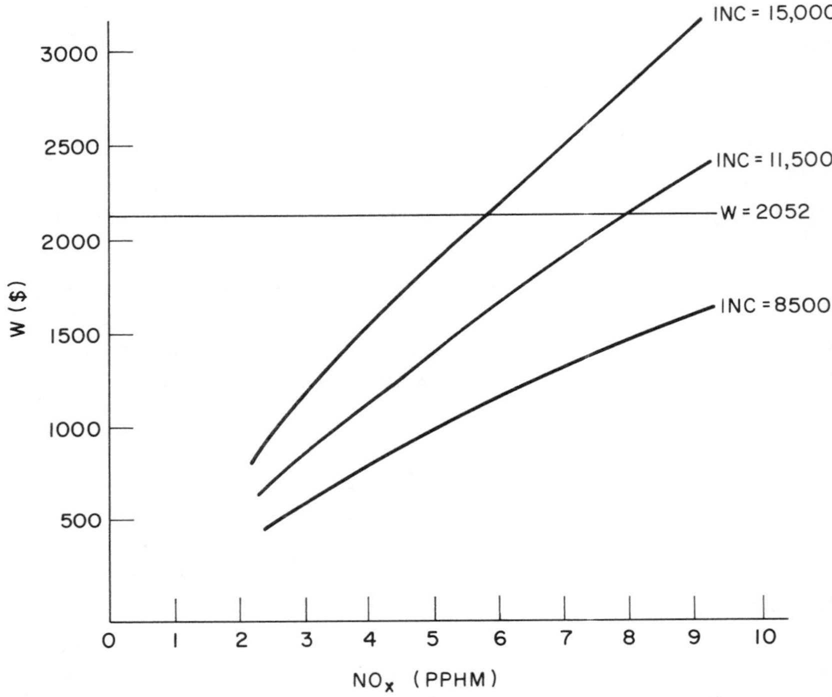

Figure 6.2 Willingness to pay for a 1 pphm improvement in NO_x concentration by income levels. (Note: pphm = parts per hundred million)

levels ranged from 3 to 9 parts per hundred million (pphm). For example, a middle-income household earning $11,500 per year would be willing to pay roughly $800 for a 1 pphm improvement in NO_x when the NO_x level was 3 pphm, but the willingness-to-pay figure would jump to approximately $2200 when the NO_x level was 9 pphm. Figure 6.2 also shows that the willingness-to-pay schedule for a marginal improvement in NO_x concentration is greater for households in higher-income groups. In this specification, the premium that high-income households are prepared to pay rises as the NO_x level increases. At low NO_x levels (3 pphm), the differential for households earning $8500 is only $200, but at high NO_x levels (9 pphm), the differential is about $700. Qualitatively speaking, it seems clear that the air-quality damage function determined from property-value data is nonlinear; that is, the marginal willingness to pay varies substantially with the level of air pollution and the level of household income.

The fact that the willingness-to-pay function slopes upward suggests that the assumption of a linear damage function will cause the benefits of a non-marginal air quality improvement to be overestimated. On the other hand, the impact of allowing for a shifting damage function (due to income changes) is not clear from the graph. To quantify the nature of the bias implicit in the assumption of a linear damage function, calculations were made of the average household benefits associated with the automobile emissions example described previously. Average household benefits were $830 (capitalized), compared to $920 when the damage function was presumed to be nonlinear. This difference suggests that property-value studies that do not take into account the nonlinearity in the air-pollution damage function are likely to overestimate benefits by 10 percent when the air quality improvement is a substantial one. It seems, therefore, that the failure to account for nonlinearities in the damage function does not seriously bias the benefits estimates obtained from previous studies.

Some Further Empirical Issues

Values or Rents? There are data problems associated with each of the variables included in a cross-section property-value study. One of the most important involves the dependent variable choice in the housing-value equation. Most property-value studies based on census tracts utilize median property value as the dependent variable.[55] The decision to examine home-owner data rather than renter data seems to be a reasonable one because these studies concentrate on the impact of air pollution on residential property values. However, because the emphasis in this paper is on the willingness to pay rather than the change in property values, there is no reason why the analysis should not be expanded to test whether the tastes of renters differ from those of home owners.[56] Doing so would introduce new issues about the measurement of the dependent variable (gross rent or contract rent?) and the choice of census tracts to be included in the sample. Therefore, to facilitate comparison with most other studies (which do not examine rental data), the matter will not be pursued empirically in this paper.

A related issue in studies that focus on home owners' willingness to pay is whether value (the capitalized stream of annual rentals) or annual rent is the appropriate choice for a dependent variable. In principle, rent is the correct measure because current rent reflects the market's implicit valuation of the current levels of air pollution as well as other housing attributes, whereas capitalized values measure the market's expectation about future levels of air quality as well as other housing and neighborhood characteristics. In fact, there is solid empirical evidence suggesting that the expectation of an improvement in neighborhood quality causes annual rents to be discounted at a low rate relative to the capitalization of annual rents in neighborhoods that are deteriorating in quality.[57] This may cause the air-quality benefit estimates obtained from property-value studies to be overestimated. To understand why, recall that the marginal willingness to pay for clean air is measured by the change in property values associated with a small change in air quality. To the extent that improving neighborhoods (in which property values provide overestimates of annual rents) have low air quality, the property-value response to a change in air pollution will be overstated. The conclusion concerning bias follows from the empirical fact that neighborhoods in downward transition tend to be located in areas in which air pollution concentrations are high.

The arguments just presented suggest that a careful study of the rental-housing market for which monthly rental data are available should tell something about the nature of the bias involved when property-value data are used. However, there are practical problems with rental data that would leave any such direct comparisons suspect. First, such an analysis would implicitly make the questionable assumption that the tastes of home owners and renters are identical. Second, there are measurement problems relating to the inclusion or exclusion of utilities and payments for services in the choice of either contract rent or gross rent to measure. Finally, rents in Boston and Cambridge (which contain many of the census tracts in the Boston SMSA sample) may have been distorted by the presence of rent control. Although these arguments suggest that serious discrepancies may exist when property-value and rental studies are compared, the results of a study by Anderson and Crocker suggest that these discrepancies may not be important.[58] Using data for Washington, D.C., Kansas City, and St. Louis, Anderson and Crocker found that benefit estimates obtained from rental data are roughly consistent with benefit estimates obtained from property-value data (they are lower for Washington, D.C., and St. Louis but higher for Kansas City).

Omitted Variables One of the serious criticisms of the property-value literature is that the effect of air pollution on property values overestimates benefits because the air pollution variable is highly correlated with important omitted variables, such as the level of noise, proximity to industry, and other measures of neighborhood quality. Because it is difficult to obtain measurements for these omitted variables at the census-tract level, it is not possible to settle this troublesome issue.[59] However, two types of experiments do allow for some crude quantitative guesses as to the magnitude of the problem. First, Harrison and Rubinfeld reestimated the housing-value equation after omitting intentionally some important accessibility and neighborhood characteristics.[60] Because pollution concentrations are higher in areas closest to locations with good accessibility, the deletion of the two accessibility variables reduced the measured impact of pollution on housing values. The pollution coefficient then reflected both the disadvantages of greater pollution concentrations and the advantages of greater accessibility. The deletion of the accessibility characteristics lowered the estimated marginal willingness to pay and the estimated total benefits by 43 percent. In a separate experiment, one of the most important neighborhood variables (a measure of the class status of the neighborhood residents) was omitted. Deleting this variable tends to credit the pollution variable with some of the neighborhood disamenities that result from a lower-class population. The deletion increased the estimated marginal willingness to pay by 26 percent and the estimated average benefits of a nonmarginal improvement by 25 percent.

The second experiment was to consider the correlation between air pollution and those neighborhood variables included in the property-value specification.[61] Because air pollution and included variables tend to be reasonably highly correlated, one might argue (given a correct specification) that the only sure conclusion about the impact of air pollution can be obtained by replacing the original variable with that portion of the air pollution variable that is uncorrelated with the other housing characteristics in the equation. This conservative estimation procedure was used by both Ridker and Henning and by the National Academy of Sciences. Ridker and Henning found that their most conservative estimate of marginal willingness to pay was just over one-third of their most favored estimate.[62] Likewise, the National Academy found that their most conservative estimate was less than one-quarter of their most reasonable estimate.[63]

Market Segmentation Most property-value studies implicitly assume that it is correct to estimate a single hedonic housing equation for the entire urban area. However, a number of authors have argued that it would be proper to view the urban housing market as a distinct set of housing submarkets, in which the submarkets are defined on the assumption that groups of individuals will (by choice or by restriction) only purchase or rent certain types of housing in certain geographical locations.[64] Unfortunately, the aggregative census-tract data do not allow for the detailed specification of housing submarkets. However, in order to get some crude estimates of the sensitivity of the benefit estimation procedure to the submarket issue, average benefits were calculated when the housing market was stratified on the basis of household income (3 categories), accessibility to employment (2 categories), and on household social status (2 categories). In each case, our estimate of the average benefits of a nonmarginal change in air pollution fell substantially, by 28 percent when the income stratification was used, by 41 percent when the accessibility stratification was used, and by 10 percent when the market was stratified on the basis of social status.[65] These calculations are admittedly quite crude, but they do suggest that the housing-submarket issue is an important one.

Damage Function Sensitivity In order to test further for the sensitivity of the shape of the air-pollution damage function described in figure 6.2, several experiments were tried. First, the willingness-to-pay equations that are associated with the alternative functional forms of the housing-value equation were reestimated. The shape of the damage function (and the resulting benefit estimates) was quite sensitive to the form in which the pollution variable entered the housing-value equation. As a consequence, the possibility that the damage function is linear could not be ruled out with great confidence, although the results suggest strongly that the function does shift with income changes. In this case, an average benefit estimate of $1013—22 percent higher than the results implicit in figure 6.2—was obtained. On the other hand, one cannot rule out the possibility that the damage function is much steeper than the one depicted in figure 6.2, in which case the average benefits are $592—29 percent lower than the most likely case.

In a second experiment, the willingness-to-pay equation was reestimated by including in the equation other housing attributes thought to be complimentary with air quality and by allowing the NO_x elasticity to vary with income.[66] Although the respecification led to changes in the elasticities on NO_x and INC, the impact on average benefits estimated in the automobile emissions example was not substantially affected.[67] (Most benefit changes were on the order of 1 percent.) These results make it clear that the shape of the damage function is quite sensitive to the form of the hedonic housing-value equation but that it is largely insensitive to changes in the exact functional form of the damage function itself.

Conclusions

This paper has critically examined the market approaches to estimating the benefits of improved urban air quality from both a conceptual and an empirical perspective. Of the two market-oriented approaches, the property-value approach appears more promising than the wage-differential approach. The wage view has, at best, a weak conceptual foundation and faces serious empirical problems that are not likely to be easily resolved. From a theoretical viewpoint, the property-value approach is more promising. When correctly interpreted, property-value studies can provide conceptually correct estimates of the willingness to pay for clean air.

The empirical evidence presented in this paper indicates that perceived marginal air pollution damages increase with the pollution level and increase with income. However, estimates of the shape of the damage function as well as the benefits associated with a nonmarginal improvement in air quality are quite sensitive to the specification of the property-value or housing-value equation. Plausible specifications of the housing-value equation may reduce aggregate benefit estimates by as much as 60 percent below the results obtained if one assumes a linear damage function that is identical for all households. In addition, the omission of variables that are highly correlated with air pollution, the possible existence of housing submarkets, and the problems inherent in the use of median property-value data all suggest that most property-value benefit estimates may be even more biased. Some specification errors and choice of functional forms support the view that property-value-related benefits are underestimated, the evidence strongly suggests that most benefit estimates obtained from property-value studies are too high, perhaps by a factor of two or more.

The results of the study appear to be somewhat discouraging, especially if one expects the property-value approach to provide an itemized list of benefits associated with each of the major pollutants in the urban area. The problem of distinguishing among the effects of various pollutants does not appear likely to be resolved with the use of housing-market information although other information may permit reasonable guesses. However, from a more realistic point of view, the results are much more promising. The estimates of housing-value-related benefits (using the air pollution proxy) obtained in the air quality example are within an order of magnitude of one another. For application to particular policy questions, the variability in the benefit estimates figures may be small relative to the uncertainty that surrounds the other parameters of a policy decision. For example, if one were attempting to determine whether the costs of instituting an inspection-maintenance scheme to reduce air pollution were greater than the benefits from improved air quality, the uncertainties associated with estimating the "true" costs of the program might be even greater than the uncertainties associated with estimating the "true" aggregate willingness to pay for the improvement in air quality.

In conclusion, despite the rather serious limitations of the property-value approach, such an approach provides a useful addition to other air-pollution-reduction benefit studies. If correctly carried out, property-value studies measure a portion of the benefits associated with air quality improvements that are not likely to be taken into account in health, vegetation, and other cost studies. Further research directed towards the problem of overlapping benefits, the question of how to impute benefits to individual pollutants, and the statistical problem of how to place reasonable bounds on the property-value estimates is likely to be worthwhile.

Notes

This paper represents a review and extension of work undertaken by the author and by A. Mitchell Polinsky and David Harrison, Jr. All statistical analyses were performed on the NBER Center for Computational Research's TROLL system, with the assistance of Laxmi Rao. The author wishes to thank A. M. Freeman, D. Harrison, and A. M. Polinsky for their helpful comments.

1. See Helen Ingram, Chapter 2 of this volume, for details.

2. See, for example, Ronald G. Ridker and John A. Henning, "The Determinants of Residential Property Values with Special Reference to Air Pollution," *Review of Economics and Statistics* 49 (May 1967): 246–257. For a general overview of the property-value studies, see Thomas D. Crocker, *Urban Air Pollution Damage Functions: Theory and Measurement* (Washington, D.C.: National Air Pollution Control Administration, June 1971); Thomas Waddell, *The Economic Damages of Air Pollution* (Washington, D.C.: Environmental Protection Agency, May 1974); Jon P. Nelson, *The Effects of Mobile-Source Air and Noise Pollution on Residential Property Values* (University Park, Pa.: Institute for Research on Human Resources, Pennsylvania State University, April 1975); and V. Kerry Smith, *The Economic Consequences of Air Pollution* (Cambridge, Mass.: Ballinger, 1976).

3. See, for example, William Nordhaus and James Tobin, "Is Growth Obsolete?" in *Economic Growth* (New York: National Bureau of Economic Research, 1972), pp. 12–13, 49–54; and John R. Meyer and Robert A. Leone, "The Urban Disamenity Revisited," in Wingo and Evans, eds., *Public Economics and the Quality of Life* (Baltimore: Johns Hopkins, 1977). For a more general analysis of wage differentials, see Irving Hoch, "Climate, Wages and the Quality of Life," in Wingo and Evans, eds., *Public Economics and the Quality of Life* (Baltimore: Johns Hopkins, 1977).

4. In most health studies, statistical estimates of the effect of air pollution on morbidity and mortality are used to compute increases in health costs and decreases in earning capacity that result from increases in pollution levels. See, for example, Lester B. Lave and Eugene P. Seskin, "Air Pollution and Human Health," *Science* 169 (21 August 1970): 723–733. For more general discussions of our knowledge concerning the effects of air pollution on human health, see Lester B. Lave and Eugene P. Seskin, *Air Pollution and Human Health* (Baltimore: Johns Hopkins, forthcoming); William R. Ahern, Jr., "Health Effects of Automotive Air Pollution," in Henry D. Jacoby et al., eds., *Clearing the Air: Federal Policy on Automotive Emissions Control* (Cambridge, Mass.: Ballinger, 1974); David Harrison, Jr., *Who Pays for Clean Air: The Cost and Benefit Distribution of Federal Automobile Emission Standards* (Cambridge, Mass.: Ballinger, 1975), Appendix A; Brian J. L. Berry

and Frank E. Horton, *Urban Environmental Management* (Englewood Cliffs, N.J.: Prentice-Hall, 1974), Chapter 4; Waddell, *The Economic Damages of Air Pollution;* and Ronald G. Ridker, *Economic Costs of Air Pollution* (New York: Praeger, 1967), Chapter 3. In other cost studies, attempts are made to analyze the physical effects of increased air pollution (on structures, for example) and to place dollar values on the costs involved. See, for example, Richard L. Salmon, *Systems Analysis of the Effects of Air Pollution on Materials* (Kansas City, Mo., Midwest Research Institute, January 1970). For a more general overview of these cost studies, see Berry and Horton, *Urban Environmental Management,* Chapter 4; Waddell, *The Economic Damages of Air Pollution;* and Kenneth A. Small, ''Air Pollution and Property Values: Further Comment,'' *Review of Economics and Statistics* 57 (1975): 105–107. Other benefit estimates have been obtained from opinion surveys of those residing in highly polluted areas. See, for example, Andris Auliciems and Ian Burton, ''Perception and Awareness of Air Pollution in Toronto,'' Working Paper Series No. 13, Natural Hazard Research Series, University of Toronto, 1970; for a general overview of studies in the area, see Berry and Horton, *Urban Environmental Management,* Chapter 4, and Waddell, *The Economic Damages of Air Pollution.* Additional surveys have been made by examining the outcomes of litigation in which plaintiffs sued for damages resulting from occupational hazards. See, for example, U.S. Senate, Committee of Public Works, *Air Quality and Automobile Emission Control,* vol. 4 (Washington, D.C.: U.S. Government Printing Office, September 1974) (hereafter cited as U.S. Senate, Committee on Public Works, *Air Quality and Automobile Emissions Control*).

5. Lester B. Lave, ''Air Pollution Damage: Some Difficulties in Estimating the Value of Abatement,'' in Allen V. Kneese and Blair T. Bower, eds., *Environmental Quality Analysis* (Baltimore: Johns Hopkins, 1972), pp. 213–242.

6. Waddell, *The Economic Damages of Air Pollution,* p. 35.

7. U.S. Senate, Committee on Public Works, *Air Quality and Automobile Emissions Control,* p. 412.

8. Waddell, *The Economic Damages of Air Pollution,* p. 2.

9. To reiterate, nonmarket benefit estimation techniques are not discussed here, nor is there a treatment of those benefits that are unlikely to be quantifiable. For example, it would be extremely difficult to quantify the benefits obtained by residents throughout the United States when air quality in one urban area is improved. These ''nonresidents'' might be better off because their options concerning the possibility of future visits or job location moves to the urban area with improved air quality would be altered.

10. See Donald Dewees, Chapter 7 of this volume, for a corresponding argument on the cost side.

11. This assumption has important income distributional consequences because market prices are dependent upon the existing distribution of income. In addition, attempts to determine implicit market prices presume that market distortions unrelated to air quality are not present. To the extent that transportation congestion and other

externalities are not fully accounted for, the use of property-value data to estimate benefits is likely to lead to biased results.

12. This is not the only possible measure of willingness to pay (or consumer surplus), but it is a convenient one.

13. The total damage function corresponds to the textual definition of total willingness to pay.

14. A different estimate of benefits would result if the willingness-to-pay function were calculated by examining the housing market after the environmental improvement took place and asking households how much they would pay, given the new set of prices that they face, in order to avoid an increased level of air pollution.

15. Case 1 is often said to describe a *linear* damage function because *total* damages are linearly related to the level of air pollution.

16. This assumes that the marginal willingnesses to pay are equal at the *original* level of air pollution.

17. For a technical discussion of the conditions under which demand is identified, see Sherwin Rosen, "Hedonic Prices and Implicit Markets," *Journal of Political Economy,* 82 (January 1974): 35–55. In a more general context, he argues (p. 51) that "if buyers are identical, but sellers differ, . . . single cross-sectional observations trace out compensated demand functions."

18. The discussion that follows relies on A. Myrick Freeman, "On Estimating Air Pollution Control Benefits from Land Values Studies," *Journal of Environmental Economics and Management,* 1 (May 1974): 74–83; Rosen, "Hedonic Prices and Implicit Markets"; A. Mitchell Polinsky and Steven Shavell, "Amenities and Property Values in a General Equilibrium Model of an Urban Area," *Journal of Public Economics,* 5 (January–February 1976): 119–129; A. Mitchell Polinsky and Daniel L. Rubinfeld, "Property Values and the Benefits of Environmental Improvements: Theory and Measurement," in Wingo and Evans, eds. *Public Economics and the Quality of Life* (Baltimore: Johns Hopkins, 1977); and David Harrison, Jr., and Daniel L. Rubinfeld, "Housing Values and the Willingness to Pay for Clean Air," *Journal of Environmental Economics and Management,* forthcoming.

19. Housing or property values are the present discounted or capitalized value of the stream of annual housing expenditures. The analysis contained in this paper considers the market for home owners rather than the market for renters.

20. To simplify, it is assumed initially that work-place choices are given.

21. Because residential property values are utilized, the health and aesthetic benefits associated with work-place location (and shopping location) are not captured by this approach.

22. Of course, there will be equivalent conditions for each of the housing attributes.

23. Estimation of the willingness-to-pay function presumes that the "demand for clean air" can be identified econometrically, that is, distinguished statistically from the supply of clean air. This assumption presents no difficulty if the level of air pollution is assumed to be fixed in the cross-section analysis but does cause difficul-

ties if air pollution adjustments are built into the model. This issue has troubled a number of authors, including Freeman, "Estimating Air Pollution Control Benefits," pp. 74–79, and Small, "Air Pollution and Property Values: Further Comment." Small argues that the demand curve is unlikely to be identified within a single urban-housing market. He feels (p. 106) that "a much more plausible assumption is that we are observing a single housing market, in which the consumer faces an entire price schedule, and that the variety observed in the choices made by people of a given income class reflects not different constraints but different preferences."

24. This problem is explicitly considered by John Muellbauer, "Household Production Theory, Quality, and the 'Hedonic Technique,'" *American Economic Review,* 64 (December 1974): 977–993, who argues that the marginal willingness to pay at each location will depend upon the utility level attained, as well as tastes (and possibly income). To the extent that it is so, Muellbauer concludes (p. 980) that "at the very least, careful attention should be paid to cross-sectional disaggregation. As far as possible, markets should be broken into segments based on commodity groupings which make it likely that their consumers have similar MRS and these segments should be studied separately." Note also that when Y is included in the willingness-to-pay function the implicit demand function is uncompensated, because money rather than real income is held constant.

25. For further discussions of these and other related issues, see Matthew Edel, "Land Values and the Costs of Urban Congestion: Measurement and Distribution," *Social Science Information,* 10 (1971): 9–12; Lave, "Air Pollution Damage: Some Difficulties in Estimating the Value of Abatement," pp. 213–215, 234–236; Thomas D. Crocker, "The Measurement of Economic Losses from Uncompensated Externalities," in William R. Walker, ed., *Economics of Air and Water Pollution* (Springfield, Va.: National Technical Information Service, 1969), pp. 192–193; Small, "Air Pollution and Property Values: Further Comment"; and Robert J. Anderson and Thomas D. Crocker, "Air Pollution and Residential Property Values," *Urban Studies* 8 (October 1971) 171–173.

26. U.S. Senate, Committee on Public Works, *Air Quality and Automobile Emissions Control,* p. 237.

27. The National Academy of Sciences study analyzed data for Los Angeles as well as Boston. They found average benefits per household to be lower in Los Angeles than in Boston, despite the greater pollutant reductions resulting from the imposition of controls. This phenomenon may be due in part to taste differences.

28. For representative examples of these studies, see Ridker and Henning, "Determinants of Residential Property Values," and Anderson and Crocker, "Air Pollution and Residential Property Values." For detailed surveys of a larger set of empirical studies, see Waddell, *The Economic Damages of Air Pollution;* Nelson, *The Effects of Mobile-Source Air and Noise Pollution on Residential Property Values;* and Smith, *The Economic Consequences of Air Pollution.*

29. Ridker, *Economic Costs of Air Pollution,* p. 25.

30. For further discussion of this issue, see, for example, Freeman, "Estimating Air Pollution Control Benefits"; Small, "Air Pollution and Property Values: Further

Comment''; David Pines and Yoram Weiss, ''Land Improvement Projects and Land Values,'' *Journal of Urban Economics,* 3 (January 1976): 1–13; and A. Mitchell Polinsky and Daniel L. Rubinfeld, ''The Long Run Effects of a Residential Property Tax and Local Public Services,'' *Journal of Urban Economics,* forthcoming 1978.

31. Polinsky and Rubinfeld, ''Property Values and the Benefits of Environmental Improvements.''

32. Polinsky and Shavell, ''Amenities and Property Values.'' For other discussions on this subject, see Yitzhak Oron, David Pines, and Eytan Sheshinski, ''The Effect of Nuisances Associated with Urban Traffic on Suburbanization and Land Values,'' *Journal of Urban Economics,* 1 (October 1974): 391–394, and Pines and Weiss, ''Land Improvement Projects and Land Values.''

33. Polinsky and Shavell, ''Amenities and Property Values,'' and Polinsky and Rubinfeld, ''Property Values and the Benefits of Environmental Improvements.''

34. See Oded Izraeli, ''Differentials in Nominal Incomes and Prices Between Cities,'' Ph.D. dissertation, University of Chicago, 1973.

35. U.S. Senate, Committee on Public Works, *Air Quality and Automobile Emissions Control,* p. 244.

36. Ibid., p. 244.

37. This model is spelled out in much greater detail in Polinsky and Rubinfeld, ''The Long Run Effects of a Residential Property Tax and Local Public Services.''

38. It might appear that the presence of labor-market adjustments invalidates the willingness-to-pay approach described earlier. However, the property-value approach does not depend upon the assumption that all the impact of the air quality improvement is felt in the land market. It depends solely on the assumption that existing *differentials* in property values in urban areas correspond to differences in air quality. Only to the extent that complex labor-market conditions might mask the true relationship between property value and air quality differentials is the property-value approach suspect.

39. See Polinsky and Rubinfeld, ''Property Values and the Benefits of Environmental Improvements,'' for an explanation of the impact of an environmental improvement on wages and property values in a closed model of an urban area.

40. For details see Paul N. Courant and Daniel L. Rubinfeld, ''On the Measurement of Benefits in an Urban Context: Some General Equilibrium Issues,'' *Journal of Urban Economics,* forthcoming 1978.

41. The results to be described are discussed in greater depth in Harrison and Rubinfeld, ''Housing Values and the Willingness to Pay for Clean Air.'' The Boston data base is particularly interesting because it was used (along with Los Angeles data) by the National Academy of Sciences in their study of the benefits of automobile emissions controls (see U.S. Senate, Committee on Public Works, *Air Quality and Automobile Emissions Control*). Roughly comparable results were obtained in a recent study of the Washington, D.C., area in Jon P. Nelson, ''Residential Choice, Hedonic

Prices, and the Demand for Urban Air Quality,'' *Journal of Urban Economics,* forthcoming.

42. Excluding tracts containing no housing units or comprised entirely of institutions, the Boston sample contains 506 census tracts.

43. On the other hand, one can argue that most neighborhood characteristics, including air pollution, are meaningful to households at an aggregated, rather than a micro level.

44. For technical details on this matter, see R. D. Larsen, ''A Mathematical Model for Relating Air Quality Measurements to Air Quality Standards'' (Washington, D.C.: U.S. Environmental Protection Agency Report AP-89, November 1971).

45. Ridker and Henning, ''Determinants of Residential Property Values,'' used sulfation data from forty-one monitoring stations to create an air pollution index that takes on one of eight distinct values. However, their particulate data came from only sixteen monitoring stations.

46. Two meteorological models, one for the stationary sources and one for the automobile emissions and other area sources, are used to translate emissions by zone into pollutant concentrations by zone. These meteorological models take into account information on wind speed, wind direction, and atmospheric stability. In addition, the stationary-source model includes information on the stack height of stationary sources. Both meteorological models assume that pollutant levels are linearly related to emissions, that is, that pollutants are nonreactive. For further details, see Gregory D. Ingram and Gary F. Fauth, *TASSIM: A Transportation and Air Shed Simulation Model,* Final Report to the U.S. Department of Transportation (Springfield, Va.. National Technical Information Service, May 1974).

47. Waddell, *The Economic Damages of Air Pollution,* pp. 132–133. For other discussions of the interactions among pollutants, see Robert U. Ayres, ''Air Pollution in Cities,'' *Natural Resources Journal* 9 (January 1969):1–22, and Jack G. Calvert, ''Interactions of Air Pollutants,'' in National Academy of Sciences, Proceedings of the Conference on Health Effects of Air Pollutants prepared for the U.S. Senate Committee on Public Works (Washington, D.C.: U.S. Government Printing Office, 1973).

48. U.S. Senate, Committee on Public Works, *Air Quality and Automobile Emissions Control.*

49. The structural attribute variables are the average number of rooms in owner units and the proportion of owner units built prior to 1940. The neighborhood variables are the black proportion of the population, the proportion of the population that is lower status, the crime rate, a large-lot zoning variable, the proportion of nonretail business acreage in the community, the full-value property-tax rate, the pupil-teacher ratio, and a dummy measure of the amenities associated with the Charles River. The accessibility variables were a measure of the weighted distance from five employment centers and an index of accessibility to radial highways. For further details, see

Harrison and Rubinfeld, ''Housing Values and the Willingness to Pay for Clean Air.''

50. The t-statistic on the pollution term was 5.6. Of all the explanatory variables, only the weighted distance-accessibility variable, the property-tax variable, the class-status variable, and the crime variable contributed more to the explained variation of 81 percent in the log-linear version of the equation.

51. This is a capitalized value. It would be equivalent to a figure of $161 per year if, for example, a 10 percent discount rate were applied. The willingness-to-pay estimate seems high because a 1 pphm improvement is greater than the average improvement implied by the 1970 Clear Air Act.

52. The estimates of 1990 NO_x concentrations were obtained by substituting the emissions characteristics of the 1990 controlled automobile fleet for the 1970 fleet emissions figures. However, the actual 1990 concentrations will differ from the calculated ones because the number of households, number of cars, travel characteristics, stationary-source emissions, and other characteristics of the urban area will change between 1970 and 1990.

53. The National Academy of Sciences study attributed one-third of the benefits to reduced automobile emissions, but their results were extremely sensitive to the particular equation specification chosen..

54. The estimated NO_x elasticity was .87 and the estimated INC elasticity was 1.00. Note, however, that the latter represents the elasticity of willingness to pay with respect to income. If we were to use this equation to calculate the NO_x income elasticity, we would find that air quality is a superior good, that is, that the NO_x income elasticity is greater than one.

55. Examples of such studies are Ridker and Henning, ''Determinants of Residential Property Values''; Anderson and Crocker, ''Air Pollution and Residential Property Values''; and Nelson, *The Effects of Mobile-Source Air and Noise Pollution on Residential Property Values*.

56. This test would most likely necessitate a different choice of census tracts for inclusion in the model because many suburban tracts contain very few rental-housing units.

57. For a detailed discussion of this issue, see George E. Peterson, Arthur P. Solomon, Hadi Madjid, and William C. Apgar, Jr., *Property Taxes, Housing, and the Cities* (Lexington, Mass.: D. C. Heath, 1973), p. 25.

58. Anderson and Crocker, ''Air Pollution and Residential Property Values.''

59. Nelson, *The Effects of Mobile-Source Air and Noise Pollution on Residential Property Values,* include a noise measure in his property-value study and still found a significant impact of air pollution on property values.

60. See note 49 for descriptions of these variables.

61. For example, the correlation between NO_x and the accessibility variables is .83

and .61, in absolute value, while the correlation between NO_x and the class status variable is .59.

62. Ridker and Henning, ''Determinants of Residential Property Values,'' p. 254.

63. U.S. Senate, Committee on Public Works, *Air Quality and Automobile Emissions Control,* pp. 237, 288.

64. For details, see Mahlon R. Straszheim, ''Hedonic Estimation of Housing Market Prices: A Further Comment,'' *Review of Economics and Statistics,* 1 (August 1974): 404–406. Muellbauer, ''Household Production Theory, Quality, and the 'Hedonic Technique','' argues for stratification because of complications arising from the willingness-to-pay calculation.

65. Details of how these calculations were made appear in Harrison and Rubinfeld, ''Housing Values and the Willingness to Pay for Clear Air.''

66. For example, when the number of rooms was included in the willingness-to-pay equation we found that the NO_x elasticity fell to .81, the INC elasticity fell to .78, and the rooms elasticity was .84. However, average benefits were largely unchanged.

67. Adjustments were also made for heteroscedasticity and simultaneity (in the NO_x variable) but the outcome was not very different.

Comment

A. Myrick Freeman III

Introduction

Rubinfeld has presented us with an excellent guide to the property-value and wage-differential approaches to estimating the benefits of air quality improvements. The paper will be very useful both for those who are not directly involved in property-value studies but who wish to get an overview of both the principles and the problems and for those who are involved in doing property-value studies or use their results in policy decisions. For this latter audience, Rubinfeld's discussion of the conceptual and empirical problems will be particularly valuable.

Rubinfeld presents two principal conclusions. The first is that the property-value technique is useful, more so than the wage-differential technique. The property-value technique is sound conceptually and theoretically. And a number of studies using different sources of data, different time periods, and different cities have all shown significant relationships between property values and air pollution levels.

The second conclusion is that there are still serious problems of specification and measurement that reduce the reliability or accuracy of specific numerical benefit estimates. Estimating benefits involves two steps: the estimation of a hedonic or implicit price function for attributes of housing, including neighborhood effects and air pollution; and estimating the demand function or willingness to pay from the implicit price data. Rubinfield has presented evidence that the latter step is not particularly sensitive to alternative specifications. But the first step, the hedonic price function, is sensitive to the specification of the model and the empirical techniques used.

Rubinfeld concludes that "most benefit estimates obtained from property-value studies are too high, perhaps by a factor of two or more." This assessment apparently is based primarily on the specification and measurement problems, rather than on the possible double-counting and overlap problems that Rubinfeld discussed in the introduction. In reviewing the results presented here and in Harrison and Rubinfeld,[1] it appears that Rubinfeld has given substantial weight to the statistical experiments described in his paper in arriving at this estimate of the magnitude of bias. But to put this uncertainty regarding the magnitude of property-value benefits into perspective, Rubinfeld argues that the uncertainty may be small relative to other uncertainties in the overall policy problem and that, in effect, some information is better than none. I would agree. And I think that Rubinfeld's paper provides the basis for the further work that is needed to reduce the magnitude of uncertainty regarding property-value benefits.

Market Approaches—an Overview

Rubinfeld limits his discussion of market approaches to the property-value and wage-differential techniques. But I think it would be useful to consider the possibility of other approaches based upon market-revealed data. I can only briefly sketch out a line of reasoning in the space allotted. Let us assume agreement on an individualistic definition of welfare gain or benefit based on some version of consumer surplus (compensating variation or equivalent variation). Broadly speaking, there are three ways to obtain estimates of welfare change for policy purposes. The first is to infer welfare measures from market behavior. Basically this method involves estimating demand curves and calculating areas under them. The second is to ask people, that is, to induce them to reveal directly their monetary equivalents of welfare changes. I would include voting schemes, bidding games, and survey questionnaires under this category. The third is to substitute politically determined value judgments for the welfare measures directly or indirectly revealed by individuals. For example, the National Academy of Sciences assumed a value of $200,000 per death avoided in calculating benefits of automative air pollution control.[2]

In principle, there is a variety of ways by which market data can be scanned in an effort to infer the demand for air quality. The appropriate empirical technique depends, in part, on the way in which air quality affects individuals' utility and on the assumptions made about the form of the utility function. Air quality changes may effect individuals only indirectly by changing the productivity of firms. Then benefits can be tracked down by examining price and quantity data in output markets and returns to specialized factors in input markets.

If air quality enters individuals' utility functions directly rather than by affecting prices, incomes, and budget constraints, there are still several possibilities. For some forms of utility function, air quality could enter as an argument in the demand functions for private goods. Then, if a complete set of private good demand functions is estimated, it may be possible under some circumstances to solve for the utility function and for the demand for air quality. On the other hand, if the utility function is separable in air quality, then all private-good demands are independent of the level of air quality. Air quality changes have no effect on observed market variables. Market approaches are inapplicable.

Other assumptions can be made about degrees of separability and the nature of substitution and complimentary relationships between air quality and some private good. These have been analyzed by Karl-Goran Maler.[3] Benefit measures based on defensive expenditures or costs of alternatives, such as cleaning or medical care are based on some assumption about substitutability between air quality and these categories of expenditures.

Finally, where air quality is a characteristic of a private good purchased by individuals and where individuals can choose among a range of alternative private-good models with varying air quality, the hedonic-price approach can be utilized. The property-value approach fits here. It requires that the utility function be separable in the private good for which air quality is a characteristic. Otherwise, quantities of other goods must be used as explanatory variables in the hedonic price function.

How does the wage-differential approach fit into this framework? It requires that one assume that utility is a function not only of goods consumed but also of some bundle of job characteristics. One such characteristic would be the air quality level of the urban area in which the job was located. The model must specify a list of other relevant job characteristics. In order to apply the hedonic technique, one must assume that the utility function is separable in these characteristics. Considering the variety of urban characteristics that might be included in properly specified model, I am not sure that any meaningful separability restrictions can be imposed on the utility function for purposes of estimating the hedonic wage equation. With this doubt in mind, I can concur in Rubinfeld's statement, "the wage view has, at best, a weak conceptual foundation and faces serious empirical problems."

Conceptual Issues

I have two comments to offer under this heading. First, after defining benefits as the area under an individual's demand or marginal willingness-to-pay curve (summed over all individuals), Rubinfeld suggests that this measure may be an underestimate if household mobility is taken into account. The argument is that the area under the demand curve captures benefits at the existing location, but if people move, it must be to capture further welfare gains. I think this argument is not correct. According to the standard welfare theory, the area under the demand curve does measure the welfare gain, provided that no other *prices* change. Individuals may alter the quantities purchased of other attributes; demand curves for other attributes may shift back and forth because of complimentary and substitution relationships; but these changes are irrelevant for benefit estimation provided that the prices of other attributes do not change. If, however, the implicit prices of other housing attributes do change, then we have the well-known problems of evaluating welfare gains with multiple price changes.[4] If other prices do change, it is not clear a priori whether evaluating the area under one demand curve leads to an underestimate or overestimate.

My second comment stems from a recent paper by Zeckhauser and Fisher.[5] They analyzed the way people respond to externalities in terms of averting behavior, that is, the steps that people take to avoid or mitigate the adverse effects of externalities. When averting behavior is present, it can help the analyst to identify and measure benefits of air quality improvement. For example, averting behavior may be viewed as a substitute for improvements in air quality. With perfect substitutability, expenditures on averting behavior provide an accurate measure of the benefits of an equivalent improvement in air quality.[6] People may attempt to avoid air pollution by purchasing residences in clean-air areas. It is this behavior and its impact on housing prices that make possible the use of the hedonic technique discussed by Rubinfeld.

Nevertheless, averting behavior can complicate and hinder the process of benefit estimation. For example, if there are several forms of averting behavior that are substitutes among themselves, they must all be identified and included in the model on which the benefit-estimation technique is based. If some forms of averting behavior are left out of the model, the model is incomplete and benefit estimates derived from it will be biased. There need not be a problem in the case of the property-value technique if all forms of averting behavior are reflected in or can be captured by those attributes of housing that are included in the hedonic price function. For example, if air conditioning and corrosion-resistant house paints are modes of averting behavior, then the hedonic price function must include terms for these attributes of housing. This modification is possible, in principle, if the data base consists of prices and characteristics for individual housing units, but it is a disadvantage of the census-tract data used by Rubinfeld and most other researchers that they cannot capture the attributes related to some forms of averting behavior.

Empirical Issues

Rubinfeld's discussion of empirical issues is excellent. It is extremely valuable to have this analysis of the effect of model specification on the magnitude of benefit estimates. But there is one question that Rubinfeld touches on that I would have liked to have seen given more discussion. This is the question of "market segmentation" first raised by Strazheim.[7] Strazheim argued that the urban housing market really consisted of a series of separate, compartmentalized markets with different hedonic price functions in each. To support the segmentation hypothesis, Strazheim showed that estimating separate hedonic price functions for different *geographic* areas of the San Francisco Bay area reduced the sum of squared errors for the sample as a whole. Rubinfeld's experiment differs from Strazheim's in two respects. First, Rubinfeld stratified the market by income and two neighborhood characteristics (accessibility and social status) rather than by geographic subarea. Second, he compared the benefit estimates rather than the statistical properties of the resulting hedonic price equations. Thus Rubinfeld's experiment is not a direct confrontation with the Strazheim segmentation hypothesis.[8]

In summary, these comments are not meant to divert attention from the overall excellence of Rubinfeld's paper. Rubinfeld and his (sometimes) coauthors, Polinsky and Harrison, have made substantial contributions to our theoretical and empirical knowledge regarding the property-value–air-pollution relationship. This paper is in the same tradition. My comments are meant primarily to demonstrate that there are still enough unsettled questions so that we can expect more contributions from Rubinfeld and company.

Notes

1. David Harrison, Jr., and Daniel L. Rubinfeld, "Housing Values and the Willingness to Pay for Clean Air," Discussion Paper D76-5, Department of City and Regional Planning, Harvard University, June 1976.

2. National Academy of Sciences, *Air Quality and Automobile Emission Control,* vol. 4 (Washington, D.C.: U.S. Government Printing Office, 1974).

3. Karl-Goran Maler, *Environmental Economics: A Theoretical Inquiry* (Baltimore: Johns Hopkins, 1974).

4. Herbert Mohring, "Alternative Welfare Gain and Loss Measures," *Western Economic Journal* 9 (December 1971): 349–368; Eugene Silberberg, "Duality and the Many Consumers' Surpluses," *American Economic Review* 62 (December 1972): 942–952.

5. Richard Zeckhauser and Anthony Fisher, ''Averting Behavior and External Diseconomies,'' forthcoming.

6. Maler, op. cit., 178–180.

7. Mahlon Strazheim, ''Hedonic Estimation of Housing Market Prices: A Further Comment,'' *Review of Economics and Statistics* 56 (August 1974): 404–406.

8. Ann B. Schnare and Raymond J. Struyk, ''Segmentation in Urban Housing Markets,'' *Journal of Urban Economics* 3 (April 1976): 146–166. Schnare and Struyk found that prices of individual attributes do vary across segments of the Boston-area housing market, but they did not include air pollution as a variable.

Comment

Lester B. Lave

Introduction

In testifying before the Senate Public Works Committee on the National Academy of Sciences study of the benefits and costs of mobile-source emission standards, Philip Handler presented evidence ''on the one hand and on the other hand.'' After a long search for a conclusion or recommendation, Edmund Muskie remarked that the nation needed more one-armed scientists. Unbiased scientists can be proud of Rubinfeld because he presents expert opinion, theory, and data that say, on the one hand, property-value and wage-differential studies have little to contribute and, on the other hand, the estimates are informative and important. I'm sure Senator Muskie would be confused.

What Do Rent Gradients Show?

In these comments, I will attempt to get rid of Rubinfeld's superfluous hand. Rubinfeld has done such a good job in critiquing the approach, I need only summarize him and add a few more qualifications. To make theoretical sense of property-value estimates, one must assume that the housing market in an SMSA is in equilibrium, that is, that property values or rentals fully reflect air pollution gradients, urban location, and so on. Of course this assumption is absurd, but random errors of, say, 5 percent will swamp any effect of air pollution. Furthermore, one must assume that utility functions are separable. Because the price of a house is the present discounted value of its future services, price embodies expectations about future levels of pollution, school quality, neighborhood quality, shoping and employment locations, and the economic health of the city. The implicit assumptions in the property-value analyses are that current levels of air pollution will prevail in the future (or at least their dispersion around the city) and that other factors are unrelated to air pollution levels; if future demand for property is related to air pollution via employment opportunities, for example, failure to include this factor in the analysis will cause the air pollution coefficient to be biased.

To get good estimates of an air pollution effect, one must have good data on property values. The only believable data are sales prices; self-reported values and assessed values are such poor surrogates for sales values as to be essentially useless. One needs data on air pollution levels in each part of the SMSA. Because few cities have such data, various techniques are used to generate "pseudodata"; often the properties of these pseudodata (such as multicollinearity) are deplored. Do these difficulties exist in the world or only in the pseudodata?

A major difficulty in interpretation is the general equilibrium nature of the problem. A new factory that fumigates what was a desirable neighborhood will cause land values to fall there and rise elsewhere. This problem is an old one in the transportation literature in cases involving freeway location. Is the measure of benefit (1) the net increase in land values, (2) the sum of land-value increases (forgetting the decreases), or (3) sum of absolute values of land changes? Strotz has an intriguing answer to this question that does not agree with the calculations of Rubinfeld's colleagues. Furthermore, Rubinfeld reminds us that wage differentials (between SMSAs) and land values should be negatively related; thus, one would have to look at labor migration and wage-rate changes and movement of capital and businesses in order to get good estimates of the effect of air pollution on land values.

The Neighborhood Effect and Spurious Correlation

These objections have been noted before. I find them overwhelming, but new studies continue to appear on air pollution and property values. Apparently, there is an empirical regularity that fascinates economists. It would seem that as long as there is an empirical regularity, we will have new studies. It strikes me that there is a good analogy between this property-value work and the early work on the association between air pollution and health. The latter empirical regularity was dismissed because the air pollution–health association was deemed to be spurious. That is, urbanization is known to lead both to air pollution and to higher mortality rates; thus the correlation between air pollution and health might be spurious. In any case, these early investigators implicitly ascribed all of the "urban component" in the mortality rate to air pollution when they contrasted polluted and clean areas (urban and rural areas).

Rubinfeld and his colleagues have identified a neighborhood effect in property values. There is nothing unexpected in this finding; casual observation suggests that a given piece of property would sell for more in a desirable neighborhood than it would in an undesirable one. I suggest that researchers have ascribed the neighborhood effect to air pollution and thus committed the same error as the early epidemiological investigations of air pollution and health. Rubinfeld is not aware of this objection, and his empirical work makes a serious attempt to deal with it, but I don't see the sort of causal modeling and testing that would be necessary to isolate the component of the neighborhood effect that is caused by air pollution. Thus, I find the empirical regularity as unsatisfying as the theoretical assumptions and data used to get the estimates; I remain a skeptic. I don't have any faith that the particular estimates accurately measure the effect of air pollution on land values; even if they did I wouldn't know how to interpret them in relation to estimates of the effects of air pollution on health, materials life, and so forth.

Air Pollution and Health

Lest I seem a skeptic about the value of air pollution abatement, I will add a few words about "nonmarket approaches." One must first estimate the causal relationship between air pollution and some type of physical damage (health, cleaning, materials life, vegetation, animals, and aesthetics) and then estimate the value of this damage. Both stages are difficult, but the first is by far the more difficult. Eugene Seskin and I have spent almost a decade attempting to unravel the causal relationship between air pollution and mortality.[1]

The empirical regularity between air pollution and the mortality rate can be seen in several annual cross-sections, as well as in daily time series and annual cross-section–time series. The result survives disaggregation by age, sex, and race, as well as by cause of death. Urbanization cannot be the source of the association because many urban characteristics are measured in the analysis and because urban areas are contrasted, rather than urban and rural areas. Furthermore, a host of factors conjectured to affect the mortality rate are controlled, statistically, in the analysis. We find that suspended particulates and sulfates exert an affect on mortality that is statistically significant and remarkably uniform across the various data sets. A 50 percent reduction in these pollutions is estimated to lead to between a 4.5 and 6 percent reduction in the total mortality rate. We believe that we have shown the relationship to be causal and thus interpret that a 50 percent reduction in these pollutants would lead to a large reduction in the mortality rate, a reduction that can be interpreted as an increase in life expectancy of about one year.

If this relationship were causal and if premature death were valued in terms of lost wages and medical costs, we find that the benefit of abating these two pollutants would be considerably in excess of the cost of abatement.

Cleaning, Plants, Animals, and Aesthetics

A nonmarket approach has also been used to estimate the benefits of air pollution abatement from cleaning, materials life, damage to plants and animals, and aesthetics.[2] Although these estimates do not seem as thoroughly supported as those for health, there are some conclusions that are worth noting. For the first four of these categories, damage appears to be slight; dollar costs total perhaps 10 percent of the value of health damage. Work on the effect of air pollution on aesthetics and the resulting dollar loss is in its infancy.[3] Particularly in a high-income society that values outdoor leisure, the value of the loss in aesthetic qualities might be far more valuable than the loss in health.

Conclusion

Rubinfeld has provided us with a valuable paper. The first half is a clear exposition of the difficulties in interpreting the estimated association between land values and air pollution. Whatever the objections to land-value studies, benefit-cost analysis has proved of immense value in public decision making about the environment. During the time EPA had to set air-pollution emission standards, studies showed that stringent abatement of stationary-source emissions (particulates and sulfur compounds) was justified by a benefit-cost analysis. Various studies showed that the evidence on mobile-source pollutants was more doubtful and cautioned that a ''go slow'' policy was necessary.[4] I believe that these studies have been influential and have served to bring evidence into a discussion that has been conducted largely on an emotional level. I cannot emphasize too strongly the preliminary nature of this evidence, but nevertheless it has had an immense and, I believe, a beneficial effect on the public debate.

Questions have evolved that are specifically required for regulation: precisely which pollutants are harmful and in what concentrations? These difficult issues require research by toxicologists, epidemiologists, atmospheric chemists and many others in the effects on health, plants, animals, cleaning, materials life, and aesthetics. This is an exciting area for research; getting a good scientific basis for public policy requires the best we can give.

Notes

1. L. Lave, "Air Pollution Damage: Some Difficulties in Estimating the Value of Pollution Abatement," in A. Kneese, ed., *Research on Environmental Quality* (Baltimore: Johns Hopkins, 1972); L. Lave and E. Seskin, "Air Pollution and Human Health," *Science* 169 (1970): 723–732; L. Lave and E. Seskin, "Health and Air Pollution: The Effect of Occupation Mix," *The Swedish Journal of Economics* 73 (1971): 76–95; L. Lave and E. Seskin, "Air Pollution, Climate, and Home Heating: The Effect on U.S. Mortality," *American Journal of Public Health,* July 1972; L. Lave and E. Seskin, "An Analysis of the Association Between U.S. Mortality and Air Pollution," *Journal of the American Statistical Association* 69 (June 1973): 284–290, 342; and L. Lave and E. Seskin, "Does Air Pollution Shorten Lives?" in J. Pratt, ed., *Statistical and Mathematical Aspects of Pollution Problems* (New York: Marcel Dekker, 1974), pp. 223–244, and in *Proceedings of the Second Annual Research Conference of the Inter-University Committee on Urban Economics* (Chicago: University of Chicago Press, 1970), 293–328.

2. F. Haynie, "The Economics of Clean Air in Perspective," *Materials Performance* 13 (April 1974); L. Lave, "The Costs and Benefits of Air Pollution Abatement," in *The Costs and Effects of Chronic Exposure to Low Level Pollutants in the Environment,* Hearings before the Subcommittee on the Environment and the Atmosphere of the Committee on Science and Technology (Washington, D.C.: U.S. Government Printing Office, 1975); and L. Lave and L. Silverman, "Economic Costs of Energy Related Environmental Pollution," J. Hollander, ed., *Annual Review of Energy,* vol. 1 (Palo Alto, Calif., 1976).

3. A. Randall, B. Ives, and E. Eastman, "Bidding Games for Valuation of Aesthetic Environmental Improvements," *Journal of Environmental Economics and Management* 1 (1974); D. Brookshire, B. Ives, and W. Schulze, "The Valuation of Aesthetic Preferences," *Journal of Environmental Economics and Management,* forthcoming (1976); and L. Lave and S. Silverman, "Setting Environmental Priorities: Amenity Versus Health?" presented at American Economic Association meeting, September 1976.

Reply to Comment

Daniel L. Rubinfeld

Lester Lave is clearly quite suspicious of the "empirical" regularity between property values and air pollution. His major point is a valid one—unless researchers can properly control for neighborhood effects, the significant air pollution coefficients in property-value equations may be the result of spurious correlation. If the correctly measured neighborhood characteristic variables were included in the equation, the air pollution term might become insignificant. Despite attempts by Harrison and Rubinfeld and others to confront this issue, Lave could be right—the outcome is still in doubt. However, in his attempt to denigrate the property-value approach, Lave has made several misleading or incorrect statements, which serve to confuse rather than clarify the methodological issues raised in the paper. In hope of clearing up some of these confusions, let me briefly elaborate some of these issues.

Lave cites a direct conflict between the theoretical sections and the empirical sections of the paper because I argue against methodological approaches that attempt to measure changes in property values and/or changes in wage rates. However, the major point of the methodological discussion is that property-value data can, in principle, be used to estimate utility functions and/or the demand for clean air. In fact, the empirical work reviewed in the paper does focus on the estimation of demand functions, not on the prediction of property-value changes. (Thus Lave's later summary of the Strotz paper is misleading because Strotz focuses on property-value changes, not on the estimation of demand functions.)

Lave makes several technical points about the demand-function or utility-function approach. He argues that *random* errors of, say, 5 percent, will swamp any effect of air pollution. In terms of the econometrics of the procedure, however, random errors pose no particular problem. Only when measurement errors are *correlated* with other included explanatory variables (or other measurement errors) or when neighborhood effects correlated with air pollution are omitted, does a serious problem arise. Lave also argues that one must assume separability of utility functions. Although such an assumption is made in much of the existing empirical work, it is not a necessary one. As Freeman points out in his comment, ''other assumptions (than strict separability) can be made about degrees of separability and the nature of substitution and complimentary relationships—between air quality and some private good.''

Finally in his technical discussion Lave argues that the ''only believable data are sales prices; self-reported values and assessed values are such poor surrogates for sales values as to be essentially useless.'' Lave fails to mention that several studies of sales and assessment data suggest that self-assessments often serve as a good proxy for sales price (see, for example, John Kain and John Quigley, ''Note on Owner's Estimate of Housing Value,'' *Journal of the American Statistical Association*, December 1972).

Comment

Frank E. Speizer

With the enactment of the Clean Air Act of 1970 came the setting of air quality standards for the primary pollutants that at that time were considered, on the basis of the available data, to be safe for human health. Since that time, a number of institutions and individuals with varied interests and motives have questioned these existing standards. They have tried to raise or lower them or to add pollutants to the already existing list.

Much of the decision-making process by which these and other standards (occupational exposure levels) are promulgated are purported to be based upon the potential human health effects of the pollutants in question. Although much of the administrative decision-making process tends not to be health related, this is the basis of the Clean Air Act, and therefore it is important to review some of the problems in assessing the health effects.

Most of the studies related to the health effects of air pollution date from about 1952, and there has been a very large increase in research activity in the last five to ten years. Air pollution episodes in the past have been associated with excess morbidity and mortality in selected population groups. The London fog of 1952 led to an excess of approximately 4000 deaths from all causes of death except automobile accidents. Similar although less dramatic events have been recorded in Europe, the United States, and Japan both before and since.

Much more difficult has been the defining of health effects in terms of morbidity and mortality from chronic exposure to air pollutants. The promulgation of standards and questions raised about changing or adding to these standards and the realization of the economic impact of existing standards or changed standards have quickened the pace of the search for documented dose-response relationship between air pollutants and health effects.

To determine human health effects means to extrapolate from animal exposure studies, to perform controlled exposure studies in both normal and possibly susceptible humans and/or to investigate the effects of "natural" experiments in which populations exposed to existing or potentially existing exposure gradients are monitored both to document their exposure and the suspected health effects over an appropriate time period.

The first two of these approaches have certain advantages. Particularly, they have the possibility of carefully defining the exposure setting and being able to make assessments before and after exposure. These approaches suffer from the distinct disadvantage that the extrapolation from animals to humans is always somewhat speculative, and controlled exposures in humans preclude the possibility of observing the effects of chronic exposure. The third approach is based upon epidemiologic investigations of the health effects of air pollutants and is the topic to which I would like to address my remarks because it is one of the areas with which I am particularly concerned at the present time.

Setting up and carrying out acceptable epidemiologic investigations are considerably more difficult than most laboratory investigations. Obviously, the investigator has very little control over what portion of the population is exposed to which agents. The researcher can only look over the natural setting, make measurements of morbidity and mortality, and hope that all the unmeasurable factors that make human society so complex occur at random in the groups being studied.

For example, when Lave and Seskin used death certificates to analyze mortality from all causes in association with suspended particular matter in approximately 100 Standard Metropolitan Statistical Areas (SMSA) in the United States, they attempted to control for social class and age in their analyses because of the known effect on mortality of these variables.[1] They had no way of knowing the cigarette-smoking patterns of the individuals or, for that matter, the smoking rates in the communities in their study. Because smoking also is associated with excess mortality, these authors had to assume either that all the communities in their study had the same smoking rates or that the smoking rates in the communities studied were randomly distributed and thus would not contribute in a systematic way to the results obtained. In this particular study, the effect of this potentially important covariable must remain undetermined.

Besides the difficulties in measurement of covariables that might be associated with the outcome, there are significant problems in making the actual measurements of outcome. Measuring outcome becomes very complex as soon as one attempts to assess anything other than how many people have died. Death, of course, is generally easy to determine because there is a death certificate for almost everyone who dies, and one has only to count them. Beyond counting certificates, there are many problems. Although there is a relatively uniform system for classifying and coding death-certificate diagnoses, clearly physicians are taught no systematic way (if they are taught at all) to fill out the cause-of-death section of a death certificate. There are examples in the literature that point to effects of changing fashion in diagnosis having an influence on what is written down on death certificates.

The place of death or last residence, which is supposed to be recorded on the death certificate, can be used as an indicator of where the decedent lived. It does not specify length of time in a particular locale, nor does it generally indicate enough about where the person lived to allow us to use the information directly to determine the ambient air pollutant exposure the individual endured. For that information one must turn to alternative sources of data, such as regional air quality data.

The problems in using alternative sources of data for air pollution index of exposure can be demonstrated by describing some data obtained in Boston. The Boston SMSA includes the greater metropolitan Boston area. This area is made up of approximately 2.75 million people and stretches from the center of Boston approximately twenty miles due north, about nine miles west, and about five miles south. This SMSA is represented by the Air Quality Control Region designated as Boston. The air quality station is located in the downtown portion of the city of Boston. We had the occasion to study sixteen different sites within the downtown Boston area, all within two miles of the official Boston site, and we were able to demonstrate that for all pollutants being measured, except sulfur dioxide, the levels recorded were highly dependent upon very local conditions, including traffic patterns and the micrometeorology of the particular site.[2] The importance of this discovery is that it points out how terribly crude the existing ongoing regional air quality data must be for estimating the exposure of large population groups.

Time does not permit getting into the difficulties of measuring outcomes other than death. Various sorts of morbidity studies that have attempted to document health effects both retrospectively and prospectively suffer from many of the difficulties to which I have already alluded. It has been difficult to obtain standardized measures of outcomes, whether they be responses to questionnaires, self-completed diaries, hospital diagnoses, pulmonary function tests, or the like, on sufficiently large populations that can be accurately defined as to how representative they are of total communities. Furthermore, the existing technology for assessment of exposure is, for the most part, cumbersome and expensive to operate, does not take into account the potential for varied human activity, and is not designed to measure interactions of the variety of pollutants that are likely to exist in any given atmosphere. To make matters even more complicated, particularly in the search for chronic effects, there does not exist an extensive data bank of exposure data that can be tapped to reconstruct past exposure. Necessarily, therefore, answers to dose-response relationship questions will have to come from carefully designed prospective epidemiologic studies.

I recognize that what I have presented suggest a rather dismal view of the epidemiologic approach to gathering data about health effects of air pollutants. My intent was not to discourage but to try to give some insight into the problems involved in attempting to assess the health effects of air pollutants as they relate to standard setting. In concluding, I would like to say that well-designed studies are underway and they have the potential for providing the data needed to make informed policy decisions. These studies are terribly expensive in both personnel time and money, and the performance of sufficient numbers of investigations will require increased numbers of appropriately trained environmental engineers and epidemiologists. At this point, these people appear to be in short supply. To train them will take money and time, but with the nature of the costs of the policy decisions that have to be made about air pollution, it seems to me that it makes good economic sense to invest the small fraction of the cost of cleaning up the environment on technicians who will be able to determine whether the cleanup has done any good.

Notes

1. L. B. Lave and E. P. Seskin, "Air Pollution and Human Health," *Science* 169 (1970): 723.

2. W. Burgess, L. DeBeradinis, and F. E. Speizer, "Exposure to Automobile Exhaust III: An Environmental Assessment," *Arch. Environ. Health* 26 (1973): 325. The pollutants measured included sulfur dioxide, nitrogen dioxide, carbon monoxide, total hydrocarbons, mass respirable particulates, and lead. Standard wet chemistry techniques were used for all measurements.

7 The Costs and Technology of Pollution Abatement

Donald N. Dewees

Introduction

The Clean Air Act of 1970 presents some of the difficulties faced by decision makers in a wide range of pollution control situations, in that both emission and ambient air quality standards were enacted before accurate estimates of the cost of meeting those standards were available and, in fact, before it was known whether it would be technologically feasible to meet those standards at all. It is rare that a legislative or regulatory authority will know the costs that must be incurred to meet its standards or regulations at the time those regulations are promulgated. In some cases, available estimates may be accurate to plus or minus 20 percent. In other cases, any estimate would be subject to an error of a factor of 2 or 3. And in a disturbing number of cases, pollution control decisions must be made before adequate control technology has been proven. Even when cost studies have been completed, the results frequently omit information that economists would regard as essential to sensible decision making.

The purpose of this paper is not to fill the lacuna in cost estimates with new estimates of actual costs for controlling pollution from various sources. The aim instead is to explore what cost data are needed for effective decision making and then to determine what portion of these data needs may be actually satisfied by available data in several specific problem areas. It is often the case that existing data can be far more useful than is first suspected if the proper questions are addressed to the data. In addition, many cost studies could be far more useful if small changes were made in the methodology of estimating those costs.

The next section of this paper will discuss the kind of data needed for effective pollution-control decision making. It will also review costing methodology in general. Following sections of the paper will look at the available data for controlling particulates, sulfur dioxide, and automobile emissions. In each case, an attempt will be made to extract from available studies and data the information needed to choose sensibly among pollution control policies. In addition, suggestions will be made for improving cost studies in the future.

Particular attention will be paid to the problems raised by technological progress in pollution control because costs are most uncertain when the technology either has not been developed or is changing rapidly. Furthermore, the adoption and design of a pollution control program may have a great effect on the rate of technological progress. If there is old technology that is well understood, costs may generally be estimated to reasonable accuracy with minimum effort.

Needs and Methodology of Pollution-Control Cost Estimates

The cost information needed to select pollution control policies depends upon the objectives and methods of the decision making process itself. It is sometimes suggested, and occasionally legislated, that a pollutant must be controlled to the maximum extent that is technologically feasible or that the best available technology be applied to its control. If this requirement were interpreted literally, it would mean that an existing pollution control device, even if outrageously expensive, must be used. Such a standard implies that we are prepared to spend unlimited amounts of money to control pollution even if it diverts resources from other socially desirable activities, such as disease prevention, education, and production of basic necessities, including food and housing. This idea is clearly nonsense. In a society that must satisfy its needs from limited resources, we must decide how to allocate those resources among many competing needs. No single goal, no matter how meritorious, can command unlimited quantities of these resources.

Thus common sense leads to a decision-making process in which the benefits of pollution control are somehow balanced against the costs. Furthermore, for most pollutants, pollution control is not an all-or-nothing matter. Usually it is possible to incorporate technologies that provide several different levels of control for a single pollutant. Because it is necessary to select the proper degree of control, it is necessary to compare the costs of different levels of pollution abatement. We must ask whether the incremental reduction in pollution achieved by going one step further is worth the incremental cost of that step. In short, we must calculate marginal costs of different levels of pollution control. The formal economic specification of an "efficient" degree of pollution control is one in which the marginal cost of abatement is just equal to the marginal benefits of abatement.

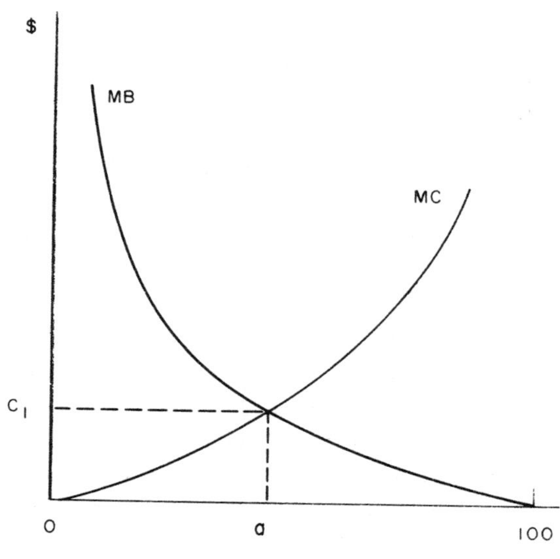

Figure 7.1. Marginal costs and benefits of abatement.

Textbooks on environmental economics[1] demonstrate this principle with diagrams such as figure 7.1. Marginal costs of pollution control are shown as an increasing function of the degree of abatement, which assumes that the first pollution control measures may be relatively inexpensive, but once the inexpensive technologies have been exhausted, further abatement at high levels of pollution control will be more expensive per unit of abatement. The marginal benefit of abatement curve is shown as a falling line that reflects great benefits from abatement at high pollution concentrations and declining benefits as the environment becomes less and less polluted. If abatement were carried to some point to the left of q_1 in figure 7.1, the marginal benefits from further abatement would be greater than the marginal costs, and social welfare, measured as the minimizing of pollution control costs and pollution damage, would be improved by increasing the degree of abatement. If abatement were carried out to a point to the right of q_1, then the cost of the last unit of abatement would be greater than the benefits, and welfare would be improved by reducing abatement. Only at point q_1 can welfare not be improved by some further change in the pollution level. Clearly, therefore, selection of an efficient degree of abatement requires the estimation of an aggregate marginal abatement cost curve like that shown in figure 7.1.

Once an appropriate degree of control has been selected for a pollutant in an area, the burden of cleanup must be allocated among many sources. The economist's solution to this problem is that a given degree of environmental improvement should be achieved at the least possible cost. If there are many sources, the least-cost means of reaching a given level of environmental quality will be achieved if each source reduces its emissions until the marginal cost of abatement is equal for all sources.[2] If each source could be represented by an individual marginal abatement cost curve and if all sources in the area did not abate until their marginal costs were equal, total abatement costs could be reduced by shifting one unit of abatement from a high-cost firm to a low-cost firm. Thus, when the objective is to minimize the cost of achieving a given total emission rate from a set of sources (this will be a sensible objective only when the environment in question is perfectly mixed so that the effect of a unit of discharge from each source is identical),[3] the marginal cost of abatement at each source must be equalized. To do so required either a policy that will automatically equalize marginal costs (such as an effluent charge) or knowledge of the marginal abatement cost curve for each source so that the efficient degree of abatement can be specified for each. Thus, in some cases it may be desirable to know the marginal cost of abatement curve for individual sources, as well as the aggregate curve for all sources in an area.

Most attempts to determine costs of pollution control use one or both of two standard methodologies. The first methodology is the engineering or simulation method. In this method, engineering estimates are used to determine the capital and operating cost of each component in a pollution control system. Frequently, these engineering estimates will produce a single cost figure for a particular pollution control device for a particular source. If the technology of abatement is reasonably well understood, equations may be generated to represent the pollution-control cost function. In such cases, one need only insert the parameters for a particular pollution source to calculate the cost of controlling that source under closely specified conditions.

The other method of pollution-control cost estimation is the statistical or econometric method. Here an equation is developed relating total costs for a pollution source to a number of variables, including the degree of pollution control. Accounting data from a number of existing sources are collected, and regression analysis is used to fit the equation to the observed data. The estimated parameters of the equation will show the cost of pollution control as a function of whatever independent variables have been included.

Each of the above costing methodologies presents serious problems. Economists tend to be suspicious of the engineering method because there is no empirical proof that it provides accurate results for an industry or set of sources.[4] The estimated costs are likely to be determined for a particular source: a new plant with a certain technology and a given rate of product output. It is very difficult to generalize such results to an industry that includes new and old firms using a variety of production technology and having a wide range of output rates. Because the costs are estimated for a single plant, it is impossible to compare them with actual experience in an industry. Rarely are the cost functions sufficiently flexible to allow consideration of the many factors that vary from one source to another and may have a profound influence on costs of pollution control.

Economists are generally more comfortable with econometric studies that sift through historical data to identify the impact of pollution control upon costs for different sources. In this method, at least one can argue that the data come from actual historical experience rather than from theoretical ideas of how costs behave. Unfortunately, it is rarely possible to complete a satisfactory econometric cost study while important pollution control decisions are being made. Cross-section analysis requires data on a large number of sources that experience different pollution control regulations, yet the best time to do the economic analysis is before regulations are imposed rather than afterward. A good time-series study showing changes in cost as pollution control requirements varied requires a long series of annual data. It is rare that an industry or firm possesses a long record of sufficient cost and pollution-emission information to allow such a study. Although some econometric studies of pollution costs have been completed,[5] the number of such studies available for a variety of sources and pollutants is not substantial. Furthermore, such studies should be expected to emerge only in cases in which the pollution control technology is stable and not subject to rapid change.

Because many decisions must be made before an accurate marginal cost of abatement curve can be estimated, it is useful to consider what use may be made of more limited cost information. Suppose first that the only available data described the cost of a particular control technology for a particular degree of pollution control. In this case, there is no marginal abatement cost curve at all, only a single point on the total cost curve. Still, with a description of the reduction in pollution and the associated cost, it should be possible to decide whether that specific pollution reduction is better than no reduction. In short, one should be able to make the all-or-nothing decision whether to adopt the control for which data exist. A single cost estimate, however, does not allow any evaluation of alternative degrees of abatement.

Another frequent situation is that we may have vague information about the cost of achieving one or two different levels of abatement, and some impression of the shape of the marginal cost of abatement curve. Such information, even if not precise, may be combined with information about the shape of the marginal benefit curve to allow sensible policy decision. Suppose, for example, that marginal costs are not known with certainty but are believed to be relatively constant over a wide range of abatement. Suppose further that the marginal damage function exhibits a sharp drop at some particular level, representing a threshold effect of the pollutant. If marginal abatement costs are believed to be between the high and low levels of the dropping marginal benefit curve, then it would be reasonable to establish an environmental quality standard just on the safe side of the benefit threshold. For a well-behaved cost function, this is the most likely point of intersection between marginal costs and marginal benefits.

Suppose, on the other hand, that marginal benefits are thought to be reasonably constant over a wide range of pollution levels because of the absence of threshold effects or because thresholds occur at different levels for different affected species or populations. In such a case, one would look for a steeply rising portion of the marginal cost curve. Establishing ambient quality standards in the vicinity of the sharp rise in marginal costs most likely would cause marginal costs and marginal benefits to equalize in this case. If both marginal costs and marginal benefits are thought to be relatively constant over a wide range of pollution levels, then social welfare will not be very sensitive to the precise degree of pollution control adopted, and criteria other than strict economic efficiency could be applied in seeking an environmental quality standard.[6]

This analysis is intended to suggest that the concept of equating marginal costs and benefits may still be useful in cases in which neither the marginal benefit nor the marginal cost curve is precisely specified. General ideas of shapes and magnitudes may be sufficient to identify superior policies. In fact, one could interpret the "best available technology" requirement as an economically efficient requirement if (1) marginal benefits were reasonably constant over a wide range of abatement and (2) ''best available technology'' were interpreted as meaning the technology that lies just below the sharply rising portion of the marginal cost curve.

The discussion of costs so far has assumed that the world is static, so that the diagram of marginal costs and benefits shown in figure 7.1 is the same for all years. Even in such a simple case, if we start from uncontrolled emissions and desire to move to point q_1, it is necessary to design a transition policy to move from one point to the other. Should all sources be controlled as of next year, should only new sources be subject to controls and old sources allowed to emit at high rates until they are scrapped, or should some other transition policy be adopted? Some method must be developed for analyzing the costs of different adjustment paths.

Furthermore, there is no reason to believe that marginal cost and benefit curves will be stable over time. Population grows continually, incomes are likely to rise, and therefore the benefits of pollution control are likely to change over time, probably shifting upward. Some pollution control technology is old and stable, but in other areas the technology may develop rapidly and cause the marginal cost curve to shift downward over time. If both marginal benefits and marginal costs may move in the future, one's cost-benefit analysis should take into account these future changes. One method is to calculate present discounted values of streams of both benefit and costs for some specified future time.

Still other complexities arise if the choice of pollution control policies is likely to influence the shifting of cost or benefit curves. For example, the establishment of pollution standards either might drive pollution costs up by creating excess demands in the pollution control industry or might drive costs down by spurring technological advances in that industry. If one expects that policies will substantially affect costs, then it is necessary to develop several scenarios of future policies and technologies for evaluation. Present values of net benefits can be calculated for each different policy development and these alternatives compared. An example of this procedure is the study of the costs and benefits of automobile pollution control.[7]

The arguments of this section can be briefly summarized. When inquiring about pollution control costs it is best to seek an aggregate marginal cost of abatement schedule covering all conceivable ranges of pollution control. In the absence of such a well-defined curve, several points on the marginal cost curve would be useful. If several points cannot be achieved, at least an idea of the relative magnitude of these marginal costs and the shape of the marginal cost curve would be essential to basic decision making. If costs may change over time, then it may be necessary to calculate present values of costs (and benefits) for several scenarios of control policies over some reasonable time in the future.

Particulate Emissions

A major source of particulate emissions in North America is the burning of coal in electrical generating stations. All coal contains ash, often from 5 to 10 percent by weight. When the coal is pulverized and burned, some of this ash is discharged with the combustion gases into the smoke stack. A large coal-fired electrical generating station could discharge over 10 tons per hour of particulates in the absence of controls.

A variety of technologies is available for collecting particulates before they can be emitted into the atmosphere. A survey of this technology was published by the Environmental Protection Agency in 1969.[8] The two primary methods for particulate control from coal-fired power plants are mechanical collectors and electrostatic precipitators. Both of these technologies have been in use for several decades. Their costs are therefore well-known and change little from year to year. Although there may be technological progress in designing and operating these collectors, such progress is not likely to be a significant factor in cost or performance over a five-year period.

Handbooks are available describing how to design particulate controls, and a number of cost studies have been published. In fact, the EPA report devotes twenty-seven pages to discussion and analysis of the costs of particulate collection. The third page of the chapter presents a theoretical cost curve with exactly the shape we would expect: marginal costs increase as the degree of abatement increases. Surprisingly, however, the report does not include a single empirical curve showing the relationship between collector efficiency and cost. Nor is there any equation in the chapter showing annual or other costs as a function of collector efficiency. Although the report recognizes immediately the importance of marginal costs for pollution-control decision making, marginal costs are never presented. Instead the equations and graphs show cost as a function of the volume of exhaust gas to be treated and a host of other variables.

Marginal cost data can be extracted from the report, however. Both tables and graphs show costs for three different levels of collector efficiency. The cost of wet scrubbers, for example, is presented for efficiencies of 75, 90, and 99 percent. Dividing the differences in annual cost by the differences in percent control, one can discover that, between 75 and 90 percent control, the marginal cost of abatement in annual dollars per percent abated is \$760 per percent. Between 90 and 99 percent control, the marginal cost is \$5756 per percent. Thus, the report permits calculation of marginal costs for two arcs, which show both the level of marginal costs, and that marginal costs are rising rapidly for wet scrubbers in that efficiency range. The cost data for electrostatic precipitators are graphed for 90, 95, and 99.5 percent abatement efficiency. A careful reading of the graphs and some brief calculations show that, between 90 and 95 percent efficiency, the marginal cost is \$5000 per year per percent abated, whereas between 95 and 99.5 percent, the cost is \$6670 per percent abated.

Even these calculations are not sufficient for performing a cost-benefit analysis. Benefits will be known, if at all, on the basis of some dollar saving as a function of a change in ambient air quality, that is, a change in the particulate density in the atmosphere. A dispersion model would be required to convert changes in source emissions to changes in ambient air quality. Even then, the changes in emission rates would be in terms of grams, pounds, or tons per hour. The marginal cost as a function of percentage abatement would have to be adjusted by the uncontrolled rates of emission converted to dollars per gram, pound, or ton controlled. Only then could one sensibly compare costs among sources or costs with benefits.

Much more complete information is provided by Watson.[9] Watson estimates econometric cost equations for mechanical and electrostatic particulate control devices using actual power-plant cost data and from that develops a cost-minimizing control cost function. Table 7.1 shows the cost of each type of collection over a range of collector efficiencies. The total cost is calculated directly from the equation, and the average cost is simply total cost divided by percentage efficiency. Marginal cost is calculated from the first derivative of the cost equation and is therefore a point estimate for the stated efficiency. The marginal cost data are plotted in figure 7.2.

A striking feature of the data in table 7.1 and figure 7.2 is the declining marginal cost of mechanical particulate collection as a function of collector efficiency. For this particular technology, the rising marginal cost curve that is invariably assumed seems inapplicable. If Watson's cost function is correct, there must be come economy of scale that more than overcomes any tendency toward declining marginal product with greater and greater abatement. Mechanical collectors, however, are limited by technological barriers to 80 or 85 percent efficiency. For higher efficiencies, electrostatic precipitators must be used. Reassuringly, the marginal cost curve for the electrostatic precipitator shows a sharply increasing cost beginning between 80 and 90 percent efficiency. Thus a combined cost curve using mechanical

Table 7.1 Particulate Control Costs (millions of 1967 dollars)[a]

Efficiency (%)	Total Annual Cost ($ × 10⁶)	Average Cost ($ × 10⁶/%)	Marginal Cost ($ × 10⁶/%)
Mechanical collectors			
10	1.53	0.153	0.047
20	1.94	0.092	0.038
30	2.30	0.077	0.034
40	2.63	0.066	0.032
50	2.94	0.059	0.031
60	3.24	0.054	0.030
70	3.53	0.050	0.029
80	3.81	0.048	0.028
$MC^1 = 106,089.92(E_m^{-.62}) + 21,130$			
Electrostatic precipitators			
50	3.57	0.0714	0.043
60	4.00	0.0667	0.045
70	4.47	0.0638	0.048
80	4.99	0.0623	0.057
90	5.64	0.0627	0.080
92	5.81	0.0632	0.092
94	6.01	0.0640	0.110
95	6.13	0.0645	0.125
96	6.27	0.0653	0.146
97	6.43	0.0663	0.181
98	6.64	0.0677	0.248
99	6.96	0.0703	0.439

$$MC^1 = 897,055/\sqrt{\mathrm{Ln}\,\frac{100}{100 - Ep}} \times \frac{1}{100 - EP} + 21,130$$

Source: Equations 7 and 14 in William D. Watson, Jr., "Costs and Benefits of Fly Ash Control," *Journal of Economics and Business* 26 (1974):157–168.

[a]Although costs are presented to three significant figures, any estimate of the absolute cost of a given installation would only be accurate to ±20 percent or so. We may have a high level of confidence about the shape of the cost curve, but a less accurate idea of its location.

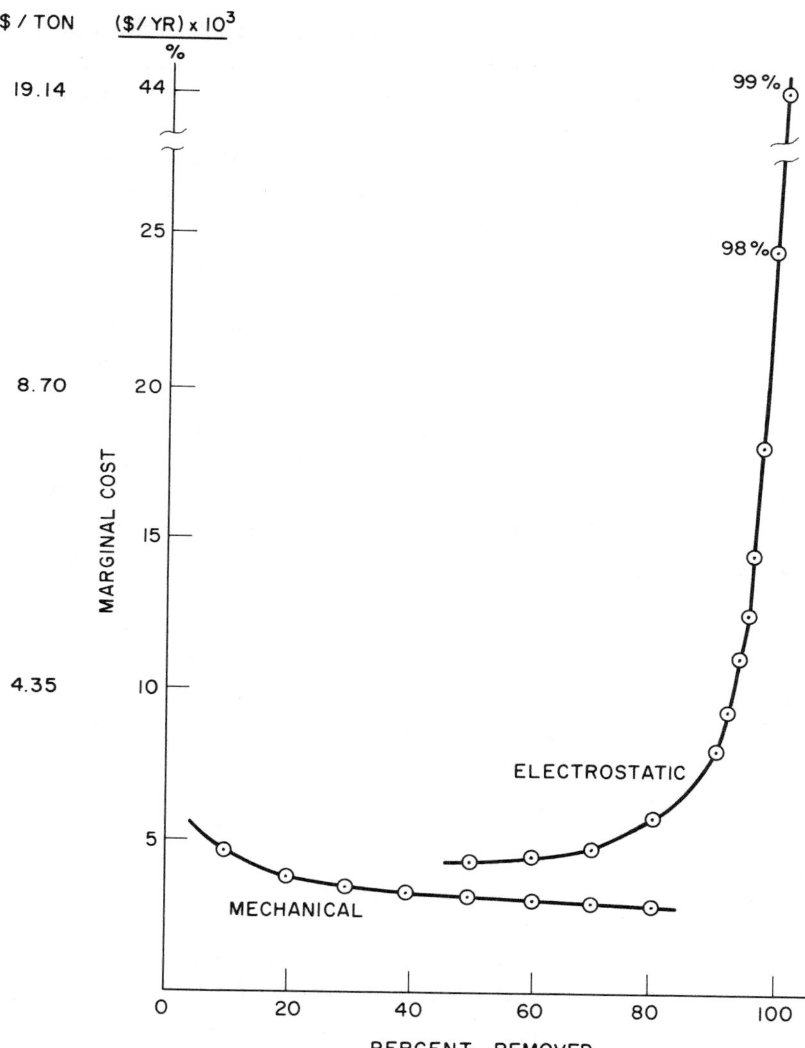

Figure 7.2. Marginal particulate control cost (based on *design* efficiency).

collectors for low efficiencies where they are dominant and electrostatic precipitators for high efficiencies where the mechanical collectors are inoperative would show a cost curve that declined initially and then rose asymtotically to the 100 percent line.

What inferences may be drawn from the data in the EPA study and the Watson study? First, a priori assumptions about well-behaved rising marginal cost curves must be somewhat modified. Although there is a portion of the marginal cost of abatement curve for particulate control that rises rapidly with the degree of abatement, there is also a portion in which marginal costs are declining. It is not proper, therefore, to assume that a marginal cost of abatement curve will rise continuously from zero to 100 percent abatement. In fact, marginal costs of the dominant technology appear to be falling or constant between zero and 80 percent abatement. If the marginal benefit of abatement curve were rapidly falling with abatement, it would be quite possible to have an intersection in the declining portion of the marginal cost curve. Of course, if the marginal benefit of abatement were reasonably constant over all pollution levels, then only the rising portion of the marginal cost curve would be relevant.

Caution must also be used in drawing inferences from these data about absolute cost levels for controlling a specific source or group of sources. The cost data in table 7.1 are for a specific set of more than half a dozen parameters describing the characteristics of the pollution source. All authors agree that control costs can vary widely from one source to another, even when they are of similar size, because of differences in fly-ash disposal costs, available space for installing the control devices, the duty cycle or load factor of the source, and a number of other variables. Costs may differ by factors of 2 to 6 because of such considerations, especially when the devices are being installed on existing power plants. Thus the data in table 7.1 are useful for establishing the rate of increase of marginal cost with increasing collector efficiency, but they indicate actual costs only for the specific conditions for which they are derived. Reasonably accurate cost estimates would require separate calculations, based on the proper values of each parameter, for every source or perhaps every group of sources. Because cost functions and data are available, such calculations need not be enormously costly, but they would require gathering some data about every source to be controlled.

Finally, the efficiency levels in all published data are design efficiencies. Measurement of the *actual* efficiency of fly-ash collectors is quite costly and thus rarely undertaken. If operation and maintenance of the boiler and collector are not of a high level, actual efficiencies may fall well below those reported. This factor is crucial at high efficiency levels because a drop in efficiency from 99 to 98 percent doubles the quantity of pollution emitted.

The method of pollution control discussed here could be classed as tail-end treatment. There is no attempt to reduce the generation of the pollutant, only to collect the pollutant once it has been created. In many cases, the least-cost means of pollution control would include some tail-end treatment and some process change to reduce pollution generation. In the case of particulate control from coal-fired generating stations, however, the cost of control is a small percentage of total generating costs. Other changes in the coal combustion process are sufficiently costly that tail-end treatment is almost certain to be the least-cost means of reducing particulate emissions.

The costs reported here are for the technology existing at one point in time. Particulate control, however, has been practiced for decades, and both technologies under consideration are well established. There may be improvements in the design and operation of these collectors, but there is no reason to expect rapid changes in cost effectiveness. Thus costs calculated for a specific installation and specific time can be expected not to change substantially over a five- or perhaps ten-year period, except for general changes in the price level. In short, future cost projections ignoring technological progress would not be seriously in error over a significant time span.

Watson's figures for particulate control costs are all expressed in dollars as a function of percentage abatement. Marginal benefit figures, however, would almost certainly be in terms of dollars per ton controlled, the same units in which an effluent charge would be expressed. Costs per percent can be converted to costs per ton by dividing through by the total tons of particulates generated per year. In Watson's sample calculations, the utility generates 2.3 million tons per year of particulates. This figure can be used to create the second scale on figure 7.2, which shows marginal abatement costs in dollars per ton controlled. In 1967 dollars, 99 percent particulate control with electrostatic precipitators costs almost \$20 per ton. This figure suggests the magnitude of effluent charge necessary to achieve that degree of control from a specific generating station.

Finally, we should step back from the cost calculation for a single source to consider the shape of the abatement cost curves for a number of sources in an area. Would the marginal abatement cost curve for an entire air shed have the same shape as the curve for an individual firm? Suppose that the pollution control policy under consideration is an effluent charge or some other means that will achieve a given reduction in emitted tons at the least possible cost. As such a policy began to control emissions it would first attack those sources that had the lowest control cost: new sources and large existing sources with favorable conditions for the installation of control devices. As the total controlled tonnage increased, it would be necessary to apply increasing controls to low-cost sources and to begin to control high-cost sources. Thus the aggregation of many individual marginal cost curves, each of which might have the shapes suggested in figure 7.2, should result in an aggregate marginal cost curve that would be continuously rising over its entire range. Any economies of scale that create a falling marginal cost curve for the single source might not exist among many sources. In short, aggregation of many sources may yield a total marginal cost curve that is better behaved, that is, one more like the traditional, rising marginal cost curve shown in figure 7.1.

When real problems involving a number of sources in a metropolitan area or a larger air shed are considered, the cost-benefit analysis described here becomes considerably more complex. Only if the atmosphere is perfectly mixed (which is unreasonable for an area of any size) is it possible to draw a single marginal cost and marginal benefit curve for the area. In a realistic case, the benefits of abating one pound from each source varies by source, and a dispersion model must be used to relate emissions at points to pollution effects at other points. An example of the use of a dispersion model for automobile emissions in a metropolitan area is given in the study published by the National Academy of Sciences, Coordinating Committee on Air Quality Studies (hereafter called the CBC Study).[10]

Sulfur Dioxide Control

Compared to the control of particulates, the technology for sulfur dioxide control from stack gases is still in its infancy. A number of different technologies are available for stack gas scrubbing, but experience to date has not identified any single technology that is clearly superior to the others. Many pilot plants have been installed around the world to test the operating capability and performance of these systems. The Sulfur Oxide Control Technology Assessment Panel[11] (hereafter referred to as SOCTAP) reported eleven full-scale stack gas desulfurization facilities in the United States as of spring 1973. Most of these were small generating stations of less than 200 megawats capacity, so they could be regarded as pilot plant or test operations.

Initial installations of sulfur dioxide control devices have shown a variety of operating problems. Some of the processes deposit unwanted materials inside the boiler itself. Others involve a buildup of limestone or other materials on portions of the cleaning devices, which must be periodically shut down for cleaning and maintenance. It has been difficult to achieve the high level of operating reliability with these sulfur dioxide control systems that is customary for electrical generating stations themselves. Thus the industry will remain wary of massive installations until satisfactory reliability can be proven.

The other major problem with sulfur dioxide stack-gas cleaning is its cost. In 1969, the U.S. National Air Pollution Control Administration[12] estimated the cost for 95 percent efficient sulfur dioxide control using lime/limestone wet scrubbing in a 200 megawatt unit at about .5 mill per kilowatt hour in operating cost plus a capital investment of $13 per kilowatt of capacity. In 1972, Slack[13] estimated the cost for a similar unit to be .75 to 1.5 mills per kilowatt hour in operating cost and $40 per kilowatt in capital investment. In 1973, SOCTAP estimated the costs of six different sulfur oxide control technologies. Their estimates are presented in table 7.2. The investment per kilowatt of capacity ranges from $17 to $65, and the operating costs range from .6 mill to 3 mills per kilowatt hour. At the time these costs were estimated, they represented a large fraction, ranging from 20 percent upward, of the total cost of electricity generation. Thus although particulate control costs may be trivial compared to total generating costs, sulfur dioxide control costs are substantial.

Not only do the costs presented by SOCTAP and other studies include a wide unexplained variation, but they also provide no reasonable basis for constructing a marginal cost of abatement curve. Five of the processes in table 7.2 are described as achieving up to 90 percent removal efficiency. Because there is no variation in abatement efficiency, no estimate of the influence of efficiency on total cost can be made. Only the first process, dry limestone injection, offers lower efficiency and lower costs than the others. Yet the range of costs and efficiency is such that it is impossible to determine whether marginal costs change significantly between dry limestone injection and the other processes.

It is possible that the figures in table 7.2 will continue to reflect the state of technology in that only two different levels of abatement will be possible. If this is the case, then accurate cost calculations might be made for a specific application and marginal costs calculated for moving between no abatement, the dry-limestone abatement level, and the 90 percent abatement level. The existing data, however, provide little information on which to base such marginal cost estimates.

It is also possible that over time a wider variety of technological possibilities will emerge. Perhaps processes will be developed that can achieve 40 to 60 percent abatement at lower cost than the 80 to 90 percent

Table 7.2 Comparisons of SO_2 control process systems.

Processing Method	Investment Costs[a] for Coal-Fired Boilers ($/kw)	Annual Costs[b] (mills/ kw-hr)	So_2 Removal Efficiency (percent)
1. Dry limestone injection	17–19	0.6–0.8	22–45
2. Wet lime/ limestone/$Ca(OH)_2$ slurry scrubbing	27–46	1.1–2.2	80–90
3. Magnesium oxide scrubbing	33–58	1.5–3.0	90
4. Monsanto catalytic oxidation (add-on)	41–64	1.5–2.6	85–90
5. Wellman–Lord Process (soluble sodium scrubbing with regeneration) .	38–65/kw	1.4–3.0	90
6. Double-alkali process	25–45	1.1–2.1	90

Source: Sulfur Oxide Control Technology Assessment Panel, *Final Report of the Sulfur Oxide Control Technology Assessment Panel to the Federal Interagency Committee of Evaluation of State Air Implementation Plans* (Washington, D.C.: U.S. Environmental Protection Agency OST, 1973), Table 4-1, p. 45.

[a] Generally, where a cost range is indicated, the lower end refers to a new unit (1000 Mwe), while the high end refers to a 200 Mwe retrofit unit. Costs include particulate removal.

[b] Assumptions: Costs calculated at 80 percent load factor; fixed charges per year 18 percent of capital costs. No credit for sulfur recovery.

technologies. If such developments occurred, then a marginal cost of abatement schedule could be developed and sensible choices made among different abatement levels.

These cost estimates were all based on extrapolations from recent installations or the design of new installations soon to be operational. They therefore represent strict engineering cost estimates. An attempt to produce a generalized cost function has been made by Burchard,[14] who used data from a number of cost studies for sulfur dioxide scrubbing systems and developed equations to represent costs under a variety of conditions. Although it is not clear that Burchard's equation is actually fitted by regression techniques to the existing data, he does use his equation to reestimate the cost of the actual facilities in his input data and finds that his cost estimates are within 15 percent of the original estimates. His cost equation is:

$$A_t = (U_sQ + 8760SU_a)L + L_o + MLC_t + R_c(D_sF_sF_r + D_aF_aF_r)$$
$$(HD_o)(1 + I_c)(1 + I_u)$$

(For definitions of the symbols used, see table 7.3.)

A notable feature of Burchard's equation and data is the tremendous range in most of the cost variables, many of which vary by at least a factor of 2. The major contribution of this cost function is to reconcile the variety of cost estimates for different scrubbing installations, which vary enormously in parts because of the tremendously varied conditions of plant size, fuel sulfur content, byproduct, disposal costs, and a number of other factors.

In May 1975, another study of flue gas desulfurization costs attempted to reduce the range of unexplained variation in the cost estimates.[15] Whereas the reported cost of existing systems varied from $33 to $197/kw, adjustments to exclude particulate control costs, to convert all costs to 1975 dollars, and to look only at lime/limestone systems reduced the cost range to between $50 and $87/kw, with an average of $70/kw. That study also developed cost models for stack gas scrubbing and found that their cost estimates for similar plants. Their model plant estimates are summarized in table 7.4.

The major contribution of this study is that it confirms that much of the previously unexplained variation in scrubbing costs can in fact be explained. The wide range in reported or estimated costs for different pilot plants is shown to be attributable in large part to a set of explanatory variables, including whether the installation is at a new plant or a retrofit, the flue gas flow rate, the SO_2 removal requirement, the degree of redundancy in the scrubbing system, the need to control particulates, and the difficulty of sludge disposal. It should be possible, therefore, to estimate scrubbing costs within 10 to 20 percent accuracy for a single plant, given sufficient data about that plant and using the lime/limestone process.

Although none of the data so far provide a real basis for constructing a marginal cost curve for an individual facility, Burchard suggests that a sensible marginal cost curve for a geographic area can be constructed. Because the many sulfur dioxide sources within an area could incorporate substantial variations in many of the cost parameters in his equation, the cost per ton of sulfur dioxide removed could be expected to vary enormously from one source to another. If a pollution control policy were adopted that could be expected to control sources in order of increasing cost, then marginal and average costs could be expected to rise as the policy target control rate increased. In fact, Burchard presents a curve representing an estimate of the percentage of coal-fired capacity in the United States that could be retrofitted with scrubbers at different cost levels. This curve (figure 7.3) shows slowly

Table 7.3 Definition of Symbols for Burchard's Equation

Symbol	Description	Units	Range of Values
A_t	Total annualized costs	$/year	10^5–10^7
U_s	Scrubber utility cost	$/SCFM-yr	0.60–0.80
Q	Flue gas rate (design)	SCFM	2.10^5–3.10^6
S	Sulfur removal rate (design)	tons/hour	10^3–10^5
U_a	Alkali handling utilities and raw materials cost	$/tons-S	18–68
L	Load factor (yearly average)	—	0.30–0.80
L_o	Operating labor cost, including overhead	$/yr	100,000–300,000
M	Maintenance as fraction of capital investment, including overhead	—	0.05–0.10
R_c	Capitalization rate based on total investment capital	—	0.12–0.25
C_t	Total capital cost	$/kw	20–80
D_s	Scrubbing direct cost based on a new 500 Mw, 4 scrubber application		
	Without part. control	$/kw	6.5–12
	With part. control	$/kw	10.5–15
D_a	Alkali handling direct cost for a specific process, based on 5 tons S/hr		230–635
D_o	Other direct costs	$\frac{\text{\$/kw other direct costs}}{\text{\$/kw process direct costs}}$	0.10–0.20
F_s	Gas flow rate and scrubber configuration adjustment factor		0.8–1.4
F_r	Retrofit difficulty factor	—	1.0–1.5
F_a	Sulfur rate scale factor	—	0.5–2.0
I_c	Contractor indirect cost	$\frac{\text{\$/kw contractor indirect cost}}{\text{\$/kw total direct cost}}$	0.25–0.50
I_u	User indirect costs	$\frac{\text{\$/kw user indirect costs}}{\text{\$/kw total direct + contractor indirect}}$	0.12–0.25
S_r	Sulfur rate	lbs/kwh	0.003–0.07

Table 7.4 PEDCo Cost Estimates (January 1975 dollars)

Model Plant Characteristics	Capital Costs Limestone	Capital Costs Wellman-Lord	Annual Costs Limestone	Annual Costs Wellman-Lord
250 Mw capacity	$/kw	$/kw	mills/kwh	mills/kwh
Retrofit				
3.5% S	81	122	5.2	6.8
0.6% S	74	94	4.5	5.3
New				
3.5% S	66	95	4.2	5.1
0.6% S	59	70	3.3	3.9
500 Mw capacity				
Retrofit				
3.5% S	70	114	4.3	5.8
0.6% S	65	88	3.7	4.6
New				
3.5% S	58	90	3.6	4.7
0.6% S	53	67	3.1	3.6
1000 Mw capacity				
Retrofit				
3.5% S	69	104	4.2	5.3
0.6% S	64	80	3.6	4.4
New				
3.5% S	57	86	3.5	4.2
0.6% S	52	64	3.0	3.4

Source: Process Cost Assessment, Report to the Office of Planning and Evaluation (Washington, D.C.: U.S. Environmental Protection Agency, May 1975), p. 3-2.

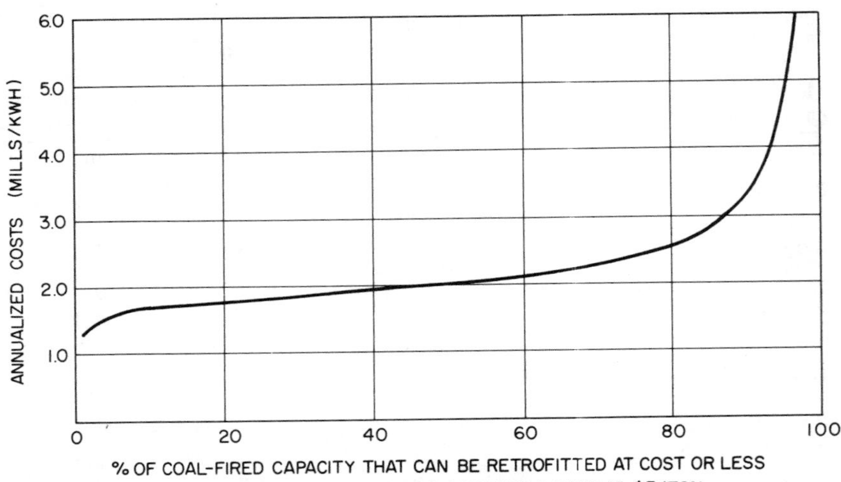

Figure 7.3. Distribution of annualized costs over utility population, limestone scrubbing. (Source: J. K. Burchard, "Some General Considerations of Flue Gas Scrubbing for Utilities," in *Proceedings—Sulfur and Utility Fuels: The Growing Dilemma* (New York: McGraw-Hill, 1972), p. 120)

increasing marginal costs from zero to 75 percent of capacity and sharply increasing marginal costs beyond 75 percent.

We might infer from this information that sulfur dioxide control at a particular source can be achieved only at one or two different abatement levels with current technology. There is consequently no basis for estimating a marginal cost of abatement curve for a single source. There is no empirical evidence to suggest what shape such a curve would have if it existed. It does appear reasonable, however, that an aggregate marginal cost of abatement curve for a geographic area containing a number of kinds of sources should display increasing marginal costs if the pollution control policy is one that would begin with the least-cost sources and move upward on the cost curve. A less efficient policy, of course, might display constant marginal costs.

The above data also provide some basis for estimating the magnitude of sulfur dioxide control costs. The point cost estimates all display a variation in cost by a factor of 2 or more; thus any estimates derived from these figures must necessarily be quite imprecise. Burchard's equation and the PEDCo study, however, provide an opportunity for reducing the unexplained variation in cost by specifically including a variety of cost factors.

Thus it may be possible to estimate control costs within 15 or 20 percent rather than a factor of 2.

The issue that has not been illuminated by these data is technological progress. Methods for sulfur dioxide removal from stack gases have been known in principle for some time, but only during the last decade have large-scale installations been made that can lead to the development of improvements and cost reductions in this technology. If policies are adopted that encourage or force the installation of large numbers of sulfur dioxide scrubbers over the next decade or two, it would be reasonable to expect that research and development would lead to substantial improvements in these processes. Maximum efficiencies should rise and costs should fall.

The problem is that there is no way to predict accurately the rate at which this progress could occur. The economic theory of technological progress is not sufficiently developed to permit quantitative predictions of the rate of development of a specific technology for which historical experience is available. We can only assume that changes will take place and will be related to the effort expended in making them.

In a situation in which current control costs are high, the possibility for technological progress is substantial, but progress will depend heavily on research efforts aimed at this specific problem. It is particularly important, therefore, to evaluate alternative regulatory policies with respect to their ability to encourage technological progress. This appears to be a case in which the effluent charge has special advantages. It would not be reasonable to demand that old and new facilities install 80 percent effective controls within a specific short time horizon, because of the unknown capacity of the capital goods industry to supply the equipment and the possibly very high cost of requiring all sources to meet a uniform standard. However, an effluent charge could be levied that would have an entirely predictable maximum impact on the firms, based upon current emission rates and the magnitude of the charge. If the charge set were about the average cost of current technology for abatement, then only those firms whose costs were lower than average would proceed with abatement devices at once. All firms, however, would have a continuing incentive to support the research and development that would lead to abatement techniques that were less expensive than the effluent charge. Thus a maximum impetus for technological progress could be achieved with a predictable impact upon the industry. Furthermore, while technology was under development, the price of electricity would rise to reflect the social cost of the damage done by the pollution. This price increase would reduce the quantity of electricity demanded and therefore the damage done in the meantime. Finally, the emissions and abatement data generated by the collection of such a charge over a period of time would provide accurate information about the cost of controlling sulfur oxide emissions. Instead of the costly and confusing congressional inquiries as to the "technological feasibility" of pollution control, observation of the rate at which orders were placed for control equipment would indicate the current costs of the technology.

All of the discussion so far has concerned the cost for tail-end treatment—the cost of removing sulfur dioxide from stack gases. It is also possible to reduce sulfur dioxide emissions by removing sulfur from the fuel before it is burned, in other words, by burning a low-sulfur fuel. Depending upon market conditions, in some cases it may be less expensive to purchase low-sulfur coal or to desulfurize oil than it is to install stack gas scrubbers. Thus cost estimates based upon stack gas scrubbing alone are likely to overestimate actual costs incurred in a cost-minimizing abatement program for an area or a country.

Automobile Emissions

Much of the theoretical analysis of economic issues on pollution control assumes that the decision maker will have available individual or aggregate marginal cost and marginal benefit of abatement curves for the relevant pollutant. The decision maker presumably determines the range of environmental quality where marginal costs and marginal benefits are roughly equal, identifies a policy that will lead toward that range, calculates income distribution effects, regional economic dislocations, and other secondary effects, and then chooses the policy that in the long run, if not in the short run, will lead to the optimal degree of pollution abatement. In this process, the level of pollution control is one dictated by economic efficiency, and the means of achieving it are designed to do so at least cost.

The extent to which real-world decision making differs from this theoretical model may be seen by examining automobile pollution control in North America. In 1970, federal new-car emission standards were set for 1975 and 1976 with no knowledge whether technology would be available to meet those standards. After the standards had been established, a committee was funded to determine the technological feasibility of meeting the standards and to report to Congress on that feasibility. This committee, the Committee on Motor Vehicle Emissions (CMVE) made several reports between 1972 and 1975. It periodically assessed the "feasibility" of meeting specific standards by specific dates and incidentally calculated some expected costs of meeting those standards. When costs appeared high or feasibility seemed questionable, standards were postponed for one or more years. At present, the standard originally legislated for 1976 has been postponed beyond 1978 and may well be postponed by several more years.

We will examine the automobile pollution case by presenting the data that were available at different times to the congressional decision makers. In this way we can see what information was available at a specific time and can compare the evolution of that information with the data that would be useful for sensible economic analyses. This procedure is adopted not as a criticism of the CMVE, but rather to illustrate the difficulty of gathering necessary information and the even greater difficulty of determining exactly what questions should be addressed by the data-gathering body.

The Cost of Meeting a Standard

The Environmental Protection Agency was directed by the Clean Air Act Amendments of 1970 to request the National Academy of Sciences to establish the Committee on Motor Vehicle Emissions (CMVE). The CMVE was to report on the technological feasibility of achieving emission control standards established by the Clean Air Act. The CMVE interpreted "technological feasibility" to mean "that an emission control system capable of meeting the standards set for the three major pollutants can be developed, designed, produced in large numbers, and maintained in service, all at reasonable cost."[16] Thus there was a recognition by the committee that the concept of feasibility incorporated some consideration of monetary costs.

For its 1974 report, the committee undertook a study of the technology that would be required to meet future emission standards and the cost of manufacturing each component in that technology. The result of these calculations was the first column of figures in table 7.5. These figures show the increase in sticker price as compared to an uncontrolled car for meeting each of the legislated standards through 1973, the originally legislated 1975 standard (which became the interim 1976 standard), and the original 1976 standard, which was later postponed to 1977 and again to 1978.

The CMVE recognized that sticker price increases do not reflect the total cost of emission controls. They identified poor driveability (stalling, hesitation, poor operation when cold) as a unquantifiable cost of emission controls and noted that increased maintenance and fuel costs were important. Estimates were made of the cost of several alternative technologies for meeting the final 1976 standard (here referred to as the US77 standard), including operating and maintenance costs. Although their estimates suggested that fuel economy penalties of 10 to 30 percent might be suffered as a result of

Table 7.5 CMVE Cost Estimates

Emission Standard	February 1973[a] Cum. Total Price	November 1974[b] Cum. Total Price	Life Total All costs
Precontrol	0		
1958	18	0	0
1970	26	0	0
1972	40		
1973 3.0, 28, 3.1 (HC) (CO) (NO_x)	100	51	557
I75 1.5, 15, 3.1		123	265
I76 0.4, 3.4, 2.0	259	193	431
US77 0.4, 3.4, 0.4	393	377	531

Sources: CMVE 1973, Table 5-3, p. 94; CMVE 1974, Table 6-4, p. 89.
[a]In 1972 dollars.
[b]In 1974 dollars.

the 1976 standards, there is no explicit presentation of annual or lifetime costs for each of the several standards presented in table 7.5. Perhaps this failure to focus on total costs can be explained by the fact that the CMVE was not supposed to evaluate which of the standards was most economical, but rather to estimate the cost of meeting the legislated set of standards.

In any event, there is no marginal cost of pollution abatement calculation among the various standards, nor are there sufficient data presented in the report to make such a calculation. Even if one had a measure of total lifetime costs for vehicles meeting the different standards, one would need an index of emissions to calculate marginal costs per percent reduced. Because the automobile legislation regulates hydrocarbons, carbon monoxide, and oxides of nitrogen, and reduces these three pollutants by different amounts in each standard, deriving a single emission index is not a trivial matter. Furthermore, the basis for measuring emission rates changed in the early 1970s, so early standards cannot be compared with later standards on a direct basis. Needless to say, it would be impossible to derive a marginal abatement cost curve from the data in the 1973 CMVE report. Still, that report did give some idea of the probable future cost of meeting a specified set of standards, and because the increase in sticker price, even for the most stringent standard, was no more than 15 percent of the price of a new car, the reader might be reassured that costs were not overwhelming.

Another CMVE report was issued in November of 1974.[17] This report calculated the cost increment for meeting a set of standards over the cost of producing a 1970 vehicle. Although it was not stated clearly, there seemed to be an implication that the technology available in 1974 would meet 1968 or 1970 standards at little or no cost. Thus the 1974 CMVE estimates may be taken as a reasonable approximation of the cost increase over an uncontrolled car. The numbers in the second column of table 7.5 show that sticker price increases for the 1973 and interim 1976 standards had decreased significantly from the estimates made in February 1973, despite the change from 1972 to 1974 dollars. When the change in the value of the dollar is adjusted, the price of the 1977 standards has also fallen.

The 1974 CMVE report includes increases in sticker prices calculated from a manufacturing simulation program by Ebner of Boston University and estimates of the change in lifetime operating and maintenance costs. Whereas early CMVE reports treated vehicle life as 50,000 miles, because that was the period during which legislation required vehicles to meet the standard, the 1974 report recognizes that the average North American automobile lasts for almost 100,000 miles. Thus, lifetime costs are based on a ten-year, 100,000-mile life. The computation of these lifetime costs, in the third column of table 7.5, shows that operating and maintenance costs are not trivial; in fact, they are greater than the increase in sticker price for all except the 1977 standard. Particularly noteworthy is the tremendous operating cost increase for the 1973 vehicle—$500 out of $557—attained primarily to the increased fuel consumption for vehicles that had to meet the reduced nitrogen oxide standards in that year. To ignore fuel costs for this computation would be to ignore 90 percent of the total cost of meeting the standard. Although the fuel penalty had been recognized in the previous CMVE report, it was not well integrated into the cost presentation and therefore was probably not reflected in decision making until the publication of the 1974 report.

The procedures for calculating the 1974 CMVE costs were developed jointly by the Committee on Motor Vehicle Emissions and the Committee on the Costs and Benefits of Automobile Emission Control, another committee of the National Academy of Sciences. The published reports contain some differences in technology and costs, but they are generally consistent. It is therefore possible to use the data presented in both the CMVE report and the Cost Benefit Committee (CBC) report.[18]

Table 7.6 shows in the fourth column the total cost over the lifetime of an intermediate-size vehicle of meeting the legislated emission standards on the schedule that existed in late 1973 or early 1974. The standard is identified in column 1. The third column in table 7.6 is an index of pollution abatement. The index is the same one used by Dewees.[19]

For each standard, the percentage by which each pollutant has been reduced from its uncontrolled level is calculated. A simple average of these three percentages (lead is ignored) is calculated as the net percent controlled (column 3). The assumption that a given percentage reduction in each pollutant should be weighted equally is obviously arbitrary, and a different index could be derived by a different weighting function. Column 5 of table 7.6 shows the average cost of abatement, which is the lifetime cost divided by the net percent controlled.

The last column of table 7.6 shows the marginal cost of abatement from one standard to another. It is calculated as the change in lifetime cost divided by the change in net percent controlled. The striking feature of this marginal cost column (which is presented graphically in the CBC Study,

Table 7.6 Lifetime Costs of Standards on Schedule (with tax)

Standard[a]	Year	NP Controlled (%)[b]	Total Lifetime Cost[c] (1974 $)	Average Lifetime Cost/NP[d] ($/%)	Marginal Lifetime Cost per NP[e] ($/%)
1968	1968	17	18	1.06	1.06
1970	1970	23	26	1.13	1.33
1972	1972	37	275	7.4	17.8
1973	1973	52	753	14.5	31.9
I75	1975	63	265	4.2	−44.4
I76	1976	80	473	5.9	12.2
US77	1977	94	844	9.0	26.5

[a]The name given to the standard in the CBC study.
[b]Net percent (NP) controlled, assigning equal weight to HC, CO, NO_x.
[c]From CBC, Chapter 2, Table 2-9, p. 64.
[d]Lifetime cost divided by net percent (NP) controlled.
[e]Change in lifetime cost divided by change in net percent (NP) controlled.

Figure 2-1, p. 69) is not only that it fails to rise continuously with increasing degrees of abatement, but also that it is sharply negative for the interim 1975 standard. It highlights the $500 reduction in lifetime cost between the vehicle meeting the 1973 standard and the vehicle meeting the interim 1975 standard. Naturally such a cost reduction allocated over an increase in pollution control yields a cost saving for that pollution abatement. This saving is greater than any possible error in the cost estimates, and is therefore significant.

Have we then found a case in which more stringent pollution control results in substantial cost savings? Is cleaner really cheaper? The answer obviously must be no. The drop in costs between 1973 and 1975 is a direct result of the adoption of catalysts to render ineffective pollutants generated by the engine. In 1973, carburetion and timing were adjusted to reduce oxides of nitrogen, an adjustment that greatly reduced the fuel economy of the vehicle. In 1975, the engine could be adjusted for reasonable fuel economy and power by using catalysts to burn up the pollutants. The tremendous fuel savings that this measure allowed far more than offset the cost of the catalysts themselves. In short, technological progress, not a more stringent standard, reduced the cost of pollution control in 1975.

It is therefore inappropriate to calculate or draw a marginal cost curve using the figures in the last column of table 7.6. A true marginal cost of abatement curve should reflect changes in total cost with variations in emission rate when all other factors are held constant. The factors to be held constant should include the menu of available technology. When we consider an emission standard for 1975, the question should be, "What would it cost to meet different standards given the technology that will be available in 1975?" It is of little interest to compare the cost of meeting the 1973 standard with the 1973 technology against the cost of meeting the 1975 standard with 1975 technology. Clearly, with 1975 technology available, if one had only to meet the 1973 standard in 1975, he would use some modification of the 1975 technology rather than the less efficient 1973 technology. It would be unreasonable to produce a 1975 car for which emission controls cost more than $265 when meeting a 1973 standard because that cost would buy the 1975 standard.

Costs as a Function of Standard and Time

The rapid change in automotive emission control technology that causes the peculiarities of table 7.6 introduces a methodological problem in cost calculation that was not found in the analysis of particulate controls. When technology is changing rapidly, it is important than any cost fighures be associated with a particular technology or a particular time. Any marginal cost calculation should be carefully designed so that only the rate of emission changes from one standard to another. Although different technology might be used to meet different standards in a given year, the menu of technology available for all points plotted on a single graph should be the same.

Consequently a marginal cost of abatement curve for automobile emissions must be associated with a particular time. Because a variety of standards have been set for different points in time, we should expect to find full information about automobile emission control costs displayed in a family of marginal cost abatement curves. Each curve could reflect costs in a particular year and different points on the curve could reflect different degrees of control possible for that year. Table 7.7 shows cost data separated both by year and by emission standard. The total cost data from table 7.6 are displayed on the diagonal of table 7.7, which represents the cost of each standard in the year in which it was scheduled to be adopted. In addition to these data, the CBC and the CMVE provided some information on technology that would be available after the date on which a standard was scheduled to appear. For example, the interim 1975 standard was met at a cost of $265 in 1975 with technology configuration number 39, a conventional engine with catalyst controls. The CMVE indicated that by 1977, this standard could be met by what it described as a ''lean burn system'' at a cost of only $134. The CBC indicated that by 1985 this interim 1975 standard could be met by a stratified charge engine at no increase in cost over a 1970 emission standard. Similar information led to the later figures for the interim 1976 and final 1977 standard costs. Table 7.7 also includes some estimated costs involving linear interpolation for years between those for which costs are known.

We can take the cost of meeting each standard at various points in time from table 7.7, combined with the percentage control reflected by each standard, to calculate marginal abatement costs in each year. These marginal costs are displayed in table 7.8. A single row in table 7.8 shows the marginal cost of meeting different standards in a given year, choosing among the menu of technology that is assumed to be available in that year. The data in table 7.8 are plotted in figure 7.4. Curves have been drawn to connect points showing marginal costs for a given year. Thus each curve in figure 7.4 is a true potential marginal cost of abatement curve for a specific year. In all cases these curves show increasing marginal costs as a function of pollution abatement. In addition, they show the downward or rightward shift of the marginal cost curve over time. In 1973, the marginal cost of control ex-

Table 7.7 Costs of Standards at Different Dates[a] (1974 dollars, gasoline @ 55¢/ gal)

	Standard[b]						
Year	1968	1970	1972	1973	I75	I76	US77
1968	$18						
1970	15 (est)	26					
1972	18 (est)	15 (est)	275				
1973	5 (est)	10 (est)		753			
1975	0 (See CBC, p. 70)	0		<265	265 (339)		
1976	0	0		<200 (est)	200 (est)	473 (#53)	
1977	0	0		<134	134 (CMVE p. 89) (LBS)	419 (est)	844 (#74)
1980	0	0		<84 (est)	84 (est)	259 (CMVE, p. 89) (LBS)	500 (CBC, p. 70) (#144,151)
1985	0	0		0	0 (CBC, p. 73)	0	0 (CMVE, p. 89) (#158)

[a] All costs after 1970 are relative to a 1970 vehicle and thus assume that the 1970 standard could be reached at no cost. When it is shown that the I75 or I76 standards become costlier, we ignore the fact that, with such technology, the 1970 standard could probably be met at a still lower cost, so the I75 or I76 standard *in fact* is not costlier, as shown here.

[b] The names of the standards (I75, I76, US77) are those used by the CMVE and CBC. I refers to an interim standard.

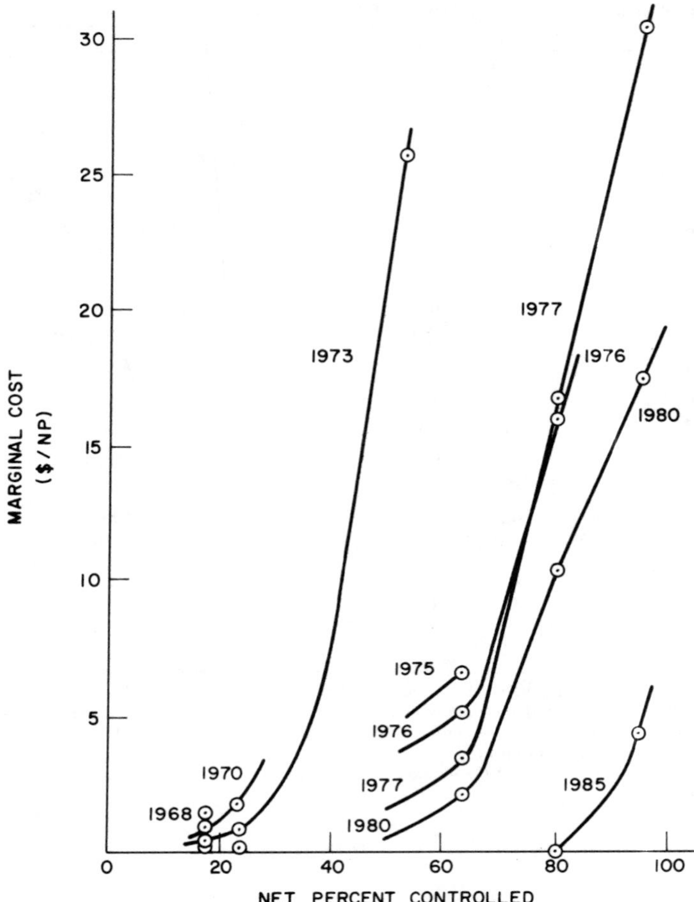

Figure 7.4. Marginal costs of standards at different dates. (Source: From table 7.8)

Table 7.8 Marginal Cost of Standards at Different Dates (1974 dollars, gasoline @55¢/gal)

Standard Net Percent	1968 17%	1970 23%	1973 52%	I75 63%	I76 80%	US77 94%
1968	$1.06					
1970	0.88 (est)	1.83				
1973	0.29 (est)	0.83 (est)	25.6			
1975	0	0		6.63		
1976	0	0		5.00	16.1 (est)	
1977	0	0		3.35	16.8 (est)	30.4 (est)
1980	0	0		2.10	10.3 (est)	17.2
1985	0	0	0		0	4.29

Source: From table 7.7.

ceeded $25 per net percent at a level of approximately 50 percent abatement. By 1977, it is expected that the $25-per-net-percent marginal cost will not be reached until the 90 percent abatement level. By 1980 or 1985, the $25-per-net-percent cost will be reached at well over 95 percent abatement. Thus these data indicate a rapid shift to the right of the marginal abatement cost over time. They also confirm a reasonably well-behaved shape for the marginal cost curve in any single year.

Because I earlier advocated what amounts to an effluent charge for automobile pollution control,[20] I am tempted to draw a horizontal line on figure 7.4 that reflects a reasonable level of effluent charge. If this line were there, one might observe the intersection of such a line with the marginal cost curves and infer the degree of pollution control that would have been achieved in different years under an effluent charge. To do so, however, would be to draw unwarranted inferences from even these data. The inference about behavior under an effluent charge would be unwarranted because we could not be certain that the technological progress displayed in figure 7.4 would have taken place at exactly the same rate if the regulatory system that has been applied were replaced by an effluent charge system.

The problem is our limited understanding of the mechanism that causes these marginal cost curves to shift downward over time. There are those who argue that the very strict federal emission standards have driven the auto companies into a frenzy of activity to try to meet the standards on schedule. Although, in fact, they have not been met on schedule, it is argued that tough standards and a threat of serious penalties are the best spur to technological progress.

Unfortunately, there is an equally compelling argument on the other side. If it appears that the standards may be more strict than technological capability can meet, there is a powerful temptation to argue for reductions or delays in the standard and to conceal evidence of progress in abatement technology. Our present regulatory system, which consists of asking the auto manufacturers how clean they can build a car and requiring them to do somewhat better, creates many incentives to move cautiously. An effluent charge system, on the other hand, would provide a continuous incentive to develop and utilize technology as rapidly as possible. There is no advantage to postponing the adoption of some device under an effluent charge system, because the magnitude of the charge depends only upon the rate of emission.

I personally do not see a way to resolve these and other conflicting theories that might explain the rate of technological advance. One can only say, however, that the form of regulation must have some impact on technological development, and therefore one could not take data that had arisen under a particular regulatory form an extrapolate those data to some other regulatory form. Far more research is necessary before we can estimate what degree of abatement would have resulted in each of the years of interest had the present regulations been replaced by an effluent charge. If I express my personal belief that a charge would have led to more rapid development, I am doing no more than expressing a faith in the principle that a steady incentive would be more productive than a game-playing situation. That article of faith, however, cannot be verified by any existing empirical information.

The interpretation of the curves in figure 7.4 is actually somewhat narrower than has already been indicated. Each curve reflects marginal costs that have been estimated to be in existence at different points in time. The menu of technology available in a given year is a function of the research and development that has been done previously. Clearly, it is thus influenced by the standards or regulatory policy that have gone before. Much of the technology that is now being applied to automobile emission control has been understood at least in principle for decades. The stratified-charge engine, fuel injection, the diesel engine, lean-burn engine, and other devices that are currently being employed to regulate the combustion process more precisely were known in the 1920s and 1930s. Why have the manufacturers now removed these devices from the shelf, dusted them off and installed them in current vehicles? More important, how long does it take to do it? It may be that development of an entirely new pollution control system is a risky business that might take decades. However, designing and building a stratified-charge engine to meet a particular emission standard may take only three to five years because the principles of operation of such engines are well understood. Thus, although the curves represent ''dusty'' technology, they might all have been advanced ten years if the emission control program had begun ten years earlier. To the extent that technology involved in these curves is a result of years of recent research, however, an accelerated standards program might have little effect on the year in which the technology would be ready. Thus the speed at which marginal cost curves such as those in figure 7.4 can be advanced may depend very much upon whether the necessary research is fundamental research or whether it is developmental research to apply known devices to current vehicles. Perhaps one reason why these curves have shifted so far and so fast is that there was an abundance of dusty technology waiting to be used. Thus we could not extrapolate from the automobile experience to the sulfur dioxide problem, for example, without understanding whether the technological problems of sulfur dioxide control were of a fundamental or developmental nature.

Cost Allocation to Individual Pollutants

It is also possible to calculate marginal costs for HC and CO together or for NO_x when only the pollutants in question change. Thus moving from standard C75 to I76 reduces HC and CO but not NO_x.[21] The same is true for T177 to T277. Conversely, only NO_x changes between C75 and T177 or between T277 and US77. These data allow a comparison of the cost of reducing HC and CO together with the cost of reducing NO_x. Because the CMVE does not indicate the date at which the various technologies will become available, we will compare only configurations based on the same technology, which would presumably be available at about the same time.

Table 7.9 shows changes in lifetime costs from varying pollutant levels separately. The top half of the table shows that HC-CO reductions cost $27 per percent (of original emission rates) reduced when NO_x is at 2 grams per mile, and $26 or $6 per percent reduced when NO_x is at 1 g/mile. The large

Table 7.9 Marginal Control Costs for HC-CO and for NO_x

Original Standard[a]	Reduced to	Percent Reduced (%)	Lifetime Cost Increase ($)	Marginal Cost ($/%)	Technology
HC-CO reductions					
C75 (0.9,9.0,2.0)	I76 (0.4,3.4,2.0)	4	107	27	Catalyst
T177 (0.9,9.0,1.0)	T277 (0.4,3.4,1.0)	4	105	26	Dual cat
		4	24	6	3-cat
NO_x reductions					
C75 (0.9,9.0,2.0)	T177 (0.9,9.0,1.0)	8.33	429	52	Catalyst
T277 (0.4,3.4,1.0)	US77 (0.4,3.4,.41)	5	253	51	Dual cat
		5	103	21	3-cat
		5	248	50	CCS

[a] C75 = 1975 California standard; 176 = the early interim 1976 standard; T177, T277 = two intermediate standards considered by the CBC and CMVE.

difference between the last two numbers reflects quite different technology whose availability date is uncertain.

The bottom of table 7.9 shows the cost of NO_x control to be $52 per percent reduced when HC and CO are at the 0.9, 9.0 g/mile level, and between $51 and $21 per percent when HC and CO are at the 0.4, 3.4 g/mile level. The three cost figures for moving from the T277 standard to the US77 standard are for a dust catalyst system, a three-way catalyst system and for a stratified-charge engine. They are presented in the chronological order in which it was expected that they would become commercially available when the CMVE report was written.

What inferences may be made from the data of table 7.9? It has been argued that these data show that, starting with the 1975 California standards, NO_x control is much more expensive than HC-CO control; the former imposes lifetime costs of up to $929, but the latter never costs more than $107. In this view, we should press on to the 0.4, 3.4 g level with HC and CO and focus the analysis upon how far to go in NO_x control, given its much higher cost.

Another view is obtained by focusing not on total cost for each pollutant but on the marginal cost per percent reduced. Here NO_x control is only twice as expensive as HC-CO control at all levels or for any particular technology, despite the fact that HC and CO are already controlled almost to the 90 percent level in the C75 standard and NO_x is only 50 percent below its precontrol level. Thus, if the benefits of HC-CO control and NO_x control are close to equal per percent reduced, one would not be justified in reducing NO_x below the C75 level of 2 g/mile unless the final 0.4, 3.4 g level for HC-CO were more than justified.

The separation of costs between pollutants therefore enables us to identify the crucial issue in automobile pollution control: how much NO_x control should we buy? Reducing NO_x from 2 to 1 g/mile may cost us as much as reducing HC and CO by 90 percent and the first 50 percent reduction in NO_x combined by means of 1976 or 1977 technology. Because marginal benefits of abatement probably decline with reduced emissions, the constancy of marginal NO_x costs between the two reduction steps shown in table 7.9 does not mean that willingness to go to 1.0 g implies willingness to go to 0.41 g. The two steps must be examined separately in light of the expected benefits.

The separation of marginal costs by pollutant also allows the calculation of marginal costs per pound of pollution reduction. Assuming a 100,000-mile vehicle life, 1 g/mile of emissions equals 220 pounds over the vehicle life. Thus the control of NO_x shown in table 7.9 expressed in dollars per pound abated is: \$1.95, \$1.91, \$.78, and \$1.76. Because stationary sources emit large amounts of NO_x, the crucial decision of how much automobile NO_x abatement to buy should take into account the marginal cost, in dollars per pound, of NO_x control from stationary sources. If stationary-source NO_x control costs less than \$1.00 per pound, it would be foolish to reduce automotive NO_x below 2 g/mile until most stationary sources had been controlled.

Scenario Costs

It should now be clear that it makes little sense to ask, "How much does it cost to meet the interim 1976 emission standard?" At a minimum, one should ask, "How much would it cost to meet the interim 1976 emission standard in a particular year?" But as we have just noted, that cost will depend upon the regulations that have preceded that standard. Furthermore, the costs of emission controls are not incurred all at once. When a new car standard comes into effect, the purchase price is paid in that year, but operating costs accrue over a period of time. As time passes, more and more vehicles on the road will experience the operating cost associated with the new standard. Thus the annual cost of a pollution control program will be related only in a distant way to the cost for a single car to meet a particular standard in a given year.

In addition, ambient air quality does not depend solely upon the emission rates of new cars. In fact, new cars are only 10 to 15 percent of the automobile population and account for less than 20 percent of the miles driven in a given year. Thus total emissions and ambient air quality depend upon the schedule of emission standards over a period of time. Sensible comparisons of costs and benefits, or costs and ambient quality, require examination of the behavior of an entire vehicle fleet over a period of years. This recognition led to the examination by the CBC of a set of scenarios for pollution control.

All the costs discussed so far refer to the cost of building and operating a single car of an intermediate size. To calculate the actual costs to American motorists of different schedules of emission standards, the CBC, Ebner, and the CMVE designed a series of emission standard scenarios. Each scenario consisted of a set of emission standards applied at particular dates. Common to all scenarios was a base vehicle population and new car sales figures for every year from 1970 through 1985. Sales were broken down into six vehicle size classes from subcompact to luxury. Sales by class were identical in all scenarios, so that the only difference among scenarios was the emission standard to be met in each year.

Table 7.10 shows the total costs over the fifteen-year simulation period of four different scenarios. Scenario A applied the legislated standards on schedule up to 1970 and imposed no more stringent restrictions after 1970. All other costs are presented as cost increments over the 1970 scenario cost. Scenario C applied all standards on schedule through the interim 1975 standards, with no further reductions thereafter. This scenario incurred $34.8 billion in additional costs between 1970 and 1985, compared to scenario A. In 1985 alone, scenario C cost $3.01 billion more than scenario A. Thus the scenario cost figures show the marginal cost of going from one set of standards to another.

Scenario E applied the legislated standards on schedule through the interim 1976 standards. It cost $45.6 billion more than scenario A, or $4.68 billion more in 1985 alone. Finally, scenario J applied the legislated standards on schedule through the final 1977 standard. Its cost was $68.4 billion more than that of scenario A, and almost $8 billion more in 1985 alone.

Table 7.10 shows the difference in costs of all motorists between different

Table 7.10 Scenario Costs of Various Standards 1970–1985 (billions of 1974 dollars, excess over scenario A)

Scenario (last standard)	List Price	1970–1985 Costs Maintenance	Fuel	Total	Total, 1985 Only
A (1970)	0	0	0	0	0
C (I75)	16.9	8.5	9.4	34.8	3.01
E (I76)	21.3	8.2	16.0	45.6	4.68
J (US77)	34.7	8.9	24.8	68.4	7.91

Source: CBC, Tables 2-13 and 2-14.

schedules of pollution controls. To choose sensibly between these scenarios, one would need to know the difference in ambient air quality from each period. The CBC study presented a set of curves showing total annual emissions that would result from each scenario. No final marginal cost for pollutant concentration was calculated, but it would be possible, in principle, to make such calculations and arrive at a final marginal cost for the entire set of scenarios.

In addition to examining different final emission standards, as was done in table 7.10, the scenarios were used to evaluate alternative timings for a set of standards. It was found that delaying the 1977 standard to 1980 would save $7 billion over the scenario period but would change ambient concentrations of the three pollutants by relatively small amounts. Thus it was possible to use a set of scenarios to examine the effect of postponing, or accelerating particular standards upon both costs and emissions. Obviously, these calculations required a set of assumptions about what technology would be available at what date. Thus there was implicit a set of assumptions about how the manufacturers would respond to different schedules of emission standards.

The use of scenarios allowed one more important consideration to be introduced into the cost calculation process. Rapid technological progress may be very costly as manufacturers quickly change from production of one engine type to another. Existing equipment may have to be amortized over a shorter lifetime than is normal, higher costs may be experienced in the entire industry as manufacturers attempt to tool up simultaneously for new engine systems. The capacity of the capital goods industry to supply auto manufacturers with new production facilities may be strained and capital costs thereby increased. These problems could not be addressed well in examining cost of individual vehicles. The scenario calculations, however, required assumptions about number of vehicles of each type that would be produced in a given year. Careful consideration of the capacity of the industry to modify its production and the cost of different rates of modification allowed injection of the performance of the capital goods industry into the technological progress in the scenarios. At a minimum, this procedure ensured that the scenario calculations did not incorporate unrealistic assumptions about the rate at which the industry could convert from producing one type of engine to another.

In summary, the scenario results rarely contradicted the impressions given from single-car calculations about cost of meeting alternative sets of standards. They provided far more information, however, about the pattern of costs over time, the effect of different sets of standards on aggregate emissions, and other cost factors. The scenario approach to pollution-control cost calculation seems to be particularly advantageous when substantial technical progress is expected, when that progress would depend upon the standards imposed, and when the costs are seriously divided between initial investment and operating and maintenance costs.

Choice of Technology

Studies of technological feasibility, such as those conducted by the CMVE and the SOCTAP study of sulfur dioxide scrubbing, tend to focus on identifying particular technologies that will be most economical or best suited to achieving a given pollution control objective. Indeed, the CMVE reports spend substantial time discussing, as was their mandate, the relative merits of stratified-charge engines, catalyst systems, fuel injection, and other alternatives. The results of some of these deliberations are shown in table 7.11. Three different technologies are presented for each of three emission standards. The right-hand column of the table, which shows lifetime total costs for each technology, illustrates the dramatic difference in lifetime costs for the various technologies even when meeting a single standard. The cost differences among technologies are frequently three or more times as great as the cost of the least expensive technology. Thus identification of the "best" technology would seem to be an important research task.

This was one case, however, in which the scenario calculations contradicted the single-car results. The standard E scenario, which terminates

Table 7.11 Single-Car Costs of Alternate Technologies (lifetime cost increase over 1970 vehicle, 1974 dollars)

Standard	Technology	List Price	Maintenance	Fuel	Life Total
I75	Catalyst	109	100	56	265
	Conventional	74	325	269	668
	Diesel	131	−75	−955	−899
I76	Catalyst	176	113	184	473
	Stratified-Chg.	201	13	−465	−252
	Diesel	131	−75	−955	−899
US77	Catalyst	302	75	467	844
	Fuel Inject.	359	38	152	548
	Stratified-Chg.	148	225	1591	1964

Source: CBC, Table 2-16, p. 93.

with the interim 1976 emission standard, relied heavily upon catalyst technology to meet that standard. Two variations of this scenario were run. In the first, emphasis was placed upon the stratified-charge engine as a substitute for the catalyst engine. In the other variation, the diesel engine was emphasized. In both cases the engines were introduced as rapidly as it was expected that reasonable diligence and investment by the auto companies would permit. Essentially the same limitations on the rate of investment in new production facilities were applied to both variations of the E scenario.

The result of these investigations of alternate technologies was that the stratified-charge engine scenario saved $5.49 billion over the life of the scenario compared to the catalyst scenario (see table 7.12). The diesel scenario saved $4.07 billion over the same period. This result seems perverse because table 7.10 shows clearly that the diesel is far less expensive than the stratified charge for meeting the interim 1976 standard. Yet, in the scenarios the stratified-charge engine saves more.

The reason for this peculiar result is that the stratified-charge engine achieved 38.2 percent of the new-car market in 1985 whereas the diesel reached only 11.6 percent. This gap occurred because conversion from the conventional internal combustion engine to a stratified-charge engine was

Table 7.12 Scenario Costs of Alternate Technologies to Meet I76 Standard (billions of 1974 dollars, compared to scenario E)

Technology	List Price	Maintenance	Fuel	Total	Market Share (1985)
Catalyst	0	0	0	0	
Stratified-Charge	0.445	0.475	−6.30	−5.49 (saving)	38.2%
Diesel	0.34	−0.285	−4.08	−4.07 (saving)	11.6%

not terribly expensive. Only cylinder heads and carburetors had to be seriously modified. Thus a given investment in new production facilities would allow a substantial conversion within the limited time.

The diesel engine, however, is different from top to bottom. All engine production lines would have to be changed, not merely those for cylinder heads and carburetors. The production of fuel-injection equipment would have to be enormously increased over present levels. Thus the same investment in new facilities yielded little more than one-quarter of the production capability for this new technology by 1985. Because the diesel market share was so small, its cost savings were not as important as those of the stratified charge.

Out of this experience comes a potentially useful rule of thumb. The diesel saves much more money per vehicle because it is radically different from the spark-ignition engine, but the stratified charge is only modestly different. The very reason for the diesel's attractiveness is also a barrier to its rapid introduction. Radical improvements involving radical changes may be either very expensive or very time-consuming or both. Thus, as between a revolutionary and a nonrevolutionary improvement, a given rate of investment may yield roughly equal improvement in overall performance of the fleet. During the transition period, it makes little difference which technology is emphasized. Only in the very long run, after all production has shifted to the new technology, will the advantages of the more radical change be fully realized. This is not to say that the concern of the CMVE or of SOCTAP to identify a "best" technology is misplaced. Rather it suggests that choice of technology may be of secondary importance for determining short-run costs of environmental improvement. Although this statement cannot be true for all technology, it may be true for leading contenders.

Summary

What can be inferred from the cost information available for automobile pollution control? First, with regard to the shape of marginal cost of abatement curves, it seems clear that they tend to be upward sloping over an important part of the abatement range. With regard to the level of abatement costs, they clearly change substantially over time. Thus the identification of marginal cost curves with specific times is crucially important.

Perhaps the most important, the automobile case shows the emptiness of the concept of "pollution control costs" compared to the concept of "marginal costs per unit of abatement at a particular time." Calculation of a single set of costs can really only determine whether the standard for which those costs are calculated will be "too expensive for us." Calculation of marginal costs of abatement at various points in time will permit one to consider carefully the design of a sensible policy. And where technological progress may be expected to relate to the policies themselves, cost calculations for different series or scenarios of emission standards can be of crucial importance to highlight the true magnitude and timing of costs over a period of substantial policy advance. It must be remembered, of course, that all of these calculations are only estimates and may be subject to substantial errors. The presentation of three significant figures in a table does not indicate the accuracy of a measurement but rather the expected value of a probability distribution with a wide range.

Finally, these calculations identify NO_x control as the most expensive portion of the auto-emission-control program, and therefore setting NO_x levels is the most important decision. This decision cannot be made sensibly without simultaneously considering the cost of controlling stationary NO_x sources.

Conclusions

What generalizations can we make after reviewing such different fields of pollution-control costs estimation? It is clear that, when decisions have to be made, the data will frequently be inadequate to perform the kinds of analysis that would lead to the best possible decision making. Up to a point, there is little that can be done about this problem. Frequently the desired cost information will develop only as a result of whatever policy is chosen, so that postponing policy decisions until costs are available may be to postpone policies forever.

Several generalizations about costing can be drawn from the above review, however. First, one should not expect to go to any existing literature and find empirical estimates of marginal cost curves for particular sources. Most available cost estimates seem to be point estimates or, occasionally, estimates of the costs of achieving several different degrees of pollution control. Construction of a marginal cost of abatement curve will usually require combining data from several studies and undertaking a variety of calculations. However, only in cases in which experience is insufficient to yield any cost estimates or in which all the technology achieves the same level of abatement will it be impossible to derive a marginal cost of abatement curve from existing data.

A second conclusion is that the assumption of an increasing marginal cost of abatement curve seems to be empirically valid at least for significant ranges of abatement level. One must be prepared, however, to find constant marginal costs over significant ranges of abatement levels. And one should not base important decisions on an untested assumption that there is no portion of the marginal cost curve that is declining. One may not want to operate in that portion of the cost curve, but it may exist.

Another principle is that total, average, and marginal costs of abatement are likely to vary tremendously among sources, depending upon factors such as size, age, and location. Scattered estimates of abatement costs that do not provide details on these variables cannot be related to each other and are nearly worthless. In many cases, however, it will be possible to discover tables or functions that indicate the range of variation according to a number of important parameters. Thus it would be possible to estimate abatement costs for a variety of sources, although such calculations will inevitably be expensive.

It is clear that estimates of abatement costs are particularly difficult when technological progress has taken place or is anticipated. Here one must be very careful to isolate the time element from the degree-of-abatement element and estimate marginal abatement cost curves separately for different points in time or menus of technology. Furthermore, when abatement policies are expected to have a significant influence on technological progress, it would be most useful to develop a set of scenarios with different schedules of policy and the resulting associated costs. The scenario approach may be particularly important when it is desired to estimate the separate impact of capital and operating costs over time or to ascertain the portion of the industry that might be affected in a given period.

Above all, unless one is talking about a single source, abatement costs depend upon abatement policies. Substantial variations in cost among sources may cause the identification of policies that will achieve the desired abatement at least cost to be vitally important to the total cost of abatement.

Notes

1. Allen V. Kneese and Blair T. Bower, "Causing Offsite Costs to be Reflected in Waste Disposal Decisions," in Robert Dorfman and Nancy S. Dorfman, eds., *Economics of the Environment* (New York: W. W. Norton, 1972), p. 137.

2. Ibid., pp. 135–154.

3. Donald N. Dewees, Carol K. Everson, and William A. Sims, *Economic Analysis of Environmental Policies* (Toronto: University of Toronto Press, 1975), Chapter 2.

4. Paul MacAvoy, "Comment," in Edwin S. Mills, ed., *Economic Analysis of Environmental Problems* (New York: National Bureau of Economic Research, 1975).

5. J. R. O'Connor and J. F. Citarella, "An Air Pollution Control Cost Study of the Steam Electric Power Generating Industry," *Journal of the Air Pollution Control Association* 20 (1970): 283–288; and Paul A. Downing, *The Economics of Urban Sewage Disposal* (New York: Praeger, 1969).

6. Martin L. Weitzman, "Prices vs. Quantities," *The Review of Economic Studies* 41 (October 1974): 477–491.

7. National Academy of Sciences, Coordinating Committee on Air Quality Studies, *Air Quality and Automobile Emission Control,* vol. 4, Senate Committee on Public Works (Washington, D.C.: U.S. Government Printing Office, 1974), Chapter 2 (hereafter cited as the CBC Study).

8. U.S. Environmental Protection Agency, *Control Techniques for Particulate Air Pollutants* (Research Triangle Park, North Carolina, January 1969).

9. National Air Pollution Control Administration, U.S. Public Health Service, U.S. Department of Health, Education, and Welfare, *Air Quality Criteria for Sulfur Oxides* (Washington, D.C.: U.S. Government Printing Office, January 1969).

10. CBC Study, Chapter 3.

11. Sulfur Oxide Control Technology Assessment Panel, *Final Report of the Sulfur Oxide Control Technology Assessment Panel to the Federal Interagency Committee of Evaluation of State Air Implementation Plans* (Washington, D.C.: U.S. EPA Office of Science and Technology, 1973), APTD-1569.

12. National Air Pollution Control Administration, U.S. Public Health Service, U.S. Department of Health, Education, and Welfare, *Air Quality Criteria for Sulfur Oxides* (Washington, D.C.: U.S. Government Printing Office, January 1969).

13. A. V. Slack, H. L. Falkenberry, R. E. Harrington, "Sulfur Oxide Removal from Waste Gases," *Journal of the Air Pollution Control Association* 22 (1972): 159–166.

14. J. K. Burchard, ''Some General Considerations of Flue Gas Scrubbing for Utilities,'' in *Proceedings—Sulpher and Utility Fuels: The Growing Dilemma* (New York: McGraw-Hill, 1972), pp. 91–124.

15. PEDCo-Environmental, *Flue Gas Desulfurization Process Cost Assessment,* Report to the Office of Planning and Evaluation, U.S. EPA, May 1975.

16. National Academy of Sciences, *Report by the Committee on Motor Vehicle Emissions* (Washington, D.C.: National Academy of Sciences, 1973), p. 1.

17. National Academy of Sciences, *Report by the Committee on Motor Vehicle Emissions* (Washington, D.C.: National Academy of Sciences, 1974).

18. CBC Study.

19. Donald N. Dewees, *Economics and Public Policy: The Automobile Pollution Case* (Cambridge, Mass.: MIT Press, 1974).

20. Ibid.

21. National Air Pollution Control Administration, U.S. Public Health Service, U.S. Department of Health, Education, and Welfare, *Air Quality Criteria for Sulfur Oxides,* (Washington, D.C.: U.S. Government Printing Office, January 1969), Table VI-4.

Comment

John B. Heywood

I find two elements in the Dewees paper especially welcome. First, his pragmatic approach to estimating the costs of pollution control—there are many practical suggestions for making the fullest use of the limited cost data likely to be available. Second, he has introduced one of the important dynamic elements of the process—technological change—into his economic analysis in notable contrast to many previous treatments of this topic. In my discussion, I want to develop further some of the issues that technological change introduces. I am going to use the automobile problem as my example.

We can distinguish between two types of technological change: first, change in an established technology as massive development resources are applied to improve product performance to meet new requirements or to meet the same requirements at reduced cost; second, the development of new technology that promises to do the same job as the established technology, only better or cheaper. For example, conventional internal combustion engines with catalysts have now become the established technology; the stratified-charge engine is a new technology.

Let me emphasize here that it is only through the application of massive resources in an engineering development process that these changes occur. The notion that new automobile engine technologies, such as the stratified-charge engine, have been waiting on the shelf for many years is false. While the concepts may have been known for some time, they have been examined only as research and exploratory development ideas. They have not, until recently, started into the extensive engineering development phase that really determines whether they are attractive market prospects or not.

In his analysis of auto emission control costs using the NAS CMVE data, Dewees illustrates the effect the introduction of new engine technologies, which the CMVE judged to be more efficient than the conventional engine technology, can have on these costs. However, though the substitution of new technology would occur over a period of time and would be constrained by the conversion of production facilities, the dynamic processes by which the existing and the new technology develop are not incorporated. Primarily because of this omission, an impression of precision in these costs is left, and it is this impression that I want to challenge.

Consider the total operating cost in cents/mile of a succession of new-model vehicles, at a fixed emission standard, over a period of ten years. The total operating cost here includes initial cost, fuel costs, and maintenance costs. Given existing fuel economy standards, the fuel costs (assuming a fixed fuel price) are likely to change the most. Figure 1(a) illustrates the model used by NAS and described by Dewees. The costs of the conventional technology A and the new technology B are evaluated today. The dynamic element is brought in by estimating the earliest introduction date D for the new technology. But figure 1(b) shows more realistically how operating costs with the conventional engine technology are likely to decrease over time. There is, of course, some uncertainty in these projections as indicated. Figure 1(c) shows how the cost of a new engine technology, say, the stratified-charge concept, is likely to change. The uncertainty is greater because the concept is less well known and defined. Development may increase costs if design changes required in the engineering development impair operation characteristics. Eventually, costs will decrease as development improves the product.

If we overlay figures 1(b) and (c) , as in (d), we see the difficulties. Note that the time scales for change in the established technology and for the introduction of new technology are likely to be comparable. These time scales are also comparable to time scales for the developing control strategies and achieving impact. Thus, all these dynamic processes must be included. Note especially that the uncertainties in the costs of established technology and of the new technology and the difference between costs of the existing and the new technologies are all comparable in magnitude.

I am convinced that a distinction should be made between cost estimates for established technologies and cost estimates where new, not fully developed technologies are compared with each other and with an established technology. For example, for conventional engines with catalysts, a number of careful studies are converging on reasonable estimates of costs of meeting different emission standards. It is important to note, however, that these have been static costs; the fuel penalties have been calculated relative to existing, less controlled vehicles and do not, therefore, incorporate the progress that would have occurred in the absence of emission standard changes. Uncertainty still remains, of course, and an important question for discussion is: How do we carry the imprecision of these cost estimates with the actual costs themselves as they are used in the political decision-making

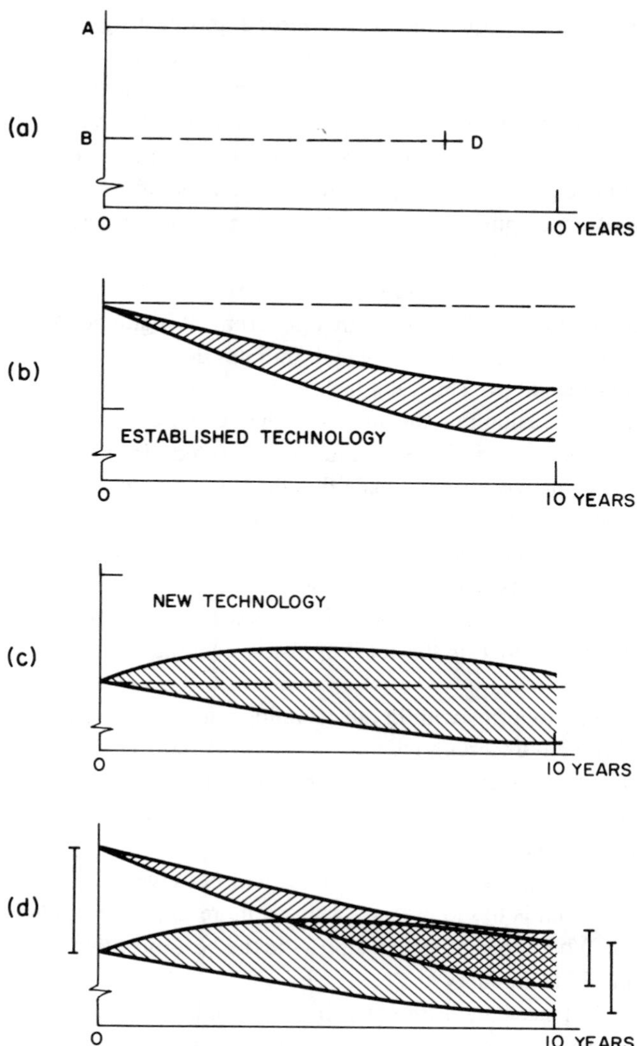

Figure 1. Total vehicle operating cost (cents/mile)

process? (A detailed point here: quoting costs to three significant figures does not help—drastic rounding of the numbers should be done.)

But with new technologies we should be brutally frank about the difficulties. While we can make rough cost estimates, it is usually the *difference* between new and established technology costs that are important. We often cannot estimate that difference with any precision at all. This difference depends on the rate of development of the established technology, when development of the new technology commences, and the rate of development of the new technology once development has started. As Dewees points out, we do not understand the impact that government regulation has on the industrial decision-making process, which initiates development and allocates development resources. Thus, the lack of precision in the CMVE's estimates makes their conclusion that the stratified-charge and diesel engines are cheaper ways to meet the emission standards unwarranted. They might be, but an equally plausible case can be made that they will not be. The uncertainty in the inputs to the calculations is too great, to permit such a definitive conclusion. Such calculations can only define what initial cost and efficiency targets these engines would have to meet to be competitive. The automobile industry disagrees with the NAS CMVE cost differentials. Other independent studies (such as that by the Jet Propulsion Laboratory[1]) have reached different conclusions.

What we have to come to terms with in our planning is that cost estimates of this type carry with them uncertainties often as large as the potential gains and penalties. It is for this reason that many of the alternative engines for automobiles command only limited development resources; they are not demonstrably attractive enough.

Improving the methodology, though worthwhile, will not give more precise answers: the input data is too imprecise. Thus, a more important question for discussion is: What types of systems provide the best incentives for progress with new or old technology?

However, in some instances the differences in marginal abatement costs between different sources are so great that uncertainties in the cost of the new technology and its effectiveness (there are considerable uncertainties there, too) are much less important. Figure 2 shows an aggregate marginal cost curve for NO_x.[2] Different control options have been added together in order of increasing marginal cost. Two automotive strategies are shown shaded. What matters in this context is the substantial change in slope along the curve. Moderate control of new utility boilers is four to five times as cost effective as light-duty vehicle emissions reduction from 2 to 1 g/mile and twenty-three times as effective as LDV emission reduction from 1 to 0.4

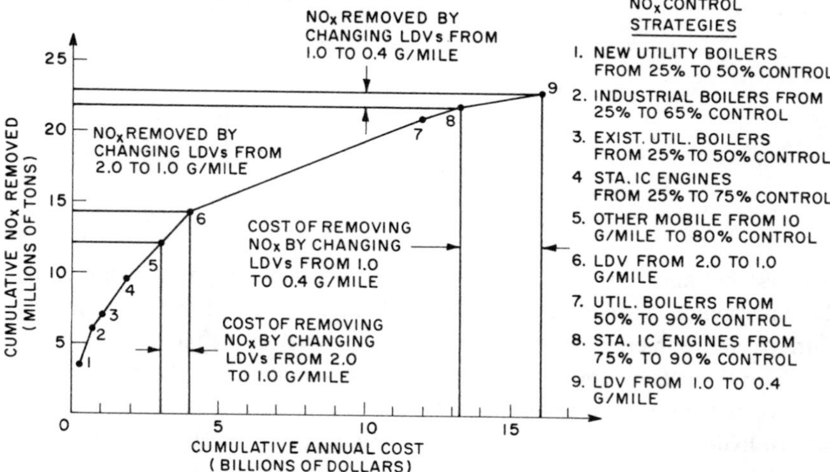

Figure 2. Cost vs. tons of NO_x removed for control strategies in year 2000. Note: Baseline emissions = 34.1×10^6 tons in 2000 and are based upon LDV standard 2.0 g/mile, other mobile 10 g/mile, 25% control of all stationary-source emissions.

g/mile. Despite uncertainties, the cost implications of different emission standards are clear. Yet apparently this information has modest impact on congressional actions.

As Dewees so clearly states, it is this type of information, carrying with it estimates of the uncertainties involved, that we should provide in the hope that more rational political decisions will result.

Notes

1. Jet Propulsion Laboratory, *Should We Have a New Engine: An Automobile Power Status Evaluation,* a report in two volumes, 1975.

2. Energy Resources Council, Task Force Report, *The Report by the Federal Task Force on Motor Vehicle Goals Beyond 1980,* vol. 2, September 1976.

Comment

Adel F. Sarofim

Among the important contributions in Dewees's paper are the illustration of how marginal costs can be inferred from literature data for some important current environmental problems and the delineation of the gaps and uncertainty in the data available. The paper includes examples of the costs of control of emissions of particulates and sulfur oxides from stationary sources and of hydrocarbons, carbon monoxide, and nitrogen oxides from automobile sources. Of necessity in a discussion of such breadth, all the pertinent background cannot be presented and there is consequently a danger that important practical constraints may be overlooked in seeking a simple methodology for cost-benefit analysis. Some specific comments on the factors that complicate the regulation of emissions from stationary sources follow.

Particulates

Dewees chose the case of the control of particulates to illustrate a well-established technology for which costs are well known and for which marginal costs can be derived and projected with confidence. These statements are valid while focus of attention is on the emission of total particulates. There is a growing awareness, however, that adverse health affects attributable to particulates are largely a consequence of the smaller particles in the 0.2 to 5 μ size range, that are preferentially deposited in the upper respiratory tract.

Significant effort is being expended currently on both the better definition of the health effects of particulates and on the development of control technologies that can efficiently collect submicron particles. It is possible that these efforts may lead to new standards and new technologies that will significantly alter the conclusions drawn in the section on particulates.

Dewees provides an interesting example in figure 7.1 of how marginal costs can decrease with increases in collector efficiency. Such curves are sensitive to the assumptions made and would show a more dramatic decrease if attention were limited to particles in the critical micron-size range, because mechanical precipitators are ineffective in collecting smaller particles and would show a relatively higher cost on this basis. The problem of assessing the costs of control of particulates may be further complicated in the future by the recognition that potentially hazardous trace elements concentrate preferentially in the smaller size ranges.[1] A major additional consideration is that the effects of particulates and sulfur oxides cannot be separated and that particle size and composition strongly influence the potentiation effect of particulates on sulfur oxides.[2] Thus, although the discussion by Dewees is a fair assessment of the current technology for the control of total particulates, it is far from clear that projection into the future can be made with confidence in view of the increasing concern about the role of particle size and composition on health effects.

Sulfur Oxides

Dewees selects flue gas desulfurization as an example of an emerging technology for which costs are difficult to estimate and for which estimated costs escalate with time. He provides estimates for scrubbing costs from different studies and makes the important point that many of the variations in costs can be explained if allowance is made for the differences in features of the plants for which costs were developed. Emphasis is placed on the difficulty of constructing a marginal cost curve because data have been supplied only for two levels of control, about 40 percent for dry-limestone injection and about 90 percent for other processes. It can be reasonably argued that costs for different degrees of abatement have not been developed because the 90 percent abatement level has been selected to meet existing emission standards for sulfur oxides with a wide margin of safety. More important, development of marginal cost estimates for one technology may be premature as long as there is still a major question as to which of several alternative control strategies is preferred. There are proponents, for example, for the use of high stacks to disperse sulfur oxides before they reach ground level, an alternative that would be economically attractive if there were no questions regarding long-range effects of sulfur oxides, such as acid rains and low levels of sulfates. The complexity of the problems of making decisions on the abatements of sulfur oxide is reviewed in detail in a report prepared by the National Academy for the Committee on Public Works.[3] The report provides an interesting cost-benefit analysis of alternative emissions control strategy[4] but properly notes that the validity of the conclusions drawn from the studies is severely limited by the uncertainty in the critical variables.

In focusing on the uncertainties in deriving marginal costs for flue gas treatment, Dewees may cause the reader to lose sight of the major uncertainty in other variables that must be considered in any decision on the abatement of sulfur oxide emissions.

Notes

1. R. L. Davidson et al., *Environ. Sci. Tech.* 8 (1974): 1107; J. W. Kaakinen et al., *Environ. Sci. Tech.* 9 (1975): 862; and D. H. Klein et al., "Pathways of Thirty-Seven Trace Elements Through Coal-Fired Power Plants," *Environ. Sci. Tech.* 9 (1975): 973.

2. M. O. Amdur, "Toxicological Guidelines for Research on Sulfur Oxides and Particulates," paper presented at the Fourth Symposium on Statistics and the Environment, National Academy of Sciences, 3–5 March 1976; K. Biersteker, "Sulfur Dioxide and Suspended Particulate Matter—Where Do We Stand?" *Environmental Research* 11 (1976): 287; and M. O. Amdur, J. Boyles, V. Ugrs, M. Dubriel, and D. W. Underhill, "Respiratory Response to Guinea Pigs to Sulfuric Acid and Sulfate Salts," in *Sulfur Pollution and Research Approaches,* a symposium held at Duke University, 27–28 May 1975, in press.

3. Committee on Natural Resources, National Academy of Sciences, National Academy of Engineering, National Research Council, *Air Quality and Stationary Source Emission Control,* report prepared for the United States Senate, Committee on Public Works, Serial 94-4, March 1975.

4. D. W. North, and M. Merkhofer, "Analysis of Alternative Emissions Control Strategies," Ibid., Chapter 13.

Comment

Richard R. Nelson

My comments will be directed mostly toward the Dewees paper, but also will be concerned with the broader question of how economists look at problems of this sort. Dewees's paper is a good example of the general art form. The problem of achieving the appropriate amount of pollution abatement is viewed as an optimization problem with well-defined benefit and cost curves. Behind the benefit curve, individual preferences are assumed to be stable and well grounded, knowable, and summable. Behind the cost curve, technologies are taken as given and known. The formal problem, then, is to produce any given level of pollution abatement at minimal cost and to pick a level of pollution abatement that maximizes the sum of benefits minus costs. Put aside, for the moment, that environmental quality is a public good. Apart from that, the problem has exactly the same form as that of determining the optimal production methods and level of production of peanut butter.

As in the peanut butter problem, the economist views the pollution abatement problem with a certain intellectual schizophrenia. It is recognized that there is an organizational problem hidden behind the maximization problem. The organizational problem stems in part from the fact that knowledge about preferences, and perhaps about technologies, is naturally decentralized and that it would be costly (perhaps even impossible) to get all that decentralized information to some central place and to process it there. The organizational problem also stems from the need to control the actions of many individuals and institutions. The economists' instinctive solution to the organizational problem is somehow to define and to establish a market. A market is viewed as at once generating the relevant information and providing the appropriate incentives. There are the standard arguments that a perfectly competitive market will generate exactly the same outcomes as those the optimization calculations would call for.

Of course the fact that environmental quality is a public good cannot be repressed. There is no way that the demand side of the market can be established on a decentralized basis. There must be a centralized mechanism to calculate benefits and to generate effective demand. Thus there is the well-known problem of somehow getting individuals to reveal their marginal evaluation schedules so that these can be summed. Or because it is widely recognized that gaining explicit revelation may be impossible, research efforts need to be mounted to calculate in some way the benefit schedule. But aside from this demand side problem, we are back to peanut butter.

I worry deeply about this way of formulating the air quality problem. I worry both about the way economists think of benefits and about costs.

While the Dewees paper was not focused on the benefit side, several other papers presented by economists in this volume have been concerned with the calculation of benefits. Both the tone of these papers and the generally derogatory remarks by economists regarding the way in which ''politics'' has entered into the policy-making process suggest that economists like to think of the calculation of benefits as a ''technical'' problem, one regarding which they have a lot to say and one that the voting, bargaining, and general hurly-burly of the political process is likely to screw up. Yet I must say the demand studies by economists have reinforced two beliefs of mine.

One of these is that there is virtually no way that payments from individuals and groups to achieve any given level can be tuned to the benefits they receive. Therefore, regarding any proposed level of expenditure on air quality some people will prefer higher expenditure and better quality and some people will prefer lower expenditure and worse quality. This is as true for a ''technically correct'' solution (one that meets Samuelson's criterion) as for one that does not. Economists like to tell graduate students about Arrow's impossibility theorem but find it convenient to forget about the problem when we pontificate on real matters of public policy. I maintain that, even were individual preferences regarding the level of air quality knowable, stable, and well grounded, reasonable people could disagree about even the meaning of the socially optimal level of air quality.

The problem is worse because it is apparent that individual preferences are hard to know, likely to be unstable, and certainly not well grounded in good understanding of the consequences of different levels of air quality. The results of attempts to measure what people have been willing to pay (for all the reasons discussed in this volume) must be viewed as subject to extremely large margins of error. Attempts by economists and other professionals to assess what people ''should'' be willing to pay are at least as shaky. We really do not know the effects of different levels of air quality on health. Even if we did, reasonable people might disagree dramatically on how much different health states would be worth. And a large portion of the benefits of cleaner air, for many people at least, are aesthetic.

None of this is to deny that the theories and empirical work of economists can contribute significantly to the public and political dialogue regarding the benefits associated with various degrees of cleaning up of the air. However, it would help enormously if economists recognized more explicitly that their contribution to the political dialogue, while potentially valuable, is limited. Further, it *should* be limited because we don't have the answer. The answer can only come out of political process.

Let me turn now to my other concern—the proclivity of economists to consider technologies as given and known. Dewees recognizes that this presumption may involve a major misspecification of the problem. His paper advances the discussion because of his explicit acknowledgment, indeed stress, that the technologies of the future are not given and are not known. Further, they will be endogenous to the policy regimes that we adopt. He provides persuasive evidence that, in fact, technologies have evolved significantly even over the last five years. In various places he gives hints that the exact way the technologies evolved would not have been readily predictable ex ante. For all these points his paper deserves praise. However, he makes several assertions that I think are debatable. More important, I do not believe that he has gone nearly far enough in recognizing the implications for thinking about policy if the evolution of technology is at once endogenous and not clearly predictable.

Dewees asserts flatly, as do several economists in this volume, that a regulatory regime involving requirements is an unhappy choice not only for standard economic reasons but also because it does not provide the appropriate incentives for the development of new technology. If requirements are set too low, there are no incentives for research and development groups to aim to do better than those standards; if they are set too high, we end up with frustration. If benefits trade-offs are not recognized in the requirements, no incentive is provided for research and development to explore trade-offs on the supply side. I believe there is something to these arguments. However, they fail to recognize two things. One is the sequential bargaining process that "standard setting" is a part of. Companies understand full well that standards originally set low may be tightened up in the future and that standards set too high will be bargained down. I am not arguing that the experience we have had under standards has been a happy one. Nor am I arguing against the proposition that a system of "prices" might not provide a better regime of incentives for research and development. All I am arguing is that the issue is not as black and white as Dewees suggests.

Second, standards do have a certain advantage as a motivator for research and development. They force attention to a problem and pose the threat of very high cost if it is not resolved in a way that prices do not. To the extent that private R and D tends to be incremental in its ambitions (and there is a lot of good evidence for that), the imposition of standards that cannot be met by incremental changes may force a loosening up in research and development thinking. Again, I am not arguing that this surely is the case, only that the issue is a complicated one.

Dewees's paper does not go far enough in recognizing the wide range of policy issues that opens up if technologies are not fixed and given but inevitably will be influenced by policies. One implication I have alluded to—the setting of standards, or of prices—must be viewed not merely in terms of how they influence actions given current technology, but also in terms of how they will influence the future evolution of technology. From this perspective, it makes sense to set standards higher than current technology can achieve or to set prices of various effluents higher than can be justified given the current marginal cost of abatement curves. But reasonable men and women will disagree on just how much higher. And it is apparent that a reasonable policy will not lock itself into constant standards or prices. Rather the policy strategy must be viewed in terms of a sequential model. I am not arguing that we now know how to specify such a model formally, much less know the numbers to calibrate such a model. However, it seems to me that this is the broad perspective that we should carry around in our minds.

Standards and prices far from exhaust the policy instruments that can be used to influence the evolution of technology. Neither Dewees's paper nor any of the other papers in this volume seems to recognize the wide range of questions relating to the nature of appropriate federal policy aimed more directly at the research and development enterprise. What kinds of R and D should the federal government itself be funding? At the least, it needs to fund enough to acquire information so that it can make plausible judgments regarding alternative ways in which technology may evolve in the future. It needs this kind of information in order to set reasonable standards and prices. To what extent should public funding of research and development go beyond this information-gathering function? Should it fund background exploratory research on different abatement technologies but stop short of moving these technologies to full operational use? This approach could be justified on grounds of both informational requirements of the government and major externalities of this kind of research and development. Should this kind of research and development be undertaken in government laboratories, in universities, or through contracts with private business firms? If the last, what should be public policy regarding the patents? Should there be compulsory licensing? What should be public attitudes toward proposals, on the part of relevant private companies, that they do research cooperatively?

I do not have the answers to these questions. I worry, however, that they were not even asked in the Dewees paper or in the other papers presented in this volume.

8 Government Policies toward Automotive Emissions Control

Edwin S. Mills
and Lawrence J. White

Introduction

The government program to control motor vehicle emissions constitutes a major effort to limit an important source of atmospheric pollutants. The costs of the program are great, but the potential benefits may also be great. The program has received an enormous amount of attention in the communications media, in government, and in scholarly publication. The program began more than two decades ago. It is much better documented than other pollution control programs. We believe the time is ripe for a definitive analysis of the program's successes and failures.

Accordingly, this paper is a critical case study of the government program to control motor vehicle emissions in the United States. We will review the history of the program, discuss its present state, analyze its strengths and weaknesses, and offer suggestions for improvement. We believe that, though the intentions behind the program have been good, the program has been poorly designed. Progress in controlling emissions has been slow, the costs have been great, and the incentives for private parties to discover and implement methods of reducing emissions have been badly distorted. We will argue that an alternative approach, relying mainly on effluent fees, would have yielded faster and cheaper progress in controlling motor vehicle emissions and continuing incentive for research and development of new and improved control devices and sources of motive power. Our suggested program, even if implemented today, could still yield substantial improvements in the present program for the 1980s.

At the beginning, it is worth listing the unique features of the motor vehicle emissions problem that differentiate it importantly from the problem of controlling stationary emissions. First, the sources of motor vehicle emissions are numerous and ownership is dispersed. There are more than 130 million cars, trucks, buses, and motorcycles on American roads, and there are about 100 million separate owners. By contrast, there are only a few thousand stationary sources of significant amounts of air pollution (for example, thermal electricity generation, metal smelting, and other industrial operations) and only a few hundred thermal electric companies and a few hundred industrial companies whose behavior must be modified by a program to control stationary sources. These differences in numbers have dramatic effects on the potential costs of administering control programs. They also affect the kinds of political support that can be generated for various control programs.

Second, motor vehicle emissions are ubiquitous and are concentrated where people live and work. Unlike, for example, thermal electric plants, the sources of the emissions cannot be relocated away from population centers.

Third, the vehicles are mostly small power sources, for which there are serious fuel and space penalties if emission control requires extra weight or bulk. They are mostly owned and operated by amateurs and receive only intermittent, sporadic, and unreliable technical care. Thus, there are likely to be large differences between the performance of emissions control devices under laboratory conditions and in the hands of owners.

Only space heating (heating of structures) rivals motor vehicles as a source of emissions with comparable numbers, ubiquity, and maintenance problems—and space heating generates only a tenth as many emissions.[1]

These differences mean that motor-vehicle emissions control policy is different from other pollution control policies technically, politically, administratively and economically. We believe that these differences justify a separate case study of the motor-vehicle emissions control experience.

The remainder of this paper is organized as follows. The first section reviews the history of the automotive emissions control program from its beginnings in Southern California in the early 1950s to the current standards for the late 1970s. The next section reviews past studies of the costs and benefits of motor-vehicle emissions control, and is followed by a critical analysis of the existing program. Then we present our proposal for an effluent fee alternative to the present program. In a separate section we present the advantages of our proposal, an estimate of its administrative costs, a discussion of the use of revenues from effluent fees, and a projection of how our proposal might have improved emissions control performance had it been adopted in 1970. A discussion of the resistance to effluent fees both in and out of government is followed by the conclusions of our study.

The History of Automotive Emissions Control Efforts[2]

The 1950s

Concern about motor vehicles as significant contributors to air pollution problems first developed in Southern California in the early 1950s. The basic process that generated photochemical smog in the area was identified in 1951 by Dr. A. J. Haagen-Smit of the California Institute of Technology. Motor-vehicle emissions were key contributors, along with industrial emissions, backyard trash incinerators and atmospheric conditions specific to the area. Emissions from industry and open burning were gradually reduced or eliminated, and vehicle emissions were left as the prime culprit. Southern California government officials began suggesting to the automobile manufacturers that they do something about emissions. The manufacturers' response was to deny that automotive emissions were a major source of smog and to declare that the entire problem of smog was very complex and needed much more careful research. In December 1953 they formed a joint committee through the Automobile Manufacturers Association to study the problem. In mid-1955 they signed a cross-licensing agreement ensuring that all manufacturers would have access to any emission control patents owned by member firms on a royalty-free basis.

This collaborative effort may have had some beneficial effect by facilitating the interchange of new knowledge about emissions control among the companies, but it surely had an overall delaying effect because it reduced each company's incentive to pursue research so as to get a jump on its rivals. Because control devices could only add to costs without making the car any more attractive to buyers, the interest of the industry as a whole lay in delaying research and discouraging the release of information to outsiders, particularly to government officials. The industry's joint interests lay in making emissions control appear impractical and costly, so as to put off the day when it would be required. The interests of the individual manufacturers were mixed. Against the joint industry interests, each manufacturer had an incentive to pursue research in order to be the first to develop a practical control system. The first firm would have garnered public relations benefits; in the event that emissions control was required and that firm had strong patents, it might be able to earn sizable profits from sales or patent licensing to other members of the industry; and even before control was required, it might achieve fleet sales to government agencies that wanted to try control devices for demonstration purposes. Chrysler achieved just such sales in the fall of 1963 and received intense industry criticism for it.[3] The incentives for the individual firm to pursue rapid research were not overly strong, but they were definitely present.

The joint industry committee and the cross-licensing agreement largely succeeded in fostering the industry's joint interest at the expense of the competitive tendencies of its members. In January 1969, the Department of Justice filed an antitrust suit against the automobile manufacturers, charging collusion in delaying the development of emission control technology. In September 1969, the suit was settled by a consent decree, without admission of guilt by the companies.[4] Some of the justice department evidence concerning the companies' joint efforts to delay research and discourage publicity concerning emission control technology was put into the Congressional Record by Representative Phillip Burton on 18 May 1971.[5]

The remainder of the 1950s produced a predictable pattern of government officials calling on the automobile industry to do something about emissions and the industry responding, "complex problem; needs more research." Government officials were unwilling to spend significant amounts of funds to generate information about the problem, yet they provided little incentive for the industry to quicken its pace. Though they were distrustful of the industry's responses, government officials could only continue to urge the industry somehow to solve "their" problem. Attitudes of government officials during the 1950s showed a lack of comprehension of the fundamental nature of the problem. Moral suasion is ineffective in dealing with pollution problems precisely because the costs are external to the agency causing the problem.

Finally, in 1959 blowby was "discovered" to be a significant source of emissions.[6] Blowby is the collection of unburned and partially burned hydrocarbons that slip from the combustion chamber past the piston rings and into the crankcase. If the vapors were allowed to collect there, they would contaminate and thin the crankcase oil. Hence, a blowby port had long been installed on cars to vent these fumes to the atmosphere. Blowby emissions accounted for roughly 20 to 25 percent of the hydrocarbon emissions of an uncontrolled car. Because the blowby port was specifically designed to allow fumes to vent to the atmosphere, because the escape of the fumes would have been obvious even to an untutored observer looking under a car while the motor was running, and because the technology to control these emissions had been known since the 1930s and had been installed on some commercial and industrial vehicles in the 1940s, it is a wonder that the discovery took so long. In any event, in 1961 the manufacturers "voluntarily" installed positive crankcase ventilation devices to eliminate blowby emissions on all new cars sold in California (they were subsequently made mandatory by California legislation), and in 1963 the devices were installed on new cars nationwide.

The 1960s

In the 1960s, the California legislature tired of relying on the auto industry's good will and slow progress and finally provided direct incentives for faster technological development. Legislation passed in 1963 required that exhaust emissions control systems be installed on all new cars sold in California one year after the state had certified that two devices were practical and available at reasonable cost. This law opened up the field to independent parts manufacturers, who, unbound by the automobile industry's joint interests, rapidly took up the challenge. In March 1964, the auto companies told the state that the 1967 model year was the earliest that they would be able to install exhaust control devices. In June 1964, the state certified four devices, all produced by independent parts manufacturers. As a result, exhaust control devices became mandatory for the 1966 model year. In August 1964, the auto companies announced that they would, after all, be able to provide exhaust control devices, of their own manufacture, for the 1966 model year.

At the national level, government interest in emissions control had been less intense in the 1950s than in California, and the companies were under less pressure to respond. Though the 1960s brought the realization that smog was not solely a Los Angeles problem, the companies were able to forestall action for a few years with their claim, ''complex problem; needs more research.'' The Clean Air Act of 1963 only directed the secretary of health, education and welfare to encourage the development of emission control devices by appointing a liaison committee to work with industry representatives. By 1965, however, the companies had already told California that exhaust control devices were feasible, and further delays were not credible. At congressional hearings in that year industry spokesmen announced that, though they still had doubts as to the wisdom of mass installation of emissions control devices, manufacturers were prepared to install exhaust control devices if required. Indeed, by this time the competitive spirit of the companies had broken loose, and each company touted its own method as the best way of controlling emissions. The 1965 amendments to the Clean Air Act directed the secretary of HEW to set emissions standards for automobiles. The secretary set standards for hydrocarbons (HC) and carbon monoxide (CO), to become effective 1 January 1968. Chrysler initially chose a system of engine modification to improve combustion and decrease emissions, while the rest of the industry chose an air pump (to increase the oxidation of the HC and CO). The former method was cheaper and was adequate to meet the standards, so it was universally adopted.[7] The cost of meeting the standards was modest, about $20 per car.[8] The secretary of HEW and later the administrator of the Environmental Protection Agency subsequently set tighter standards for HC and CO for later years, added nitrogen oxide (NO_x) to the list of controlled pollutants, and set separate controls on evaporative emissions from gas tanks and carburetors. Table 8.1 contains the exhaust standards that have been set for the various years.

The 1970s

By 1970, congressional dissatisfaction with the progress that was being made on emission control was strong. The environmental movement was at high tide, and the automobile was held directly or indirectly responsible for much of the alleged deterioration of the environment. The companies that produced automobiles were judged ultimately culpable. An aggressive attitude toward the automobile companies was clearly a popular political stance for Congress at the time.

A report by Delbert Barth and his colleagues at the National Air Pollution Control Administration,[9] a predecessor agency of the EPA, provided a criti-

Table 8.1 Exhaust Emissions Standards[a] (grams per mile)[b]

	HC	CO	NO_x
Uncontrolled car[c]	8.7	87.0	4.0
1968	5.9	50.8	NR[g]
1970	3.9	33.3	NR
1972	3.0	28.0	NR
1973	3.0	28.0	3.1
1975 (interim)[d]	1.5	15.0	3.1[h]
1975 (California)[d]	0.9	9.0	2.0
1975 (original)[e]	0.41	3.4	2.0
1976 (original)[f]	0.41	3.4	0.4

Source: NAS-NAE vol. 4, 1974, p. 57.

[a] Evaporative emissions are not included in these figures.

[b] As measured by the federal constant-volume sampling, cold- and hot-start test.

[c] Uncontrolled except for crankcase blowby control.

[d] Interim standard set by EPA 11 April 1973.

[e] Ninety percent reduction (from 1970 levels) in hydrocarbons and carbon monoxide originally mandated for 1975; currently delayed until 1978.

[f] Ninety percent reduction (from 1971 levels) in nitrogen oxides originally mandated for 1976; currently delayed until 1979.

[g] No requirement.

[h] Standard will be tightened to 2.0 for 1977.

cal impetus. Barth et al. selected 1990 as a target date and selected threshold levels of pollutant concentrations above which they believed there were significant health risks to exposed populations. They found the worst current pollutant concentration readings[10] and assumed a growth rate of vehicles between 1967 and 1990 that would yield more than double the automobile population by 1990. They then predicted that the worst pollution concentrations would more than double by 1990. They applied a rollback model—an assumed simple proportional relationship between emissions and pollutant concentrations in the atmosphere—to determine the extent to which automotive emissions would have to be reduced so as to reduce pollutant concentrations below the threshold levels.[11]

In somewhat modified form, the Barth et al. recommendations became the basis for the standards set in the 1970 amendments to the Clean Air Act. The act mandated a 90 percent reduction in HC and CO emissions (from 1970 levels) by 1975 and a 90 percent reduction in NO_x emissions (from 1971 levels) by 1976. In the 1970 amendments Congress took a tough line: earlier it had simply authorized an administrative agency to set standards; in 1970, Congress actually set standards itself. The 1970 amendments were intended to show contempt and vindictiveness not only for the auto companies but also for the federal administrative agency. Also, the 90 percent reductions were beyond the technical capabilities of the automobile industry at the time. The legislation was specifically intended to force the companies to quicken the pace of their emissions control research. The act also instructed the EPA to encourage the states to establish periodic emissions tests for cars on the road and transportation control plans for most urban areas; implementation of these provisions of the act has been hopelessly delayed in virtually all states.

Despite anguished screams that the mandated 90 percent reductions were impossible or would be extremely costly and might shut down the automobile industry, the EPA began seriously to administer the act. The act did permit the EPA administrator to grant a one-year delay in the enforcement of the 1975 and 1976 emission standards, but the auto companies' initial request in 1972 for the year's delay was denied. The decision was appealed to the courts, and the U.S. Circuit Court of Appeals remanded the case to the EPA for a new decision. In April 1973, the administrator of the EPA decided to grant a one-year delay in the original 1975 standards for HC and CO and set interim standards for 1975 (see table 8.1); in July, a one-year delay in the original 1976 standard for NO_x was granted. The only grounds permitted in the law for delaying the enforcement of the standards was technological infeasibility. Yet EPA granted the delay shortly after certifying the Honda CVCC engine as meeting the 1975 standards. It is hard to imagine that the delay would have been granted had a large U.S. company had an engine certified. If that assumption is correct, the program was used as a protectionist device. This conclusion does not depend on whether the original 1975 standards were justified.

In June 1974, in the wake of concern about the "energy crisis," Congress granted an additional year's delay in the enforcement of all standards. And in April 1975, because of fears that the catalytic converter devices that were being used to meet the interim standards might generate serious sulfur oxide emissions (EPA has since concluded that they do not), the administrator of the EPA granted yet another year's delay in the enforcement of all of the standards. The original 1975 standards for HC, CO, and NO_x are now scheduled for 1978. (For purposes of clarity in the text, we shall continue to refer to these as "the original 1975–1976 standards.")

In late 1976, Congress considered and rejected comprehensive amendments to the Clean Air Act that would have permitted additional delays in meeting the highest standards. But by the summer of 1976 the auto companies had already had to have firm designs for 1978 cars. They opted for designs that would not meet the mandated 1978 standards because they had reason to believe that Congress would grant additional delays. Thus, at the time of this writing, the auto companies are committed to manufacturing 1978 cars that it will be illegal to sell! Presumably, Congress will grant new delays in early 1977. The original, high standards may never be enforced.

Conclusions

The history of emissions control is not overly encouraging. Despite twenty-five years of government involvement, progress in reducing emissions has been slow and fitful. But EPA data do show modest decreases in ambient concentrations of HC and CO since 1970 in urban areas for which comparable records exist.

A Review of Studies of Benefits and Costs of Motor-Vehicle Emissions Control

Background

Is the program worth the costs? This question should be asked about any legislative program and is certainly necessary for one that reaches the size of the automotive emissions program. Unfortunately, the Congress did not give careful attention to this question when it passed the 1965 and 1970 amendments to the Clean Air Act. The attitude in Congress seemed to be that the nation's health was at stake and that clean air was an absolute good. Lives could be saved and sickness avoided by a reduction in pollution, and costs were a secondary consideration. It is unclear whether this lack of concern about costs reflected a belief that costs would be comparatively low (incoporating a belief that American technology, if properly challenged, could reduce the cost of anything) or a belief that costs really were irrelevant when questions of health were at stake. The 1970 amendments, in particular, show a deep hostility to cost measurement. The emphasis is on technical feasibility, to the neglect of costs.

Even if Congress had been interested in comparing costs and benefits, there were few studies that would have been helpful. There were no cost-benefit studies of the prospective programs in 1965 or 1970, and surprisingly little interest in such studies was shown by Congress or by appropriate administrative agencies. For example, a major Department of Commerce study issued in October 1967 (*The Automobile and Air Pollution: A Program for Progress*) scarcely mentioned the notions of costs and benefits and did not attempt to quantify them. The subpanel reports underlying the main study gave some attention to the question of cost-benefit analysis but, again, did not attempt one for automotive emissions. Testimony in 1965 congressional hearings generally projected moderate costs for the proposed standards and projected large health benefits from the elimination of pollution but could not really link emissions reduction with health improvements or identify reliable estimates of the monetary value of projected benefits. Modeling that would link emissions to ambient air quality, air quality to health and other effects, and health and other effects to dollar value was only in its infancy. The testimony in the 1970 hearings was little better, except that the estimates of the costs of meeting the original 1975–1976 standards were higher and wilder because the technology was unknown.

Since the passage of the 1970 amendments, information concerning the likely effects of the program has increased substantially. Between 1972 and 1975 eight major studies assessed the program and calculated likely costs and/or benefits. These studies are listed in table 8.2 in their order of appearance. Instead of summarizing each one, we shall discuss the general problems of measuring costs and benefits and include the results of each study as appropriate. An extensive review of five of the studies is in White.[12]

Costs

The costs of the emission control program are far easier to quantify than benefits. In principle, costs should include the opportunity cost of any emis-

Table 8.2 Major Studies of the Automotive Emissions Control Program (1972–1975).

U.S. Ad Hoc Committee on the Cumulative Regulatory Effects on the Cost of Automotive Transportation (RECAT), *Final Report*	February 1972	Costs, benefits
National Academy of Sciences, *Report by the Committee on Motor Vehicle Emissions*	Febrary 1973	Technology, costs, testing
H.D. Jacoby et al., *Clearing the Air: Federal Policy on Automotive Emissions Control*	1973	Policy analysis, costs, benefits, technology
D.N. Dewees, *Economics and Public Policy: The Automobile Pollution Case*	1974	Costs, alternatives
National Academy of Sciences and National Academy of Engineering, *Air Quality and Automobile Emission Control* (prepared for the U.S. Senate, Committee on Public Works	September 1974	Costs, benefits, health effects, air quality modeling, alternatives
National Academy of Sciences, *Report by the Committee on Motor Vehicle Emissions*	November 1974	Costs, technology, testing
F.P. Grad et al., *The Automobile and the Regulation of Its Impact on the Environment*	1975	Costs, benefits, air quality modeling, technology, testing, alternatives, legislative analysis
D. Harrison, Jr., *Who Pays for Clean Air?*	1975	Income distribution

sions control hardware, extra fuel costs necessitated by the controls, extra maintenance costs (including the value of the owner's time), an estimate of the monetary value of decreased consumer satisfaction if controls lead to a deterioration in car performance, and the costs of administering the program itself. The first four items should be estimated for each major variety of car (subcompact, compact, intermediate, and so on) and then combined on the basis of estimates of the composition of the national automobile population. All of the studies estimate the hardware, fuel, and maintenance costs (but none include the vehicle owners' time costs). Only Dewees[13] tries to measure the costs of inferior performance, and he could measure only the higher costs of horsepower necessitated by the controls, to the neglect of other performance losses; because of a very low estimated demand elasticity for horsepower, he finds that the loss of consumers' surplus from the implied higher price of horsepower is small. Harrison[14] adds an extra dimension by focusing on the incidence of the costs and benefits among individuals with different incomes. None of the studies mentions the costs of administering the program. Though the costs are nontrivial in absolute terms (in FY1975, all federal spending on air pollution abatement and control was $274 million,[15] but an unknown fraction of this amount went for control of stationary sources), the administrative costs are dwarfed by other costs of the program.

The cost estimates have been surprisingly consistent over time, despite improvements in and increased certainty of control technology. The early RECAT report estimated costs of the original 1975 standards at $426 per car; its estimate of the cost of the original 1976 standards was $935 per car.[16] Two and one-half years later, the NAS report provided the following estimates (which will be used for the remainder of this paper): the extra cost over a 1970 vehicle for meeting the original 1975 standards for an intermediate six-cylinder car is estimated to be $400 per car in 1974 dollars, with a $200 uncertainty band; the original 1976 standards are estimated to cost $600 per car, with a $350 uncertainty band.[17] The marginal cost of emissions control definitely increases with increasing severity of the standards, as Dewees demonstrates.[18]

The NAS-NAE report[19] expands the NAS per-car estimates to annual cost figures and projects them into the future, along with estimates of future technological changes. Because the costs consist of initial hardware and subsequent fuel and maintenance costs and because it takes time for the automobile stock to turn over, maximum annual costs are not reached until 1985.[20] The report projects annual costs of $8 billion in that year and assumes that they will remain at that level in the following years as technological and learning curve improvements offset increases in the size of the national automobile fleet. The discounted present costs for the period 1975–2010 are $126 billion.

Harrison[21] also draws on the NAS cost methodology, but he is more optimistic. Although he predicts that total cost per vehicle of meeting the original 1976 standards may be as high as $1000 for the years 1977–1980, he predicts that technological change will bring the cost per vehicle down to $450 for 1981 and following years. He estimates total annual costs at $6 billion in 1980 but only at $4 billion by 1990.

Unfortunately, these NAS-based cost estimates are probably too low. First, they neglect the costs of consumer dissatisfaction with cars that have inferior performance characteristics. Admittedly, such costs are hard to quantify. Nevertheless, they could be substantial for some models and apparently already have been for 1973 and 1974 cars. Also, they neglect the cost of owners' time and the inconvenience involved in extra maintenance. Second, the NAS cost methodology errs by comparing the fuel consumption of cars equipped with emission controls with that of an uncontrolled 1970 vehicle. Rather, the comparison should be with fuel consumption levels that would otherwise occur. For example, the NAS report indicates that catalyst-equipped cars meeting the interim 1975 standards will achieve 2 percent better fuel economy than an uncontrolled 1970 vehicle. This change counts as a $65 *benefit* of the program to be subtracted from other costs. But today, in the absence of emission controls, fuel economy would be at least as good as that achieved by the catalyst-equipped automobiles and probably better. These fuel economy gains over 1970 should not count as benefits. Even if one believed that fuel consumption improvements have come about solely because of the emissions control program (which is unlikely; most resulted from higher fuel costs after 1973), they are benefits from the past operation of the program and should not be included in calculations trying to measure the future worth of the program. They would not disappear if the program disappeared. Consequently the NAS cost estimates are too low by a sixth to a third, depending on the year.[22]

In any event, the costs of the program are sizable. And, as Harrison shows,[23] the incidence of these costs is likely to be regressive, because the standards increase the cost of automotive transportation, which absorbs a larger fraction of the incomes of poor than of rich people. Dorfman's results are similar.[24]

Benefits

The benefits of the emissions control program are much more difficult to quantify than costs. The benefits occur because of improvements in the ambient air and should include the following: delayed deaths; illnesses avoided, delayed, or made less severe; reduced or avoided physical discomfort or annoyance caused by polluted air; decreased destruction of agricultural crops; and decreased materials deterioration.[25]

The first step in measuring benefits is to ascertain changes in emissions and changes in ambient air quality. This determination involves projections of the size and composition of the national automobile fleet, projections of driving patterns, and models of the relationship between specific emissions and ambient air quality. The last step is the most difficult. The NAS-NAE report provides a summary of the problems of modeling these relationships.[26] Complex physicochemical models of air quality, which take into account geographical dispersion, time effects, meteorological conditions and specific local factors, are still in their infancy and have not been used in policy formation. Instead, rollback models, which assume a simple proportional relationship between emissions and ambient air quality, have been used. At best, these rollback models might be suitable for an inert gas such as CO. They are inappropriate for modeling the complex, nonlinear, and still incompletely understood processes whereby photochemical oxidants are generated from HC and NO_x. As the NAS-NAE report points out, improvements in these models are badly needed.

The second step is to translate improvements in ambient air quality into health and comfort improvements and materials savings. Again, the relationships are only imperfectly understood, as the summary in the NAS-NAE report indicates.[27] Toxicological evidence from laboratory experiments indicates substantial short-run health effects from very high pollutant concentrations, but the evidence for health effects at the lower concentration levels that have been or are likely to be present in the ambient air is mixed. Some researchers find effects, others do not. The existence of threshold pollutant levels is doubtful in large populations. There may be a threshold level below which a particular pollutant does not affect a healthy adult. But populations at risk consist also of children, the aged, and people with many kinds of illnesses and disabilities. Doubts about threshold effects apply equally to CO, NO_x, and photochemical oxidants.[28] These pollutants, except CO, frequently can be sensed at current levels and cause minor discomfort, such as eye irritation, but the evidence for more serious health effects is uncertain.

The second kind of evidence relating health to ambient air quality comes from epidemiological studies. These are statistical efforts to measure the health and mortality effects of everyday exposure to pollutants, either through cross-section regressions among cities or by means of detailed investigations of particular episodes of exposure to unusually high concentrations of pollutants. The strongest results have been found in the measurement of the effects of sulfur dioxide and particulates, but automobiles are not significant emitters of either. The evidence for automotive pollutants is again mixed; some researchers have found effects and others have not.

Estimates of the benefits to crops and materials from reduced pollution are at much the same state as the toxicological measurements of health benefits. The effects of high concentrations of pollutants are well known, but the effects of actual concentrations are much less understood. The estimates of actual damages that have been made have a high degree of uncertainty attached to them.

The final step is to convert the health benefits and materials savings into a dollar measure. There is little problem in converting crop and materials savings into dollars. There are more problems in trying to put a dollar figure on the delay of a death, the avoidance or reduction in severity of an illness, and the elimination of physical discomfort such as eye irritation. The NAS-NAE report puts a value of $200,000 on each death delayed but makes little effort to defend the figure.[29] It also chooses a range of $1 to $20 for the value of the prevention of a person-day of discomfort or restricted activity. One could easily provide an equally wide range for the value of delaying a death. In principle, the right way to value these things for cost-benefit purposes is to ascertain how individuals themselves value them through revealed behavior. A small literature on the subject has developed, starting with Schelling and extending through Thaler and Rosen.[30] The report notes that its figure of $200,000 is consistent with the findings of Thaler and Rosen, but it does not make any further efforts in this direction.

The report does try to measure the value of benefits from pollution abatement by examining the operation of two markets in which pollution may affect the prices that are perceived: urban housing and intercity labor markets. The urban housing model, which Ridker and Henning developed,[31] involves a cross-section regression within one metropolitan area of the value of housing in a census tract against various explanatory variables, including pollutant concentrations reported (or predicted from a separate model) for that census tract. The negative coefficients on pollutant concentrations may indicate how much people value a marginal change in pollution. The labor model assumes that real returns to labor tend to be equalized among cities. Nonmarket amenities ought to be offset by lower money wages. Air quality may be such an amenity, and higher pollution levels in one city may require higher money wages so as to make workers as well off as in other cities. The test of the model involves a cross-section (among cities) regression of wages on a number of explanatory variables, including pollutant concentrations. Again, the positive coefficient on the pollutants should indicate the value that people put on air quality, in terms of what they need to be paid to endure worse air quality or what they are willing to give up in return for better air quality.

The strengths and drawbacks of these models have been reviewed by Rubinfield in this book and also by White,[32] so they need not be repeated extensively here. Briefly, both models are vitally dependent on the assumption of an "open city" (the free migration of labor among cities in response to potential utility gains or losses equalizes utility among residents of different cities); they have simultaneous equation problems; there are probably serious serial correlation problems (the error term for one census tract or city may be related to the error term of a neighboring census tract or city); and the estimates of benefits appears to be sensitive to the exact specification of the equation.[33]

Only the RECAT report and the NAS-NAE report made explicit dollar estimates of the benefits of the program. The RECAT report estimated benefits in 1985 at $3.5 to $9.1 billion per year. Its estimates of costs come to $10.1 billion per year in 1985. It thus finds the original 1975–1976 standards not to be worth the costs.

The NAS-NAE report tries to make the full connection between emissions reduction, ambient air quality improvement, health improvement and materials saving, and dollar benefits. With suitable caveats concerning the problems of uncertainty and of double-counting, the report provides a best guess of $5 billion per year in benefits (in 1973 dollars and at 1970 income and population levels) in a likely range of $2.5 billion to $7.5 billion. Because it takes time for the automobile fleet to turn over, the full benefits will not be felt until 1985; because of higher incomes and larger urban populations, the best guess of benefits in that year is $7.2 billion; for 2101, it is $13.6 billion. The discounted benefits for 1975–2101 come to $137 billion.

The effects on income distribution of these benefits are likely to be regressive. Harrison[34] calculates only the improvements in ambient air quality that different income classes, on average, will receive. He finds the improvements to be substantially the same for all income classes. He does not attempt to put dollar values on these improvements for each income class. But Rubinfeld's evidence[35] indicates that the willingness to pay for marginal improvements in air quality rises more than proportionately with income. Thus, equal levels of physical improvement imply a regressive pattern of benefit values.

The NAS-NAE report estimates the discounted costs of the program at $126 billion and the discounted benefits at $137 billion. But, as argued previously, the costs are underestimated because consumer dissatisfaction costs are neglected and fuel consumption improvements are incorrectly compared to rates for 1970 vehicles rather than to those that would develop in the absence of emissions controls; this latter failing probably raises costs by a sixth to a third. Also, the benefits are probably overstated because the report assumes that actual cars on the road will meet the standards, whereas there is good evidence that average emissions from cars on the road exceeded the pre-1975 standards (this topic will be discussed later), and it is not at all unreasonable to expect actual emissions to exceed the original 1975–1976 standards by 20 percent or more.[36] Because the law currently specifies that only the test vehicle must meet the emissions standards, these in-use results do not constitute violations of the law. Also, the standards apply only for the first 50,000 miles of a car's life, whereas the average car lasts 110,000 miles. Deterioration in emissions after 50,000 miles may well be a serious problem, but the NAS-NAE report does not take it into account.

Furthermore, the emphasis on new and relatively new cars in the emission control program must reduce the relative cost of old cars to induce the motoring public to keep cars longer than otherwise. The result is to lessen the improvement in ambient air quality from the program. A 10 percent increase in the age of the average car when it is junked would increase emissions by considerably more than 10 percent. The overall effects of the program are likely to be regressive with respect to income distribution because the patterns of both cost and benefit incidence are likely to be regressive.

Recognizing the uncertainties and the large margins of possible error in its estimates of costs and benefits, the NAS-NAE report is agnostic as to the wisdom of implementing the original 1975–1976 standards. There are alternative strategies, however, and some offer a better balance of costs and benefits.

Alternatives

The most promising alternative recognized by the studies is a "two-car strategy": maintain the original, tough 1976 NO_x standards for the 37 percent of cars that would be sold in high-pollution areas but require only the 1973 standards for other cars. Since the marginal cost of meeting the tough NO_x standard is high, it makes little sense to require it for areas that are unlikely to have NO_x pollution problems. Under this scheme, according to the NAS-NAE report,[37] annual costs would hit a peak of only $5 billion in 1985 and the present discounted benefits of the program would be only $82 billion, which is well under the estimated benefits (which would be almost unaffected by this option). Because California currently requires stiffer emission controls than the remainder of the country, this two-car strategy does not depart radically from current practice. It is certainly well within the manufacturing capabilities of the automobile companies. The RECAT report also endorses this strategy.[38]

A second alternative considered by the NAS-NAE report involves freezing the emission standards at the interim 1975 standards. This plan would reduce the present discounted cost of the program to $41 billion. Unfortunately, it would also reduce the benefits, but the report does not provide estimates of benefit reductions, so no assessment of the option is possible.

Promotion of public transportation and/or restrictions on the use of automobiles are suggested in the 1970 amendments and by many commentators as alternative ways of reducing pollution. The studies by Dewees, Jacoby et al., Grad et al., and the NAS-NAE[39] unanimously reject these alternatives as cost ineffective compared to automotive emissions controls.

The cross-elasticity of demand between automobile use and public transport appears to be far too low for improvements in the latter to woo significant numbers of drivers away from their cars. This situation is due not to an irrational attachment of drivers to their cars but to substantial cost and other advantages of cars over alternative modes for most car users. Restraints on vehicle use are quite costly in terms of time and inconvenience and mostly just shift pollution from one area to a neighboring area. Steep downtown parking fees are thought by some writers to be cost effective, but they would be cumbersome to administer and would increase the amount of traffic passing through downtown areas.[40]

Finally, does new technology offer a low-cost, low-emissions alternative for the future? Truly different motive power technologies—turbine, steam, and electric—are still far from the production line and seem to be getting no closer. The internal combustion engine is likely still to be with us well into the 1980s and probably through the 1990s as well. The standard engine can meet the original 1975 standards through engine modifications and the addition of an oxidizing catalyst to oxidize the HC and CO emissions. Stratified-charge engines—either dual carbureted (such as the Honda CVCC) or fuel injected (still under development)—appear to offer hope of meeting the 1975 standards without catalysts and at lower overall costs. Diesel and rotary engines (such as Mazda) can also meet the 1975 standards without catalysts, but diesels have noise and handling characteristics that may discourage many American buyers and rotaries appear to have problems with seal wear and fuel economy. All of the engines are beset with serious problems that must be solved before they will meet the original 1976 NO_x standards. An additional reducing catalyst for the NO_x or a three-way catalyst to eliminate all three pollutants is necessary, and the technology for both is still relatively undeveloped.[41] A recent NAS report indicates that the research effort to develop the necessary technology has slackened,[42] probably as a consequence of the delays that have been granted and the perceived likelihood that Congress will grant further delays; we will return to this point in later sections.

Given the huge costs and benefits of the emissions control program, the social returns to the discovery of a low-cost, low-emissions source of motive power would be very high. Jacoby et al.[43] argue strongly that the federal government should undertake a massive research effort to discover such a source. This ought to be province of the automobile companies, but they have shown little inclination in the past to undertake this kind of research (perhaps the policies that we will advocate later would induce better performance in this respect). The risks are high here, and it is easy to waste large sums of money. But the returns are also high, and a large research effort financed by the federal government might be worthwhile.

Conclusions

Congress, in passing the 1970 amendments to the Clean Air Act, seems to have been unconcerned with or hostile toward questions of costs and benefits. Since then, much information has been generated on both costs and benefits. The measurement of both costs and benefits is difficult and inevitably includes uncertainties and wide margins of error. The best estimates indicate that the program originally enacted may be worthwhile, but only marginally so, and that superior alternatives—specifically a two-car strategy—definitely exist.

But are there better ways of achieving these benefits at yet lower costs? We do not believe that the current policies and methods have been even close to optimum in this respect. We now turn to our critique of current policies.

A Critical Evaluation of the Existing Program

A Description of the Current Program

As mentioned earlier in this paper, the national government has set emissions standards for automobiles according to the schedule listed in table 8.1.[44] Manufacturers must deliver representative vehicles to EPA for certification that they meet the standards before any cars of that type can be sold at retail. Failure to obtain the certification leads to a fine of up to $10,000 per car sold. No fines have been levied. Even if the penalties were within the range of the costs of emissions control, it is unlikely that an automobile company would choose this option because it would be branded as a violator of the law.

The certification process proceeds as follows: emissions are measured by the CVS-CH test.[45] The car is held (soaks) for 12 hours at 60° to 86°F. It is then started (a cold start), driven over a 23-minute cycle that simulates an urban driving pattern, then turned off for 10 minutes, restarted (a hot start) and driven for an additional 8.5 minutes. Emissions are collected from the exhaust from the time the key is turned in the ignition, regardless of whether the car starts. Since the 1970 amendments require that cars meet the standards for five years or 50,000 miles, whichever comes first, a small set of vehicles is driven 50,000 miles. The CVS-CH test is adminsitered to these cars every 5000 miles. Normal maintenance on the vehicles is permitted. The vehicle must pass the emission standards at every test. A least-squares regression line is calculated from the testing points to determine the deterioration factor per mile. Then a larger fleet is driven only 4000 miles and tested. The deterioration factor from the least-squares line appropriate to that model car is applied to the 4000-mile results to predict the 50,000-mile emissions. If the predicted emissions are below the standards, it passes the test and cars of that type can be sold; if not, the car may be tested again. All the cars in an engine family must pass this latter test before the engine family can be certified.

Once the test vehicles have been certified, there are no further requirements for testing vehicles in actual use to ascertain that they conform to the standards. The EPA will begin assembly-line audits only during the 1977 model year.[46] The 1970 act instructs the EPA to encourage the states to develop programs to induce motorists to maintain their cars so as to keep emissions low. Little has been done in this direction. Only New Jersey has annual emissions inspections for all cars. California has spot checks on the highway, although Southern California plans to institute mandatory inspection in 1977. Cincinnati and Portland, Oregon, have mandatory emissions-inspection programs. No other states or cities have mandatory programs currently operating, though a few, including Chicago, have voluntary programs.

Is the Program Well Administered?

By many of the standards of Washington regulatory agencies, the EPA's automotive emissions program appears to be quite good. EPA has been even-handed in its decisions. It would be hard to argue that the EPA has become a tool in the interests of either the manufacturers or the environmentalists. Fairness does not imply efficient outcome, as we shall argue. It is a nontrivial accomplishment, however, given the performance of some other regulatory agencies.

The CVS-CH test appears to be a reasonably good one. There has been some question as to whether it represents urban driving patterns accurately, but the NAS is satisfied that the test "is sufficiently accurate representation of urban driving patterns to be used for purposes of determining light-duty motor vehicle exhaust emissions."[47]

Yet the certification procedure has its faults. First, the more or less continuous driving for 50,000 miles greatly underrepresents the number of cold starts that a car in actual use experiences. Since additional cold starts tend to increase emissions, actual-use emissions exceed the test emissions.

Second, motorists are likely to maintain their cars less conscientiously than the recommended maintenance the manufacturers are permitted during the 50,000-mile driving period. Again, actual-use emissions will be higher than those during tests. The failure of the EPA to encourage the states to promote maintenance by car owners is a major factor here. The states could encourage maintenance either by requiring periodic preventive maintenance and repairs to manufacturers' standards or by requiring periodic emissions testing, backed by suitable incentives or requirements for repair. The former would involve component reliability and functional tests. The latter would involve system reliability. Because the CVS-CH test could not be replicated for mass testing—the time and inconvenience involved in the overnight soak and fifty-minute test required by the CVS-CH test would generate enormous costs for the owners of 100 million cars—the EPA would need to settle on a short, hot-start test that correlates highly with the CVS-CH. (Even repeat testing of the same vehicle using the CVS-CH test does not yield a perfect correlation, so one could not expect a different test to do any better.[48] A number of tests are possible, but the EPA simply has not moved actively in this direction. Simple, privately developed tests whose results have correlation coefficients with the CVS-CH test in excess of 0.95 have been available since the early 1970s. One easy solution to the correlation problem is to require that a new car pass both the CVS-CH and a short hot test. This is a slightly more stringent requirement, but the short test could reasonably be used as the basis for state-encouraged maintenance. In the absence of any action, motorists have not had and will not have proper incentives to maintain their cars, so emissions have been and will be higher than they should be.

Third, the absence of assembly-line tests has probably encouraged the manufacturers to provide "mint" models for certification. Again, this probably means that actual-use emissions exceed test emissions.

Fourth, in one respect the actual certification testing procedure may be too rigid because it requires that all cars pass the standards at all points up to 50,000 miles. Since average emissions (times the total number of cars) determine ambient air quality, it would be more logical to require only that the average emissions over the entire 50,000 miles of all cars tested (weighted by likely sales) meet the standards. But given the other, previously described elements that cause actual-use emissions to exceed the test emissions, given the opportunity for the manufacturers to retest a car that fails, and given the lack of standards beyond 50,000 miles, there is at least a moral case to be made for stricter testing procedures.

Fifth, recent in-use testing indicates that evaporative emissions from gas tanks and carburetors exceed the standards by a large factor.[49] Better tests, which correspond more closely with in-use conditions, and better devices are needed.

Finally, there has probably been too much focus on automobiles and not enough on other motor vehicles. Trucks and buses are covered by the 1970 amendments but in a less stringent fashion than cars. Section 202(b), which covers automobiles, specifically mandates the 90 percent reduction. But section 202(a), which covers trucks and buses, simply instructs the EPA to set standards for motor-vehicle emissions of any air pollutant that is a threat to public health and welfare. In the absence of the 90 percent mandate, the EPA has moved more slowly to control the emissions of these vehicles than those of cars. Prior to 1974, diesel trucks had only smoke standards and gasoline trucks had only mild HC, CO and NO_x standards. Under the current (interim) 1975 standards, automobiles can be considered to emit only 17 percent as many pollutants as would an uncontrolled car. Light-duty trucks[50] emit roughly 25 percent as much as uncontrolled trucks. Heavy-duty gasoline-engine trucks emit 47 percent. And heavy-duty diesel-engine trucks emit 75 percent.

Though the emissions from all trucks and buses in the United States, if uncontrolled, might come to only a quarter of the emissions of all automobiles, if also uncontrolled,[51] it is likely that current truck and bus emissions approach equality with current automobile emissions. Tighter truck standards are planned for the next few years. Although the proper level of truck emissions relative to automobile emissions is not obvious (except, as we shall argue, the relative levels that would result from a uniform effluent fee), it is quite likely that the marginal benefit-cost ratio for tightening truck emissions standards would have exceeded the marginal ratio for tightening automobile emissions standards in the past few years.

Along the same lines, motorcycle emissions are not controlled at all, and standards are only currently being proposed. The motorcycle fleet is only 5 percent the size of the automobile fleet, but a typical motorcycle's emissions are now three to four times those of a car that meets the 1975 standards.[52] Accordingly, motorcycles now constitute a significant source of emissions and therefore earlier action on motorcycles would have been appropriate.

How Successful Is the Program So Far?

There are three questions we can ask with regard to the possible success of the program. First, is ambient air quality improving? This, of course, is the ultimate goal of the program. But in the absence of a complete model of the determinants of ambient air quality, we cannot be sure that the emissions control program is responsible for any of the changes observed. Second, are actual in-use emissions meeting the standards or at least below the levels of uncontrolled vehicles? If so, then ambient air quality is surely better than it otherwise would have been. Finally, we might ask if progress in controlling emissions has occurred as rapidly as it otherwise might have under different policies. We leave the last question for discussion later.

Ambient Air Quality The ambient air quality with respect to CO and oxidants has improved in recent years in most metropolitan areas.[53] The data—in the form of annual averages of daily highest one-hour (or second highest one-hour) readings for metropolitan areas, or the fraction of days in which readings exceed the national air quality standards—do indicate a modest downward trend from the late 1960s through 1974. The data are incomplete, measurement techniques change over time, and it is difficult to put together continuous time series, but this does seem to be a safe conclusion. The millennium has certainly not arrived; CO and oxidants have not disappeared; but there have been modest improvements.

Have the auto emissions controls played a role in this improvement? As noted earlier, in the absence of a complete model and complete data on weather changes and on other emissions sources, it is difficult to offer a definitive judgment. There is, however, one piece of circumstantial evidence that leads us to believe that the emissions controls have played a role: ambient air quality with respect to NO_x has grown worse; the trend through 1973 was definitely upward. The emission control devices put on cars between 1968 and 1972 only reduced HC and CO emissions; they *increased* NO_x emissions, for which there were no standards, by at least 25 percent.[54] Thus the improvements in CO and oxidant ambient air quality and the deterioration of NO_x ambient air quality are quite consistent with the pattern of auto emission reductions and increases.

In-Use Emissions As noted, the current emissions control program requires only that a presale sample of vehicles meet the standards. EPA has also collected information on actual emissions from in-use vehicles but has used the data only as an input into local transportation control plans (to determine the necessary severity of the controls one needs to know the likely pattern of actual emissions). Only recently has EPA indicated that this in-use data might imply that the automobile companies were failing in their obligation to meet the standards.[55]

The in-use data are collected from small, stratified samples. Cars are selected, and owners are asked by mail to allow their cars to be tested. In return, the owners receive a substitute car for the few days involved plus a small payment. There may be some bias toward clean cars since owners of dirty cars may be ashamed to allow them to be tested and may refuse or have them serviced before delivery for testing.[56]

Tables 8.3–8.6 show the results for a set of tests performed on samples of cars in various cities. It should be noted that, for the 1968–1971 model cars, only the average value of the presale certification vehicles had to meet the standards; hence, average in-use values are an appropriate test of the predictive value of the certification process. Table 8.3 is expressed in terms of the standards and tests at the time the cars were tested; the remaining tables are expressed in terms of the 1975 CVS-CH test.

A striking pattern emerges from these tables. In every testing year and for every model year, average in-use CO emissions significantly exceeded the applicable federal standards. In some testing years, average HC emissions from some model years also significantly exceeded the federal standards. And the 1973 model cars significantly exceeded the NO_x standards. Cars in Denver produced the greatest HC and CO emissions; Denver's altitude of about one mile appears to create severe emissions problems and manufacturers were not required to make any allowance for effects of altitude during years covered by the tables. In all cases, the emissions from post-1967 cars were less than those from uncontrolled cars (1967 and earlier). Nevertheless, these results do cast serious doubt on the certification process for predicting in-use emissions.

Table 8.7 rearranges the data from tables 8.4–8.6 so that the emissions from a given model year can be tracked in subsequent testing years. A quite consistent pattern of higher emissions in subsequent testing years emerges from the table. Age and mileage definitely take their toll in increased emissions. The CAPSPAN[57] study found rates of emissions increases of 0.5–1.0 percent per 1000 miles.

For the 1972 and later models, the certification procedure required that all models pass the test. Accordingly, it is worth ascertaining how well in-use emissions corresponded with this requirement. Table 8.8 provides these results. In 1973, more than half of the 1972 vehicles in five cities failed either the HC or CO test; only 39 percent passed both. In Denver only three percent passed both. A year later, only 21 percent passed both standards in the five cities. Also in 1974, only 15 percent of the 1973 models passed the test for all three pollutants in the five cities and only 42 percent of the relatively new 1974 models passed all three standards. Once again, the certification procedure proved to be a poor predictive device for in-use emissions. As was pointed out previously, owners lack incentive to maintain cars in ways

Table 8.3 In-Use Emissions from 1968–1971 Automobiles in Six Cities[a] Based on "7 × 7" Federal Test Procedure" (in 1971)

			HC			CO		
Model Year	Number of Cars	Average Mileage	Federal Standard	Mean Emissions	Estimated Standard Deviation of Mean	Federal Standard	Mean Emissions	Estimated Standard Deviation of Mean
1968[b]	878	34,600	275 ppm	333 ppm[f]	8.41	1.5%	2.14%[f]	0.04
1969[b]	922	23,300	275 ppm	296 ppm[f]	5.2	1.5%	1.88%	0.03
1970[c]	1829	11,000	2.2 g/mi	2.83[f]	0.04	23 g/mi	35.3[f]	0.51
1970 (Denver)	352	10,100	2.2	4.25[f]	0.13	23	65.1[f]	1.56
1971[d,e]	225	5,400	2.2	2.12	0.08	23	29.0[f]	1.33
1971 (Denver)[e]	144	5,200	2.2	3.16[f]	0.33	23	45.2[f]	1.90

Source: CALSPAN Corp., *Automobile Exhaust Emission Surveillance: A Summary* (Ann Arbor, Mich.: U.S. EPA, Office of Air and Water Programs, May 1973, Tables 2–22.

[a] Denver, Detroit, Houston, Kansas City, Los Angeles, and Washington, D.C.

[b] Houston and Kansas City only; excludes Volkswagons.

[c] All except Denver.

[d] Detroit, Houston, and Los Angeles.

[e] Only engines that had "stabilized."

[f] Significantly above the applicable federal standard at the 95 percent confidence level.

Table 8.4 In-Use Emissions from 1957–1971 Automobiles in Six Cities[a] Based on 1975 CVS-CH Test (in 1972)

			HC			CO		
Model Year	Number of Cars	Average Mileage	Federal Standards	Mean Emissions	Estimated Standard Deviation of Mean	Federal Standards	Mean Emissions	Estimated Standard Deviation of Mean
1957–1967[b]	458	68,500	—	8.74 g/mi	0.36	—	86.5g/mi	1.88
1957–1967 (Denver)	97	65,100	—	10.16	0.10	—	126.9	4.92
1968[c]	84	46,400	5.9	5.54	0.77	50.8	67.8[d]	6.27
1968 (Denver)	18	42,000	5.9	7.35[d]	0.64	50.8	109.2[d]	12.37
1969[c]	89	39,500	5.9	5.19	0.45	50.8	61.7[d]	3.29
1969 (Denver)	17	38,900	5.9	6.32	0.84	50.8	76.4[d]	11.57
1970[c]	86	28,700	3.9	3.90	0.21	33.3	48.2[d]	2.66
1970 (Denver)	17	26,000	3.9	6.72[d]	0.93	33.3	94.8[d]	8.20
1971[c]	101	15,600	3.9	3.06	0.13	33.3	40.1[d]	2.44
1971 (Denver)	20	15,100	3.9	5.59[d]	0.32	33.3	88.1[d]	8.41

Source: J.Bernard, P. Donovan, and H.T. McAdams, *Automobile Exhaust Emission Surveillance—Analysis of the FY1973 Program*, (Ann Arbor, Mich.: CALSPAN Corp., July 1975), p. 60.

[a] Chicago, Denver, Houston, Los Angeles, St. Louis, and Washington, D.C.

[b] All except Denver and Los Angeles.

[c] All except Denver.

[d] Significantly above the applicable federal standard at the 95 percent confidence level.

Table 8.5 In-Use Emissions from 1966–1972 Automobiles in Six Cities[a] Based on 1975 CVS-CH Test (in 1973)

Model Year	Number of Cars	Average Mileage	HC Federal Standards	HC Mean Emissions	HC Estimated Standard Deviation of Mean	CO Federal Standards	CO Mean Emissions	CO Estimated Standard Deviation of Mean
1966–67[b]	140	69,300	—	8.67	0.59	—	93.5	3.39
1966–67 (Denver)	34	65,300	—	11.91	1.41	—	141.0	9.30
1968[c]	105	59,300	5.9	6.34	0.50	50.8	63.7[d]	3.29
1968 (Denver)	21	51,400	5.9	6.89	0.81	50.8	101.4[d]	14.36
1969[c]	110	50,900	5.9	4.95	0.31	50.8	64.2[d]	3.25
1969 (Denver)	22	46,100	5.9	5.97	0.27	50.8	97.8[d]	8.13
1970	135	37,500	3.9	5.24[d]	0.41	33.3	58.3[d]	3.30
1970 (Denver)	27	31,600	3.9	5.56[d]	0.30	33.3	87.5[d]	6.01
1971[c]	150	27,300	3.9	3.95	0.18	33.3	52.8[d]	3.30
1971 (Denver)	30	18,200	3.9	5.19[d]	0.32	33.3	80.3[d]	6.80
1972[c]	175	15,400	3.0	3.13	0.21	28.0	28.8[d]	1.84
1972 (Denver)	35	14,100	3.0	4.75[d]	0.41	28.0	80.4[d]	5.49

Source: J. Bernard, P. Donovan, and H.T. McAdams, *Automobile Exhaust Emissions Surveillance—Analysis of the FY1973 Program.* (Ann Arbor, Mich.: CALSPAN Corp., July 1975), p. 57.

[a]Chicago, Denver, Houston, Los Angeles, St. Louis, and Washington, D.C.

[b]All except Denver and Los Angeles.

[c]All except Denver.

[d]Significantly above the applicable federal standard at the 95 percent confidence level.

Table 8.6 In-Use Emissions from 1967–1974 Automobiles in Six Cities[a] Based on 1975 CVS-CH Test (in 1974)

Model Year	Number of Cars	Average Mileage	HC Federal Standards	HC Mean Emissions	HC Estimated Standard Deviation of Mean	CO Federal Standards	CO Mean Emissions	CO Estimated Standard Deviation of Mean	NO_x Federal Standards	NO_x Mean Emissions	NO_x Estimated Standard Deviation of Mean
1967[b]	68	68,100	—	8.65	0.71	—	108.3	6.44	—	4.04	0.22
1967 (Denver)	17	66,400	—	9.87	0.70	—	146.1	10.30	—	2.22	0.32
1968[c]	90	61,500	5.9	6.95	0.84	50.8	74.3[d]	5.23	—	4.94	0.26
1968 (Denver)	18	63,500	5.9	7.65[d]	0.76	50.8	97.0[d]	8.10	—	3.21	0.32
1969[c]	105	59,200	5.9	6.10	0.52	50.8	68.0[d]	3.43	—	5.17	0.21
1969 (Denver)	21	54,700	5.9	7.07[d]	0.48	50.8	104.6[d]	8.44	—	3.76	0.36
1970[c]	110	50,600	3.9	5.36[d]	0.47	33.3	65.0[d]	3.15	—	4.68	0.20
1970 (Denver)	22	45,700	3.9	6.56[d]	0.44	33.3	105.2[d]	6.76	—	3.22	0.33
1971[c]	135	36,500	3.9	4.15[d]	0.20	33.3	51.5[d]	3.02	—	4.50	0.17
1971 (Denver)	22	32,700	3.9	5.51[d]	0.29	33.3	96.9[d]	4.27	—	3.18	0.24
1972[c]	150	29,500	3.0	4.05[d]	0.29	28.0	53.7[d]	3.22	—	4.56	0.16
1972 (Denver)	30	27,500	3.0	5.40%0.35	28.0	28.0	90.5[d]	9.73	—	3.29	0.28
1973[c]	175	18,800	3.0	3.64[d]	0.18	28.0	45.4[d]	2.49	3.1	3.38 [d]	0.12
1973 (Denver)	35	14,300	3.0	4.54[d]	0.30	28.0	84.7[d]	6.98	3.1	1.96	0.15
1974[c]	50	6,900	3.0	3.03	0.16	28.0	35.5[d]	3.24	3.1	2.75	0.17
1974 (Denver)	10	5,300	3.0	4.19[d]	0.16	28.0	79.0[d]	4.55	3.1	1.81	0.26

Source: J. Bernard, P. Donovan, and H.T. McAdams, *Automobile Exhaust Emission Surveillance—Analysis of the FY 1973 Program* (Ann Arbor, Mich.: CALSPAN Corp., July 1975), p. 40.

[a] Denver, Detroit, Houston, Los Angeles, Newark, and St. Louis.

[b] All except Denver and Los Angeles.

[c] All except Denver.

[d] Significant above the applicable Federal standard at the 95 percent confidence level.

Table 8.7 Emissions by Model Year over Time (testing year)

Model Year	Mileage			HC Emissions				CO emissions			
	1972	1973	1974	Federal Standards	1972	1973	1974	Federal Standards	1972	1973	1974
1968[a]	46,400	59,300	61,500	5.9	5.54	6.34	6.95	50.8	67.8[b]	63.7[b]	74.3[b]
1968 (Denver)	42,000	51,400	63,500	5.9	7.35	6.89[b]	7.65	50.8	109.2[b]	101.4[b]	97.0[b]
1969[a]	39,500	50,900	59,200	5.9	5.19	4.95	6.10	50.8	61.7[b]	64.2[b]	68.0[b]
1969 (Denver)	38,900	46,100	54,700	5.9	6.32	5.98	7.07[b]	50.8	76.4[b]	97.8[b]	104.6[b]
1970[a]	28,700	37,500	50,600	3.9	3.90	5.24[b]	5.36[b]	33.3	48.2[b]	58.3[b]	65.0[b]
1970 (Denver)	26,000	31,600	45,700	3.9	6.72[b]	5.56[b]	6.56[b]	33.3	94.8[b]	87.5[b]	105.2[b]
1971[a]	15,600	27,500	36,500	3.9	3.06	3.95	4.15	33.3	40.1[b]	52.8[b]	51.5[b]
1971 (Denver)	15,100	18,200	32,700	3.9	5.59[b]	5.19[b]	5.51[b]	33.3	88.1[a]	80.3[b]	96.9[b]
1972[a]		15,400	29,500	3.0		3.13	4.05[b]	28.0		38.8[b]	53.7[b]
1972 (Denver)		14,100	27,500	3.0		4.75[b]	5.40[b]	28.0		80.4[b]	90.5[b]

Source: Tables 8.4–8.6.
[a] All cities except Denver.
[b] Significantly above the applicable federal standard at the 95 percent confidence level.

Table 8.8 Percentage of Vehicles that FAILED the Applicable Federal Standards in Six Cities

	1973 Testing Year				1974 Testing Year			
Model Year	HC	CO	Failed at Least One	Failed Both	HC	CO	NO_x	Failed at Least One
1972[a]	59	48	61	32				79
1972 (Denver)	86	91	97	80				97
1973[a]					54	58	55	85
1973 (Denver)					80	92	11	97
1974[a]					40	40	30	58
1974 (Denver)					100	100	10	100

Sources: J. Bernard, P. Donovan, and H.J. McAdams, *Automobile Exhaust Emission Surveillance—Analysis of the FY1973 Program* (Ann Arbor, Mich.: CALSPAN Corp., July 1975), pp. 8, 38–40; and M.E. Williams, J.T. White, L.A. Platte, and C.J. Domke, *Automobile Exhaust Emission Surveillance—Analysis of the FY1972 Program* (Ann Arbor, Mich.: U.S. EPA, Office of Air and Water Programs, February 1974), p. 6.
[a] All except Denver.

necessary to enable them to meet standards in use. Failure to meet standards in use therefore could reflect as much on owners as on manufacturers, at least after cars have been driven two or three years.

The interim standards imposed for 1975 appear to be generating similar results. Tables 8.9 and 8.10 provide the relevant figures for the 1975 model cars. Average CO emissions were significantly above the standard in five cities, and only a third of the cars met all three standards.[58] In Los Angeles, where the tougher California standards apply, most cars in use (74 percent) met the federal standards, but average NO_x emissions were above the California standards. One-year-old cars should continue to meet new car standards, regardless of maintenance, unless control devices have been tampered with.

Finally, in-use evaporative losses appear to be much greater than the standards. The standards were originally set at 6 g/test (roughly 0.6 g/mile) for 1971 (for 1970 in California) and then lowered to 2 g/test (0.2 g/mile) for 1972 and the years following. The federal testing procedure uses a charcoal cannister technique for measurement. In-use tests with a Sealed Housing for Evaporative Determinations (SHED) technique showed that 1970–1971 cars

Table 8.9 In-Use Emissions from 1975 Automobiles in Six Cities[a] Based on 1975 CVS-CH Test (in 1976)

Model Year	Number of Cars	HC			CO			NO_x		
		Federal Standards	Mean Emissions	Estimated Standard Deviation of Mean	Federal Standards	Mean Emissions	Estimated Standard Deviation of Mean	Federal Standards	Mean Emissins	Estimated Standard Deviation of Mean
1975[b]	587	1.5	1.3	0.04	15.0	22.9	0.97	3.1	2.4	0.04
1975 (Los Angeles)	35	0.9	0.5	0.05	9.0	6.6	1.16	2.0	2.4[c]	0.19

Source: Preliminary EPA figures.
[a] Chicago, Houston, Los Angeles, Phoenix, St. Louis, and Washington, D.C.
[b] All except Los Angeles.
[c] Significantly above the applicable standard at the 95 percent level.

Table 8.10 Percentage of 1975 Automobiles which FAILED the 1975 Federal Standards, in Six Cities[a]

	HC	CO	NO_x	Failed at Least One
1975[b]	28	49	20	63
1975-Los Angeles	0	9	9	26

Source: Preliminary EPA figures
[a,b]See Table 8.9

in Los Angeles had evaporative emissions of 27.2 g/test, or 2.1 g/mile; 1971 cars in Denver had evaporative emissions of 82.0 g/test, or 6.4 g/mile.[59] The Los Angeles emissions were 30 percent below the emissions of uncontrolled cars, similarly measured. Nevertheless, these results indicate that in-use HC evaporative emissions are serious and may exceed in-use HC exhaust emissions.

Overall, it is clear that in-use emissions have exceeded and are likely to continue to exceed the federal standards. This may be less true for HC exhaust emissions, but it is certainly true for CO exhaust emission and HC evaporative emissions. These excess values are not trivial; in Denver, the CO exhaust emissions have been two to three times the level of the standards and evaporative emissions have been more than ten times the standards. In-use emissions are still below those of uncontrolled cars, but the program simply is not working nearly as well as the certification procedures would lead one to believe. The EPA's lack of action to encourage better maintenance of vehicles in use must count as a serious failing in the administration of the emission control program.

Are the Original 1975–1976 Standards Worthwhile?

The best estimates of the costs and benefits of the current program come from the NAS-NAE report.[60] As we have seen, that report's best guess is that the discounted benefits come to $137 billion, whereas discounted costs come to $126 billion. We also claimed that the costs were probably understated, whereas the benefits were probably overstated. At best, then, the program probably just breaks even. The conclusion that total benefits are no larger than total costs of the program is a serious indictment. If the program chooses abatement levels at which total benefit equals total costs and if marginal benefit decreases and marginal cost increases with the percent abatement, it follows that marginal benefit is less than marginal cost at the abatement level dictated by the program. In other words, the large increases in costs exceed the small increases in benefits for the last percentage points of abatement mandated by the program. Our guess is that the program should aim at 80 to 85 percent abatement instead of 95 percent. This conclusion is independent of the criticism that the program has achieved abatement at needlessly high cost.

The report does point out that one alternative strategy would push the benefit-cost ratio substantially above 1.0. This two-car strategy provides for enforcement of the strict NO_x standards originally scheduled for 1976 only against the roughly one-third of cars sold in high-pollution areas. Since the marginal cost of meeting the NO_x standard is high, this appears to be a reasonable alternative within the framework of the existing program. Further, as Harrison demonstrates,[61] this alternative would be less regressive than the present program because most poor people live in rural areas, where the cost of emission control would decrease. The two-car strategy cannot be achieved by EPA administrative action alone, however; it would require new congressional action.

A Fundamental Critique

In the previous pages we criticized specific provisions of the 1970 legislation and some of the EPA's specific methods and procedures. Changes in these areas would improve the operation of the program. But many of these failures are indicative of the general spirit and philosophy underlying the emissions control program, namely, the notion held by Congress that the setting of specific, absolute, all-or-nothing emission standards is the best way to deal with pollution problems. It is this general philosophy that makes us most uneasy and that, we feel, has led to *slower and more costly* emissions control than otherwise could have been achieved.[62]

Congress has correctly perceived that the automobile companies will not provide emissions control voluntarily. Air pollution is a classical example of an external diseconomy in which a motorist emits pollutants and generates uncompensated costs for others but directly experiences few or none of the costs of these actions; in the absence of an incentive, few will pay the extra cost of emissions-control equipment. Congress's response to this problem has been to turn to the automobile companies and demand, "reduce the emissions of your automobiles to the following levels, or else. . . .'' The ''or else'' is an implicit threat to close down the company.

In a world of certainty, in which the technology, costs, and benefits are well known and understood, the correct set of standards could lead to optimum emissions control. But in a world of uncertainty concerning the benefits and particularly concerning the costs and the technology necessary to achieve the desired emissions control, this proposition becomes much less likely. The important aspect of policy in such circumstances is to create the proper incentives for productive research on new technology to control emissions and to distribute the burden of emissions control in a manner that minimizes total social costs. It is in exactly these areas that a policy of setting standards fails.

It is important to emphasize that the argument has to do only in part with the loss of welfare from an erroneous standard or incentive when social cost and benefit curves are subject to uncertainty. Much more important is the continuing incentive to seek cheaper and better means of production, an incentive that a market system provides. Economists might argue whether prices are at socially optimum levels at any given time in a particular manufacturing industry, but most economists agree that the market system provides powerful incentives to seek cheaper methods of production and better products. Applied to emissions control, this familiar argument provides a strong reason for economic incentives instead of standards.

Congress in 1970 wanted to push the automobile companies to undertake research and develop new technology for emissions control. It set tough standards for 1975 and 1976 that were beyond the technology available in 1970. Congress, confident in the marvels of American technology and satisfied with its ''or else'' sanction, fully expected the companies to comply by 1975 and 1976. But Congress recognized that technology was somewhat uncertain and provided the EPA with the power to grant a single one-year delay in the enforcement of the standards if it found that the technology was not available and the companies had made a good-faith effort to discover it. And, of course, Congress always retained the option of subsequently amending the legislation to delay or soften the standards. Because these options existed, however, the ''or else'' sanction to shut down the automobile companies in the event of noncompliance was, and is, simply not credible. All parties know that neither the EPA nor Congress would shut down the entire automotive operations of General Motors, Ford, or Chrysler for a significant time. This is a political fact recognized by all players in the auto-emissions-control game, but politics aside, it would not be justifiable to shut down a giant corporation because its cars missed an arbitrary emissions standard by a few percentage points.

The "brinkmanship" problem of the noncredibility of threats to shut down the industry is not an abstract, theoretical problem devised by academics. It has arisen a number of times in the administration of the emissions control program. In the spring of 1972, the Ford Motor Company revealed to the EPA that its personnel had improperly maintained its cars during the certification testing of the 1973 models. Ford had to begin testing its models again from the beginning and did not complete the certification until September, just before the new models were to be introduced. In the interim, EPA allowed Ford to ship the uncertified models to its dealers but not to sell them at retail. If most of Ford's models had failed the retesting, would EPA have refused to allow Ford to sell its cars? In an election year?

In April 1973, when the EPA granted a year's delay in meeting the original 1975 standards, the decision of the EPA administrator, William Ruckelshaus, appeared to be heavily influenced by Chrysler's total inability to meet the 1975 standards. By contrast, he estimated that 93 percent of General Motors' production would be able to meet the standards. He all but accused Chrysler of deliberately dragging its feet and, hence, of failing the good-faith requirement for justifying a delay.[63] Chrysler's expenditures on emissions control were between a sixth and a tenth of those of General Motors and Ford in absolute amounts and were a third as much per sales dollar. Chrysler had switched catalyst suppliers in September 1972, and the change apparently set back its emissions control efforts by six months. Ruckelshaus devoted four pages of the decision, plus a six-page appendix, to an analysis of Chrysler's behavior. In the end, he concluded:

> On such a record, the gravest questions as to Chrysler's compliance with the statutory requirements [of good faith] must arise. But a determination that they have not been met cannot be lightly made. . . .
>
> With regard to Chrysler, I conclude with serious reservations that the statutory requirements concerning good faith have been met. *In reaching this conclusion, I am placing decisive reliance upon the consideration that the sanction that arises from a negative finding on this issue with respect to a particular manufacturer could force that manufacturer to close down in 1975.* Such a result would not only create extreme hardship for large numbers of innocent employees of the manufacturer concerned but would also severely impact numerous suppliers of the manufacturer and ultimately the public at large. Thus, despite the very serious questions I have concerning the record as it relates to Chrysler on this point, I do not believe that Congress intended me to make a finding of bad faith in the absence of a very high degree of certainty that the acts of a particular manufacturer require such a finding. On this record, Chrysler's defense of its procurement deci-

sion and of its acts toward Engelhard [the original catalyst supplier] have raised sufficient doubt to preclude a positive finding of bad faith. [emphasis added][64]

The EPA was simply not prepared to shut down Chrysler. It is politically unthinkable to close down one of the domestic car manufacturers because its products miss an arbitrary emissions standard by a few points. Such action would hurt not only the manufacturers, the auto unions, and the thousands of parts suppliers, but also much of the motoring public.

In May 1975, a NAS report[65] concluded that there had been a slackening in the manufacturers' efforts to develop the technology to meet the strict NO_x standards. The slackening was surely a response to the three delays that had already been granted and the likelihood that, if no technology were available, Congress would have to grant further delays.

In 1976, the auto companies simply announced that the 1978 cars they were committed to manufacture would not meet the 1978 standards then mandated. This statement may have been a move in the brinkmanship game, but it was made in the knowledge that the standards had been repeatedly postponed, that Congress was considering further postponement, and that neither Congress nor the public had much sympathy left for extremely stringent and expensive standards. In any case, the auto companies, not the Congress, have decided what emissions standards 1978 cars will meet. Nobody any longer takes seriously the notion that the federal government will permit the law to be enforced as it stands in January 1977.

The program relying on standards has thus created conflicting incentives for the automobile companies. If they want to meet the deadlines, the companies must find quick technological solutions with a high probability of success, even if they are costly. Lower-cost solutions that take longer or have a lower probability of success (or perhaps only a higher variance) are likely to be neglected. At the same time, the companies realize that the EPA is unlikely to shut them down if they appear to make a good-faith effort. The good appearance is also necessary for public relations and to avoid subsequent punitive action by Congress. Accordingly, they have an incentive to slacken their costly research efforts as long as they can maintain appearances, because their bluffs are unlikely to be called.

As a consequence, we are getting an unsatisfactory research effort: the total research effort is too low and it is focused on quick, high-probability, high-cost solutions. The catalytic-converter technology chosen by the American companies to meet the original 1975 standards was considered a high-cost but high-probability method of meeting the standards. It now appears that stratified-charge engines may be a superior technology, but they were a riskier prospect in 1970. With a better set of incentives, the American companies might have spent more effort on the stratified-charge engines and on other technologies. It is not surprising that foreign manufacturers were first to demonstrate automobiles with noncatalyst technologies that could meet the original 1975 standards (the Honda stratified-charge engine, the Mazda rotary engine, and the Peugeot diesel engine). Foreign manufacturers considered the American market important but were not vitally dependent on it; they would survive if they failed to develop appropriate technologies and had to withdraw from the market for a year or two. Hence, they could afford to explore the higher-risk technologies, and they succeeded.

The delays that have been granted in the enforcement of the standards have undermined the credibility of the program. They have been granted at scattered and uncertain intervals; they have introduced uncertainty needlessly and have discouraged the best long-run research. This is simply not good policy.

The existence of simultaneous standards on a number of pollutants has also impeded research. There is a trade-off for most engine technologies between HC-CO emissions and NO_x emissions. Efforts to reduce the former frequently lead to increases in the latter.[66] In the absence of a reliable reducing catalyst or some other technology to reduce NO_x emissions, exhaust gas recirculation is the only way to reduce those emissions. But exhaust gas recirculation exacts a high penalty in fuel consumption, at least in some engines. Consequently, the NO_x standards have discouraged efforts on diesels and stratified-charge engines that could achieve low HC and CO emissions, because of the high levels of NO_x emissions they produce.

The policy of standards has also focused virtually all control efforts on the manufacturers. Incentives for motorists to maintain their cars could produce significant reductions in emissions, but they are totally lacking. In part, this gap is a consequence of the administration of the program. The EPA has not encouraged the states to develop inspection programs, nor has it endorsed the tests that would be necessary for state inspections. But the states, with the exception of New Jersey and California, have not come forward to develop inspection programs on their own, and the Congress has not pressured EPA or the states to move faster. Thus, the absence of maintenance encouragement is primarily a consequence of the philosophy that has underlain emissions control policies of the federal and state governments for more than two decades, namely, "emissions are the manufacturers' responsibility; here are the standards; the manufacturers must meet them or else. . . ."

The section of the 1970 amendments that requires manufacturers to warrant their emissions control systems for five years or 50,000 miles is another product of the philosophy. The potential difficulties surrounding this warranty provision have been described by the RECAT report, White, one of the NAS advisory panels, and Grad et al.[67] Briefly, there are no current incentives for motorists to enforce the warranty (which may be a blessing in disguise); because the entire engine is part of the emissions control system, the manufacturers might well require all engine repairs to be done at their dealers before they will honor the warranty, with disasterous consequences for independent repair shops; because each car is not currently tested as it leaves the factory, there may be some cars that could not have passed the standards when they were new and could never pass the standards; and because the CVS-CH test is not a practical test for cars on the road, a short test will be necessary, but it could lead to complications if a car could pass one test but not the other.

Furthermore, not only do motorists lack incentives to maintain their cars, but there is also nothing to prevent them from disconnecting or readjusting the emissions control devices, the act of which might *improve* the performance of the car.[68] The NAS report[69] cites recent surveys in Washington, D.C., and New Jersey that indicates that up to 15 percent of 1970–1974 vehicles had one or more emissions control devices deactivated and up to a third exhibited signs of some form of tampering.

In addition, the program has probably placed too much emphasis on automotive emissions and not enough on other sources. This is especially true if one adds the EPA's proposed transportation restriction plans to the original 1975–1976 emissions standards. We noted earlier the looser standards that have applied to trucks and buses and the lack of standards for motorcycles. Stationary sources of the same pollutants—thermal electric generation, space heating, gasoline distribution, industrial processes, and solid waste disposal—have not been restricted to a comparable degree with respect to HC, CO, and NO_x emissions.

Finally, the philosophy behind the policy of standards setting has meant a nearly total disregard for the quantification of costs and benefits and for their consideration in policy choice. Cost considerations are not part of the language of the 1970 amendments with respect to the 1975–1976 standards.[70] Though the EPA had conducted some benefit studies,[71] it is shocking that the first full-length study of costs and benefits, with rigorous efforts to quantify and compare both, was conducted only in 1974 (the NAS-NAE report), nine years after the federal policy of setting standards had begun.[72]

In sum, the philosophy underlying the emissions standards is one that stresses moral solutions over efficient ones. Perhaps that is what the body politic wants. As economists rather than moral philosophers, we can express only dismay and disagreement.

Conclusion

The EPA is a fair, even-handed agency. There has been no obvious bias in its decisions, and there have been no scandals. But many of its administrative decisions have not encouraged efficient emissions control. The larger program of setting emissions standards and the philosophy underlying the program have not generated the right incentives for rapid and properly conceived research on emissions control or a proper distribution of the burden of control. As a consequence, we have had slower and more costly reductions in emissions than need have been the case. It is ironic that the policy of direct emissions standards and controls—a policy that is usually favored by policy makers because it is supposed to yield rapid and predictable results—has instead yielded slow and tortuous results.

If alternative policies were not available and if asked whether the program is worth it, we would respond, following the NAS-NAE report, with a grudging "yes," particularly if some modifications could be made. But superior alternative policies could have been and still could be, implemented. We will advocate just such a policy in the next section.

A Practical Proposal for an Effluent Fee Program

Introduction

The goal of public policy should be to structure incentives so that the maximization of private interests serves also to maximize social interests. No serious writer believes this goal can be accomplished with respect to motor-vehicle emissions in the absence of an explicit government program for the purpose. Present policies do not achieve this goal.

The proximate goal of an auto pollution control policy should be to motivate competition in the production of clean, nonpolluting cars. It should reward those who produce clean cars and those who maintain their cars so as to keep them clean; it should penalize those who produce dirty cars or who allow them to become dirty. Our proposal for an effluent fee program achieves these proximate goals.

Effluent fees have long been proposed by economists as a desirable means to solve pollution problems, but the proposals are usually vague and abstract and pay little attention to practical detail. The proposals are rejected by policy makers and are read mainly by other economists. The members of the profession have been talking largely to themselves!

We believe our proposal is practical and workable. The remainder of this section will present our proposal in detail. Then we will discuss the advantages of our proposal.

Effluent Fees for New Cars

We propose that effluent fees for new cars be levied on all cars sold (we hope this program will be extended on a comparable basis to all other motor vehicles) on the basis of measured emissions for a test sample of vehicles. The manufacturers would submit the sample of vehicles for representative models, in advance of sales, as is done now. A small fleet would be tested for 50,000 miles; a larger fleet would be tested for 4000 miles and the results extrapolated to 50,000 miles, as is done now. Ideally, the cars would be tested for 110,000 miles, the average life, but the extra time and cost probably make this impractical. If the used-car effluent fee program described below is *not* implemented, then the present CVS-CH test can be used because it measures what is appropriate. On the other hand, if the used-car effluent fee program is implemented, the EPA will have to decide on a stationary hot test to be used for both new and used cars.

There would be no certification or absolute standards. It would be legal to sell any car, regardless of how dirty it is, provided the proper effluent fee is paid.

The *average* emissions for each model over the 50,000 miles would be used to compute the effluent fee for that model. The fee structure would be the following:

$$F = F_{\mathrm{HC}}(\mathrm{HC}) + F_{\mathrm{CO}}(\mathrm{CO}) + F_{\mathrm{NO_X}}(\mathrm{No_x})$$

where F is the total fee and F_{CH}, F_{CO}, and $F_{\mathrm{NO_X}}$ are the individual fees for HC, CO, and $\mathrm{NO_x}$. The HC emissions would include evaporative and blow-by emissions as well as exhaust emissions.

The individual fees should be set so as to equate the marginal benefits and marginal costs of abatement. As we saw previously, estimation of the marginal benefits is particularly difficult, and rough guesses must be made. It is important to emphasize that these guesses must also be made in setting standards; an effluent fee program does not add any extra burden in calculating costs and benefits. The data from the NAS-NAE report suggest that F should be about \$300 for an uncontrolled car in an area with minor pollution problems and about \$900 for an uncontrolled car in an area with major pollution problems.[73] As a realistic example, we propose the following fee schedules for the two areas:

Low pollution area:

F_{HC} = \$6.33/g/mi.
F_{CO} = \$1.15/g/mi.
F_{NO_X} = \$25.00/g/mi.

High pollution area:

F_{HC} = \$19.00/g/mi.
F_{CO} = \$3.45/g/mi.
F_{NO_X} = \$75.00/g/mi.

Tables 8.11 and 8.12 illustrate the effluent fees that would be paid on cars that met the standards that have been implemented or proposed. As can be seen, with the low fee schedule, manufacturers would find it worthwhile to build cars that met the 1970 standards; with the high fee schedule, they would find it worthwhile to meet the 1975 California standards. Furthermore, we shall argue later that if an effluent fee schedule had been implemented in 1970, the costs of meeting the various standards would be lower than under the present program and still greater emissions reductions would be likely.

Since different areas of the country have different pollution problems, different fee schedules ought to be used. One possibility would be to have the federal fee differentiated according to the destination of the car. An alternative, which we prefer, would be to have a single federal fee schedule, say, at the level of table 8.11, and to permit the states to add extra effluent fees of their own. The states could establish both a fee schedule for new cars, to be paid at the time the car is first registered in the state, and a separate schedule for cars in use (with a separate testing program, discussed below) to encourage maintenance. We would expect the states with serious pollution problems to set fees so that the discounted sum of new-car and used-car fees on an uncontrolled car would be \$600 or more; thus the combined federal and state fees would approximate those in table 8.12. States with minor pollution problems or whose citizens did not care about pollution would be free to abstain from imposing effluent fees. States that wanted to differentiate their fees by county or administrative district would be free to do so. Adjoining states that share an urban area with pollution problems would be encouraged to coordinate and equalize their fees.

Differential state fees may create an incentive for "bootlegging" new cars from low-fee states to high-fee states. To our knowledge, California has not faced any bootlegging difficulties because of its more stringent controls. Nevertheless, if it became a problem, there is any easy solution: the state effluent fee would be owed on any new or used car when it is first registered in that state. The effluent fees paid to another state would count as an

Table 8.11 Effluent Fees (Low Schedule) on Cars that Meet the Various Standards

	Fees Paid[a]				Estimated Discounted Cost of Meeting Standards[b]	Total of Fee and Cost of Meeting Standards
	HC[c]	CO	NO_x	Total		
Uncontrolled car	$100	$100	$100	$300		$300
1968 standards	56	59	125[d]	240	20	260
1970 standards	44	39	125[d]	208	30	238
1972 standards	20	32	125[d]	177	72	249
1973 standards	20	32	78	120	587[e] (280)	(400)
1975 interim standards	11	17	78	106	280	386
1975 California standards	7	10	50	67	380	447
1975 original standards	4	4	50	58	430	488
1976 original standards	4	4	10	18	630	648

[a] F_{HC} = $6.33; F_{CO} = $1.15; F_{NOX} = $25.00.
[b] Costs for 1968–1972 are from D.N. Dewees, *Economics and Public Policy:The Automobile Pollution Case* (Cambridge, Mass.: MIT Press, 1974), Appendix C. Costs for later years are from National Academy of Sciences, *Report by the Committee on Motor Vehicle Emissions* (Washington, D.C.: U.S. Government Printing Office), p. 94. The NAS figures are compared with a 1970 car, so $30 has been added to the NAS figure.
[c] Uncontrolled blowby emissions are 4.1 g/mile; controlled blowby emissions (1968 and after) are nil. Uncontrolled evaporative emissions are 3.0 g/mile; controlled emissions (1972 and after) are 0.2 g/mile.
[d] The 1968–1972 controls cause NO_x emissions to rise to about 5.0 g/mile.
[e] $557 is the NAS-estimated cost with the 1973 technology.

Table 8.12 Effluent Fees (High Schedule) on Cars that Meet the Various Standards

	Fees Paid[a]				Estimated Discounts Cost of Meeting Standards[b]	Total of Fee and Cost of Meeting Standards
	HC[c]	CO	NO_x	Total		
Uncontrolled car	$300	$300	$300	$900		$900
1968 standards	169	176	375[d]	720	20	740
1970 standards	131	116	375[d]	622	30	652
1972 standards	61	97	375[d]	533	72	605
1973 standards	61	97	232	390	587[e](280)	(670)
1975 interim standards	32	52	232	316	280	596
1975 California standards	21	31	150	202	380	582
1975 original standards	11	12	150	173	430	603
1976 original standards	11	12	30	53	630	683

[a] $F_{HC} = \$19.00$; $F_{CO} = \$3.45$; $F_{NOX} = \$75.00$.
[b] Costs for 1968–1972 are from D.N. Dewees, *Economics and Public Policy: The Automobile Pollution Case* (Cambridge, Mass.: MIT Press, 1974), Appendix C. Costs for later years are from National Academy of Science, *Report by the Committee on Motor Vehicle Emissions* (Washington, D.C.: U.S. Government Printing Office), p. 94. The NAS figures are compared with a 1970 car, so $30 has been added to the NAS figure.
[c] Uncontrolled blowby emissions are 4.1 g/mile; controlled blowby emissions (1968 and after) are nil. Uncontrolled evaporative emissions are 3.0 g/mile; controlled evaporative emissions (1972 and after) are 0.2 g/mile.
[d] The 1968–1972 controls cause NO_x emissions to rise to about 5.0 g/mile.
[e] $557 is the NAS-estimated cost with the 1973 technology.

offset up to the amount due, as is now done for state income taxes on individuals who work in one state and reside in another.

The federal effluent fees would be paid when the manufacturer shipped the car to the dealer; the state fee would be paid when the car was first registered. The manufacturer would be free to absorb or pass on as much of the fee as he saw fit. The emissions results for that model would be added to the window sticker of every car, along with a statement of the fee paid.

Assembly-line audits would supplement the prior-to-sale tests. If the audits revealed that certain models were dirtier than the presale tests had indicated (with sufficient margin allowed for the inherent variability of the tests), the manufacturer would be liable for extra effluent fees on all cars likely to be involved. This provision would discourage the practice of providing carefully engineered ''cream puffs'' for the presale tests.

To ensure proper planning of research, engineering, and design by the manufacturers, the effluent fees should be announced five years in advance and should be unalterable, except in extraordinary circumstances, after they have been announced. They should, however, be indexed to the CPI for new automobiles to take into account inflationary trends. This indexing could also be announced five years in advance. Each year a new set of fees would be announced for five years hence.

Comparable federal and state fee schedules should apply to all sources of emissions, mobile and stationary. Because the important measurement with respect to pollutants is emissions per time period—say, per year—comparability on this basis should be the goal. Thus, if an average car travels 12,000 miles per year and an average truck travels 20,000, the truck emission fees, in terms of grams per mile, during the test should be 20/12 of those for cars.[74] The stationary-source fees should be similarly calculated. Our proposed federal fee for HC, $6.33/g/mile during the test, is the equivalent of 0.007¢/g emitted during the 110,000 mile lifetime of a car (using a 5 percent discount rate); this amount would be the appropriate corresponding fee for stationary sources.

Finally, the fee schedule could be expanded to include other harmful pollutants that are presently or might potentially be emitted, such as lead compounds or sulfur oxides. Again, additions to the list should be announced five years in advance.

Cars in Use

In a perfect world, the automobile manufacturers would discover a pollution control device that would last forever, never deteriorate, require no maintenance, and be tamper-proof. Unfortunately, this situation is unlikely. All emissions control systems, actual and proposed, suffer deterioration and require maintenance for proper operation. As demonstrated previously, present programs offer no incentives for maintenance, and a car owner may even gain from degrading his system. Studies by Horowitz and Gafford and Hule indicate that an inspection-maintenance regime can yield significant decreases in emissions.[75] Accordingly, we believe that an effluent fee system on cars in use would be a valuable part of an emissions control program. It would encourage maintenance of the emissions control system by the car owner, the only one who can do anything about maintenance after the car leaves the dealer's showroom.

The states would be encouraged to establish emissions inspection-testing procedures, perhaps as part of safety inspections or as a new, separate inspection. The EPA could provide technical assistance and financial grants to the states to promote these programs.

A quick (one- to five-minute) hot-start test would be required for state inspections. Because the hot-start test could not correlate perfectly with the CVS-CH test, administrative simplicity would probably call for the same hot-start test to be administered to both new and in-use cars.[76]

The states should be encouraged to establish an effluent fee schedule, similar in form to that for new cars, to be levied on the basis of the annual test results. The states would be free to set their fee schedules for each pollutant or could forgo inspecting and taxing entirely. States without serious pollution or whose citizens did not care about pollution would be free to choose the latter route.

The fee schedule should be geared not only to emissions per mile but also to annual mileage driven. Thus, with an annual schedule of, say, $F_{HC} =$ 0.014¢/g/mile, $F_{CO} =$ 0.003¢/g/mile, and $F_{NO_X} =$ 0.057¢/g/mile, an uncontrolled car driven 12,000 miles in a year would pay $85; the same car, if it could meet the original 1975 standards, would pay only $16.[77] This schedule should provide adequate incentives for maintenance. And it would provide a strong penalty for very high emitters.

Gearing the effluent fee on cars in use to annual mileage would require that cars be manufactured with a tamper-proof odometer, or at least one that had a reliable and accessible seal that would indicate any tampering. We guess that this is not a difficult or expensive task. If it is, however, the effluent fees could exclude the mileage factor and be scaled up accordingly, though a desirable property of the fee schedule would be lost.

Every car should be given the hot test as it comes off the assembly line, before it is shipped to the dealer. Since all cars are now given a short test by the manufacturers to see that they run properly, the extra hot test should not be a significant burden. Then, the following information would be added to the window sticker of each car: the expected average emissions over 50,000 miles, as determined by the presale tests for the federal effluent fee program; the emissions at zero miles of the sample vehicles in the presale test; and the actual emissions (at zero miles) of the vehicle in question. Thus buyers would know the likely future emissions of the particular car they wanted to purchase. They could estimate its likely future emissions taxes and resale value and decide whether the vehicle was likely to be an emissions lemon.

A further word on testing is warranted. We have endorsed the concept of a hot-start test for both the new-car effluent fees and the subsequent state used-car effluent fees. There are great administrative advantages to the use of a single hot test. There would be complete consistency between the federal and state fee programs. Each new car could be tested as it came off the assembly line, and as already explained, this information would provide motorists with an idea of the likely future emissions taxes they would have to pay. The information on the window sticker of the car would be concise yet complete for the buyer's purposes. In addition, the automobile companies would learn about the likely emissions properties of each car and, when possible, would ship the high emitters to the states with the low emissions fees. We would be willing to sacrifice something in the representativeness of the test in order to gain these advantages.

There are many who believe that the CVS-CH test has such superior properties that it should be retained for the new-car test. If it were, the used-car program would face a choice: (1) the hot start test would be administered only on used cars, or (2) although the CVS-CH would continue to be the basis of the new-car emissions fee, the hot-start test would be administered to every new car as it rolled off the assembly line for future information purposes only. Option (1) would have the disadvantage that new-car buyers would have only a rough idea from the CVS-CH test (administered only to sample cars) as to the future hot-test emissions performance (and hence fees) of their cars. There will always be some cars that are low emitters on the CVS-CH and high emitters on any hot test, or vice versa; new-car buyers would be quite unhappy about buying a car they believed to be a low emitter and discovering that, despite their best maintenance efforts, it turned out to be a high emitter on hot-start tests.[78] Option (2) would provide buyers and manufacturers with desirable information about the likely future performance of the car, but it would mean posting an extra set of test results on the car's window, a practice that would probably cause confusion on the part of new-car buyers.

In sum, we would prefer a uniform hot-start test, for the reasons argued. Nevertheless, a decision to keep the CVS-CH for a sample of new cars would not make a used-car testing program impossible, and either of the alternatives just mentioned is feasible. Finally, the states would be encouraged to institute the same program for all motor vehicles and a similar one for stationary sources.

Conclusions

The program we have outlined is consistent, practical, and workable. The next section will discuss the advantages and costs of the program and the uses to which revenues collected under the program might be put. We will also offer some speculation as to the control levels and costs that would be present today if our program had been adopted instead of the current standards program.

Advantages, Costs, Revenue Use, and Some Speculations

Advantages

The effluent fee schedule we have proposed, unlike the current program, would provide the right incentives for research, low-cost production, and maintenance. It would introduce vital elements of competition and flexibility into the national program. Manufacturers and owners of clean cars would be rewarded and those of dirty cars would be penalized.

The manufacturers would no longer face all-or-nothing, zero-one deadlines for meeting standards. Instead, they would have proper incentives to plan and pursue research on emissions control in an optimum fashion, and therefore they might pursue riskier strategies if they thought the expected value of a particular strategy was positive. They would no longer have incentives to collude, to delay progress on emissions control, or to bluff or engage in brinkmanship contests with government authorities.[79] They would be rewarded for producing clean cars and for producing control devices that were durable and easily maintained. They would be able to trade off among pollutants and could choose low-cost strategies that were especially good at reducing some pollutants even if they could not reduce others. We would not have the current situation, in which difficulties in reducing NO_x emissions appear to be holding up the production of low-cost engines that are very good at reducing HC and CO emissions. Finally, the hot test applied to each car as it emerged from the assembly plant would give each manufacturer a good guess as to the likely future emissions of that particular car. Accordingly, the manufacturers would have an incentive to ship the dirty cars to states with low effluent fees (where, presumably, pollution problems were less severe or citizens did not care) and ship the clean ones to states with high effluent fees. All of these actions would be consistent with efficient resource allocation and the rapid and efficient production of new knowledge about emissions control. The program would shift companies' incentives away from political lobbying and brinkmanship, where their interests clash with those of the rest of society, to designing and building low-cost, clean cars, where their interests and society's coincide.

In accordance with the fundamental theorem of effluent fees, the proposed effluent fees would induce the greatest emissions abatement of those cars and by those manufacturers for which abatement was cheapest; it would equate marginal abatement costs among sources, as the present program does not. Furthermore, if a similar fee schedule were applied to other mobile sources and to stationary sources, as suggested in the preceding section, the program would equalize marginal abatement costs among all sources, mobile and stationary. Again, present programs do not come close to achieving this.

Consumers would have an incentive to seek, buy, and maintain clean cars. They would also become interested in additional retrofit devices if they found the devices cost-effective. The owners' incentives to maintain their vehicles would extend over the entire life of the vehicle; current pollution control policy ceases to account for a vehicle's emissions after 50,000 miles. Manufacturers would have a competitive incentive to inform consumers about the low-emissions properties of their cars and to compete on their emissions performances. Consumer publications would have an incentive to do the same. Again, we would be encouraging an efficient allocation of resources in reducing emissions.

Furthermore, the emissions fee program would probably be less regressive in its income consequences than the current program. Harrison[80] calculates that a two-car strategy would be considerably less regressive in its overall effects than enforcement of original 1975–1976 standards, because most poor people live outside the large urban areas with pollution problems and would therefore escape the high costs required for emissions controls in these areas. Our effluent fee proposal is, in effect, an *n*-car strategy, and we believe that it would have similar consequences for income distribution. Most poor people live in rural areas and states where effluent fees are likely to be lowest. There will, of course, be instances in which poor people with dirty cars will face high effluent fees or expensive repair bills. However, surely no one can still believe that the poor are not currently paying for pollution control, and the evidence by Harrison and Dorfman indicates that, on average, they are paying in a regressive fashion.[81] Despite instances of dirty cars, we believe that our program overall would be less regressive than the current program.

Costs of the Program

The administrative costs of our proposed program should not be large. The new-car effluent fee program should not cost more than the current program. The used-car program would involve added expense for the tests on all new cars at the factory and the annual inspections in the states that choose to have effluent fee programs. The cost of the factory test should be quite modest: one to two dollars for the labor involved in a two- or three-minute test and perhaps another dollar for the equipment and overhead.[82] Three dollars per car for a 10 to 12 million cars per year comes to only $30 million to $36 million.

The state inspection program cost would be more substantial. It would consist of inspection costs, repair costs to owners (less any consequent fuel savings)[83] and time and inconvenience costs to owners. At the high end, we could suppose that all states chose to have inspection systems. At present, thirty states plus the District of Columbia require annual or semiannual safety inspections, either in state inspection facilities or in commercial repair shops. In these states, the added costs for the manpower and equipment necessary for the emissions tests would be small, probably less than a dollar per car. The cost for a state effluent fee program would be hardly more than the cost for a state effluent standards program.

The extra repair costs, less fuel savings, are difficult to estimate. Because many owners normally repair and maintain their cars regularly, we must try to measure only the extra costs induced by the fee schedule. Gafford and Hule report one experience with maintenance: repairs costing $27–$37 per car yielded a 44 percent reduction in HC emissions and a 34–43 percent reduction in CO emissions on high-emissions cars.[84] Unfortunately, fuel savings, if any, are not reported.[85] Since a 2 percent gasoline savings means a $10 annual saving to offset repair costs, the figure could be significant.[86] It is difficult to predict exactly how much extra maintenance the effluent fees would induce. As a rough example, let us stick with the Gafford and Hule experience, though it did not involve effluent fees and involved only adjustments to high emitters to bring them to a set of HC and CO standards. The high emitters constituted 30 percent of the car population. Their repair costs were $27–$37 per car. If we conservatively allow a $10 offset for extra fuel economy, we are left with \$17–\$27 per car. If we average over the entire population (remembering that 70 percent did not require repairs), we get \$5–\$8 per car.

The time and inconvenience costs are equally difficult to quantify, and we have no prior measurements. Because emissions inspections could be done at the same time as safety inspections and hence the only extra time would be for the extra maintenance itself, we feel that an *average* of one additional hour of inconvenience is a conservative guess.[87] A value of \$4 for that hour seems reasonable. The total cost per car for these states, then, is about \$10–\$13 per car, or \$590–\$737 million for the 59 million cars in these states. The remaining eighteen states would have to upgrade and expand the occasional inspection facilities now in commercial repair shops or would have to develop new facilities. We estimate this cost would be, at most, \$2 per car in facilities and labor.[88] The other costs would be the same, except that an extra hour of inconvenience would probably be involved, so the total cost per car would be \$15–\$18 per car, or \$690–\$828 million for 46 million cars. The grand total for all states would be \$1.3–\$1.6 billion.

Finally, if we assume that the states with pollution problems could administer the inspections so that only cars registered in the affected counties need be tested, only 36 percent[89] of the cars in the country, or 39 million, would need inspection. At an average of \$15 per car, the cost would be \$585 million.

Even the last figure is not small, but recall that we have included owner time and inconvenience costs, which estimates of the current program's costs have not. We believe that these costs are worth the improved maintenance behavior and consequent emissions reduction that would occur.

Putting the Effluent-Fee Revenues to Use

A question that is sometimes raised in discussions of effluent fees concerns the use to which the collected revenues would be put. If new-car owners pay an average of \$100 per car, 10 to 12 million new-car sales means \$1.0 to \$1.2 billion in extra federal revenues. If 35 million cars in use pay an average of \$10 per car, the program would generate \$350 million for the states. How should these funds be used?

Earmarking is, in general, a bad fiscal principle. In principle, government budgets should be allocated to equate costs and benefits at the margin for each program. It is hard to think of a program in which earmarking of funds will achieve the optimum public resource allocation. Optimum government expenditures on municipal sewage-treatment-plant construction hardly depend on revenues collected in the motor-vehicle effluent fee program. For example, an innovation in auto emissions control technology could reduce auto effluent fee collections substantially during a few years. But that would hardly justify a cutback in government money to build municipal sewage plants. Furthermore, earmarking tax revenues removes public resource allocation from the appropriations process, and hence from scrutiny by the democratic process. Even in an aggregate sense, there is no reason for institution of an effluent fee program to increase government expenditures. It may be that other taxes should be reduced so that government expenditures are unchanged.

Nevertheless, earmarking effluent-fee revenues is a remarkably popular notion. Some people are greatly attracted to the notion that effluent fees should be used to reduce the suffering from remaining polluting discharges. This type of appropriation must be made with care to avoid loss of efficiency. The use of effluent fees from a smoky factory to compensate downwind residents for damages from remaining discharges inappropriately lowers the cost of residence there relative to other sites and results in excessive pollution damages. In practice, however, resource misallocation is not likely with auto effluent fees. In principle, there is no loss of efficiency if fee revenues are earmarked for programs on which optimum expenditures exceed earmarked funds. The federal government spends more for pollution control (most of it for waste treatment plants and research) than would be collected from a motor-vehicle effluent fee program. Careful earmarking of fee revenues for these purposes would not introduce inefficiency and would, in a general way, lessen the suffering from pollution that remains.[90]

What Would Have Happened if . . .

Suppose that in the early 1950s government officials in Southern California had tried a different strategy. Instead of alternatively pleading with and shaking their fists at the automobile manufacturers, suppose they had spent money on research, gathered more information, and in 1955 established an effluent fee schedule similar to the one we have proposed. Suppose also that only the new-car effluent fees were imposed. Is it conceivable that progress would have been any *slower* than it has been during the past twenty years?

Suppose, again, that instead of establishing standards the federal government had instituted an effluent fee program along our suggested lines in 1965. Recall that the sizable effluent fees not only would have inspired the manufacturers to engage in productive research but also could have created a sizable public fund to finance yet more research. Again, is it conceivable that progress could have been slower?

Finally, let us be more concrete and ask what would have happened if the

1970 amendments to the Clean Air Act had incorporated our effluent fee proposal instead of the mandated 1975–1976 standards. Research by the manufacturers would have been better focused. More attention would have been given to systems other than the oxidizing catalyst. Chrysler's shenanigans in 1972 would have had no effect on public policy. There is a much greater likelihood that Wankels, stratified-charge engines, and/or diesels would be on a significant fraction of new cars. We strongly suspect that new cars would be cleaner or cheaper than the cars that have been produced under the current program. Some of the expensive mistakes of the past few years would have been avoided.

Let us try to put some numbers on these claims. The NAS report[91] indicates that the 1973 standards cost, on average, $587 per car (compared to an uncontrolled car). Two years later, technology had advanced so that catalyst-equipped cars could meet the stiffer interim 1975 standards at only $280 per car. The high costs associated with meeting the 1973 standards were mostly the result of extra fuel consumption. If the effluent fee schedule had been in effect, no one would have used this costly technology. As a conservative guess, the saving would have been $350 per car. Because 11.4 million cars were sold in 1973 and 8.7 million were sold in 1974, almost all of which embodied the 1973 technology,[92] about $7 billion in extra costs could have been avoided had our effluent fee schedule been in effect.

Second, suppose that the better-focused research encouraged by our effluent fee program had brought stratified-charge engines, a fuel-efficient rotary, and/or diesels to the practical production level. Since the manufacturers would not have worried about the necessity of meeting an absolute NO_x standard, research on these engines would have proceeded faster and more productively. It would be impossible for the entire industry to change from conventional engines to these new engines overnight. But 10 or 20 percent of the 1975 models might have contained these engines. The NAS report indicates[93] that the Honda stratified-charge engine (with a likely improvement in carburetion) could meet the original 1975 standards at a cost of only $380, or $50 less than the estimate of current likely costs. It estimates that the fuel-injected stratified-charge engine meeting the original 1975 standards would save $280 compared to a conventional uncontrolled car (because of large fuel savings), so this engine would be $760 cheaper than the likely cost of meeting the standards; it would cost only $124 to meet the original 1976 standards. A diesel would *save* $473 in meeting the original 1975 standards, so it would be $953 cheaper than the likely cost; given current technology, a diesel could not satisfactorily meet the original 1976 standards.

Savings of this magnitude, even for only 10 to 20 percent of 1975 cars, would yield appreciable total annual savings. If only 10 percent of 1975 cars could have used fuel-injected stratified-charge or diesel engines, at an average saving of $850 for 830,000 vehicles, the aggregate saving would have been $705 million for that year alone. And these cars would be meeting the original 1975 standards in that year! In subsequent years, the savings would grow as the market shares for these efficient vehicles grew. The NAS-NAE report[94] shows that a gradual increase in fuel-injected stratified-charge engines to a market share of 38.2 percent in 1985 would yield a present discounted saving of $5.5 billion for the years 1970–1985, compared with a strategy relying on conventional engines with catalysts (the original 1976 standards having been suspended). We believe that this is a conservative estimate of the magnitude of savings that the better-focused research program would have yielded.

Finally, the emissions fee schedule would probably bring an improvement in benefits. First, the fee program would ensure that cars did not deteriorate in use as badly as they now seem to. Actual emissions from new cars meeting any given standard would be less than is now likely to be the case. Further, an emissions fee program could quite realistically bring appreciable reductions in emissions of existing cars. Let us again refer to the maintenance experiment described by Gafford and Hule.[95] Maintenance costing $27–$37 per car decreased the emissions of the high emitters by 4.0–7.5 g/mile for HC and by 36–65 g/mile for CO.[96] For a car driven 12,000 miles per year, an annual fee schedule of 0.014¢/g/mile for HC and 0.003¢/g/mile for CO would probably be sufficient to induce most of this maintenance, plus some additional maintenance from low emitters.

In the Gafford and Hule experiment, the average decrease in emissions for all cars—those that required repair and those that did not—was 20 percent for HC and 18 percent for CO. If we project these improvements to national levels, we obtain the equivalent of the reduction in emissions that would be brought about by three years of new cars meeting the original 1975 standards. The NAS-NAE report values this latter improvement at $1.7 billion.[97] Thus, emission fee programs on cars in use, which would cost $1.3–$1.6 billion in administrative, repair, and inconvenience costs, would have a favorable benefit-cost ratio. If the program were restricted to high pollution areas, the benefits would remain about the same, whereas costs would fall to $585 million. The program appears to be quite worthwhile.

Conclusions

An effluent fee program would have great advantages in encouraging sensible economic behavior by all actors involved in emissions control. The costs are modest, especially for the new-car emissions fee program, and the benefits substantially outweigh the costs. The program would be less regressive than the current program. And the revenues, viewed as a payment to society for the use of clean air, could sensibly be used for pollution-related purposes.

The Opposition to Effluent Fees

Background

We are not the first to propose effluent fees as a way to solve pollution problems, though other proposals are rarely as detailed. The advantages outlined here have been advanced before. Yet there is little interest in effluent fees by any of the major groups involved in pollution problems. Except for an abortive sulfur tax bill, effluent fees have never been seriously considered as a way of dealing with air pollution problems. To say that there is opposition to them may be overstating the case. They do not seem to be within the framework of arguable policies that anyone would bother to oppose. This section will discuss the attitudes of each group.

Government Employees

The EPA has sole responsibility for administering the automobile emissions program. Neither EPA nor CEQ (Council on Environmental Quality) has shown much interest in measuring benefits from environmental programs or in economic incentives. An outstanding example of this lack of interest occurred shortly after the EPA's April 1973 decision to grant the first delay in implementation of the original 1975 standards. The administrator, William Ruckelshaus, was asked to testify before the Senate Public Works Committee concerning his decision. He had gone through a confrontation with the industry, he had had to back down, and he recognized that the sanctions at his disposal were inadequate and noncredible. Yet, when asked, he showed a complete lack of interest in effluent fees.[98] He clearly had not been briefed on them. They obviously had not been actively debated before him by his staff.

The EPA currently has extraordinary power to influence decisions concerning production, investment, consumption, product design, and waste disposal for many goods and services. An effluent fee system would be a threat to this direct power. Small wonder that the EPA has given it little attention. Standards setting or enforcement justifies a considerably larger organization and permits discretionary actions such as those we have witnessed in the auto program. These discretionary actions involve the kind of exciting, tense negotiations and confrontations that are irresistable to politically ambitious people.

The Congress and State Legislatures

Legislators appear to have a deep-seated suspicion of economic incentives and of economic analysis. Most behavior is assumed to be completely price inelastic, so effluent fees would not deter polluters and sufferers would continue to suffer. And responsibility for any resulting price changes could be easily traced to the legislation. By contrast, standards and controls offer the promise of direct results; results do not depend on indirect relationships via taxes, incentives, and behavioral changes. Few seem troubled by the fact that the promise fails to coincide with reality. And the responsibility for price increases associated with the standards is more easily muddied. Congress, in particular, has adopted a hostile and vindictive attitude toward the automobile companies and has passed legislation that has had large indirect costs for consumers; but the costs are difficult for the average voter to identify, trace, and assign responsibility for.

It used to be claimed that legislators did not like effluent fees because they were "licenses to pollute." A recent scanning of congressional hearings indicated insufficient interest in effluent fees to generate even the "license-to-pollute" denunciation.[99]

Companies

Individual polluters would certainly oppose effluent fees. The fees would be harder to evade than standards. Delays would be unlikely. Polluters have much to gain from the current system.

Some enlightened business groups, notably CED, have endorsed effluent fees as a sensible way of dealing with pollution problems.[100] But there is deep anxiety that industry may get both standards and fees and thus be saddled with an extra set of taxes but no extra flexibility.

Environmentalists

Like other groups, environmentalists generally show little interest in economic incentives and economic analysis. Many environmentalists are distrustful of the for-profit private enterprise system and hence are suspicious of incentives that rely on that system. They fail to see the coincidence of private and social interests that effluent fees could bring. They fear the income equity consequences of effluent fees; they appear unaware of (or refuse to believe) the regressive nature of current programs. They want clean air, water, and land, and appear to be largely unconcerned with costs. They apparently believe, or hope, that the costs will be absorbed out of profits.

Some environmental groups have recently endorsed effluent fees on a theoretical basis, but their hearts do not seem to be in it. They still want standards.

Summary

We are left with a fundamental paradox of taxation, which applies to pollution and to other areas as well: our political system is willing to legislate direct controls and standards (under the guise of having a direct impact) that have the effect of high distorting taxes and great administrative complexity. The system is unwilling to levy simple taxes that remove distortions. Perhaps the fault is the system's inability to arrange for costless income transfers so that winners could compensate losers. Perhaps it is a deep distrust of economic motives. Perhaps too many of us still believe in free lunches.

Conclusions

More than twenty-five years have been spent studying, analyzing, and trying to control automotive emissions. The country has learned a great deal about the technology of emissions creation, pollution formation, emissions monitoring, and emissions control. Unfortunately, much less has been learned about the adoption of sensible policies. One indication of this lack of knowledge is the Energy Policy and Conservation Act of 1975, which mandates improvements in automotive fuel economy from the 1975 averages of 13–14 miles per gallon to 18 mpg in 1978, 20 mpg in 1980, and 27.5 mpg by 1985. Congress has shown the same lack of concern for costs, benefits, economic incentives and consumer tastes in this area that it has shown in auto emissions policy. Once again cars have been emphasized and trucks neglected. We confidently predict a pattern of confrontation and improperly motivated research in this area similar to that in emissions control.

The effluent fee program that we have proposed is simple and direct. It would motivate people to take socially beneficial actions at the lowest cost possible. It would encourage proper research. It would decrease the large indirect costs of bureaucratic controls and result in a pervasive emphasis on producing and maintaining clean cars.

The problem is important. The means of properly dealing with it are at hand. It is a shame they are not being used.

Notes

Support has been generously provided by MIT and by a grant from the Sloan Foundation to the Economics Department, Princeton University. We thank Paul Rampell and Jeffrey Smisek for research assistance on this paper.

1. See U.S. Department of Commerce, *The Automobile and Air Pollution: A Program for Progress,* Report of the Panel on Electrically Powered Vehicles (Washington, D.C., October 1967), p. 11. As hydrocarbon emissions from automobiles decrease, evaporation of gasoline from service stations may become a relatively important source of emissions. In 1972 there were 226,459 gasoline stations in the United States.

2. Further details on the history of emissions control can be found in F. P. Grad et al., *The Automobile and the Regulation of Its Impact on the Environment* (Norman, Okla.: Oklahoma University Press, 1975), chapter 8; H. D. Jacoby et al., *Clearing the Air: Federal Policy on Automotive Emissions Control* (Cambridge, Mass.: Ballinger, 1973), pp. 9–14; and L. J. White, *The Automobile Industry Since 1945* (Cambridge, Mass.: Harvard University Press, 1971), chapter 14.

3. See R. Nader, *Unsafe at Any Speed* (New York: Pocket Books, 1965), pp. 122–123.

4. See *U.S.* v. *Automobile Manufacturers Association, Inc.,* 1969 Trade Cases, No. 72907.

5. This is reprinted in U.S. Senate, Committee on Public Works, Subcommittee on Air and Water Pollution, *Decision of the Administrator of the Environmental Protection Agency Regarding Suspension of the 1975 Auto Emission Standards,* Hearings (Washington, D.C., 1973), pp. 445–456.

6. See Nader, op. cit., p. 120.

7. It is worth noting that the devices that reduced HC and CO emissions tended to increase nitrogen oxide emissions, for which standards were not set until 1973. HC and CO are products of incomplete combustion, whereas NO_x is a natural product of combustion.

8. See D. N. Dewees, *Economics and Public Policy: The Automobile Pollution Case* (Cambridge, Mass.: MIT Press, 1974), Appendix C.

9. D. S. Barth et al., "Federal Motor Vehicle Emission Goals for CO, HC and NO_x Based on Desired Air Quality Levels," in U.S. Senate Committee on Public Works, Subcommittee on Air and Water Pollution, *Air Pollution—1970,* Part 5 (Washington, D.C., 1970).

10. For example, the highest carbon monoxide reading was in Chicago in 1967.

11. Criticism of Barth's procedures is found in Jacoby et al., *Clearing the Air,* pp. 166–174. It is doubtful that relevant thresholds exist for large populations. In any case, the measurement procedure employed in collecting data used by Barth et al. was defective, and the calculations produced an excessively stringent threshold.

12. See L. J. White, "American Automotive Emissions Control Policy: A Review of the Reviews," *Journal of Environmental Economics and Management* 2 (April 1976): 231–246.

13. See Dewees, op. cit.

14. See D. Harrison, Jr., *Who Pays for Clean Air?* (Cambridge, Mass.: Ballinger, 1975).

15. U.S. Office of Management and Budget, *The Budget for Fiscal Year 1977, Special Analyses* (Washington, D.C., 1976), p. 302.

16. The maintenance and fuel costs estimated by the report have been discounted at 4 percent for eleven years and added to the hardware costs.

17. These include extra hardware, fuel, and maintenance. A 4 percent discount rate is used for the latter two items. Dewees (*Economics and Public Policy,* Appendix C) estimates that the extra costs of meeting the 1970 standards come to $30, so this sum should be added to the NAS estimates for the total cost of meeting the emissions standards.

18. See Dewees, op. cit., pp. 104–111.

19. National Academy of Sciences and National Academy of Engineering, *Air Quality and Automobile Emission Control,* prepared for the U.S. Senate, Committee on Public Works (September 1974), vol. 4, chapter 2.

20. The report, written in 1974, assumed that the original 1975–1976 standards would take effect in 1977–1978 and also included the costs of the interim 1975 standards.

21. Harrison, op. cit., chapters 3 and 6.

22. The size of the cost underestimate depends on the assumed mix of vehicles and their fuel consumption in each year.

23. Harrison, op. cit., chapter 6.

24. N. S. Dorfman, "Who Will Pay for Pollution Control?—The Distribution by Income of the Burden of the National Environmental Protection Program, 1972–1980," *National Tax Journal* 28 (March 1975): 101–115.

25. For other pollutants, such as particulates, decreased cleaning costs would also count as benefits, but automotive pollutants do not generate significant cleaning costs.

26. NAS-NAE, *Air Quality and Automobile Emission Control,* vol. 3.

27. Idem, vol. 2.

28. Hydrocarbons are not harmful pollutants by themselves, but they are a vital input, along with NO_x, into the production of ozone and other oxidants.

29. NAS-NAE, op. cit., pp. 351–360.

30. T. C. Schelling, "The Life You Save May Be Your Own," in S. B. Chase, ed., *Problems in Public Expenditure* (Washington, D.C.: The Brookings Institution, 1968), pp. 127–166.

31. R. Ridker and J. Henning, "The Determinants of Residential Property Values, with Special Reference to Air Pollution," *Review of Economics and Statistics* 49 (May 1967): 246–257.

32. L. J. White, "American Automotive Emissions Control Policy," pp. 240–241.

33. Comments and criticisms of the housing model are also found in A. M. Polinsky and S. Shavell, "The Air Pollution and Property Value Debate," *Review of Economics and Statistics* 57 (February 1975): 100–104; and K. A. Small, "The Air Pollution and Property Value Debate," *Review of Economics and Statisitcs* 57 (February 1975): 105–107.

34. Harrison, op. cit., chapter 7.

35. D. L. Rubinfeld, *Market Approaches to the Measurement of the Benefits of Air Pollution Abatement,* June 1976, mimeographed.

36. See National Academy of Sciences, *Consultant Report to the Committee on Motor Vehicle Emissions, Commission on Sociotechnical Systems, National Research Council, on Field Performance of Emissions-Controlled Automobiles* (Washington, D.C., November 1974), chapters 6–8. The National Academy of Sciences, *Report by the Committee on Motor Vehicle Emissions* (Washington, D.C., November 1974), seems unduly optimistic on this point. A CALSPAN report (CALSPAN Corp., *Automobile Exhaust Emission Surveillance: A Summary* (Ann Arbor, Mich.: U.S. EPA, Office of Air and Water Programs, May 1973), Sec. 1.2) indicates that HC and CO emissions tend to increase at a rate of 5 percent per 1000 miles for low mileage and 0.5–1.0 percent per 1000 miles after the engines have "stabilized."

37. NAS-NAE, op. cit., pp. 415–421.

38. U.S. Ad Hoc Committee on the Cumulative Regulatory Effects on the Cost of Automotive Transportation (RECAT), *Final Report* (Washington, D.C., February 1972), pp. 33–34.

39. See Dewees, op. cit., chapter 6; Jacoby et al. op. cit., chapter 3; Grad et al., op. cit., chapter 5; and NAS-NAE, op. cit., vol. 4, chapter 3.

40. They would be cumbersome to administer because they would have to be graded carefully by area and because they would have to be levied on all parking spaces, including the large number for which no charges are made. To the extent that they reduced driving to and from downtowns, they would divert through traffic from beltways and other routes around downtowns.

41. But Volvo may have recently made an important breakthrough in three-way catalysts. See the *Los Angeles Times,* 30 May 1976, p. I-1.

42. National Academy of Sciences, *Report of the Conference on Air Quality and Automobile Emissions, May 5, 1975, to the Committee on Environmental Decision Making,* June 1975, p. 1, mimeographed.

43. Jacoby et al., op. cit., pp. 38–47.

44. Standards for evaporative emissions—2 g of HC per test—also exist.

45. CVS-CH stands for constant-volume sampling—cold and hot (start).

46. See *The Wall Street Journal,* 21 July 1976, p. 10.

47. NAS-NAE, op. cit., p. 49.

48. For the problems of the variability in the CVS-CH test, see NAS, *Report by the Committee on Motor Vehicle Emissions,* pp. 150–162; for other tests and the problems of correlation, see National Academy of Sciences, *Feasibility of Meeting the 1975–76 Exhaust Emission Standards in Actual Use,* Panel on Testing Inspection and Maintenance for the Committee on Motor Vehicle Emissions (Washington, D.C., June 1973), chapters 3 and 4.

49. NAS-NAE, op. cit., p. 164; and Grad et al., op. cit., pp. 130–132.

50. These are trucks under 6000 pounds in weight; they are mainly pickup trucks.

51. The computations underlying this statement are in White, ''American Automotive Emissions Control Policy,'' p. 244.

52. See *The New York Times,* 26 October 1975; and U.S. Department of Commerce, Bureau of the Census, *Statistical Abstract of the United States, 1975* (Washington, D.C., 1975), p. 571.

53. U.S. Environmental Protection Agency, *Monitoring and Air Quality Trends Report, 1974* (Research Triangle Park, N.C.: Office of Air and Water Management, February 1976), chapter 4; U.S. Council on Environmental Quality, *Environmental Quality—1975* (Washington, D.C., December 1975), chapter 3.

54. See Grad et al., op. cit., pp. 125–129. Motor vehicles are the source of slightly less than half total NO_x emissions.

55. See *The New York Times,* 22 July 1976, p. 28.

56. Owners of problem cars might be tempted to have them tested if they thought it would help them to obtain valuable servicing from manufacturers. But only emissions are tested, and servicing to improve emissions control is of benefit to the owner.

57. CALSPAN Corp., *Automobile Exhaust Emission Surveillance.*

58. These results were released to the press and received a modest amount of attention. See *The New York Times,* 22 July 1976, p. 28.

59. CALSPAN, op. cit., section 2.2.3; Grad et al., op. cit., pp. 130–132.

60. NAS-NAE, op. cit., vol. 4.

61. Harrison, op. cit., chapter 8.

62. For a similar critique applied to other pollution control programs, see L. J. White, ''Effluent Charges as a Faster Means of Achieving Pollution Control,'' *Public Policy* 24 (Winter 1976): 111–125.

63. The administrator's decision is found in U.S. Senate, Committee on Public Works, op. cit., pp. 2–51.

64. Ibid., p. 43.

65. NAS, *Report of the Conference on Air Quality and Automobile Emissions*, p. 1.

66. This is primarily due to the fact that high engine temperatures discourage HC and CO emissions but encourage NO_x emissions; low engine temperatures do the opposite.

67. U.S. Ad Hoc Committee on the Cumulative Regulatory Effects on the Cost of Automotive Transportation, op. cit., pp. 36–37; L. J. White, "The Auto Pollution Muddle," *The Public Interest* 32 (Summer 1973), pp. 97–112; NAS, *Feasibility of Meeting the 1975–76 Exhaust Emission Standards in Actual Use*, pp. 79–88; and Grad et al., op. cit., pp. 372–374.

68. Mechanics are forbidden to deactivate the devices, but owners are not.

69. NAS, *Report by the Committee on Motor Vehicle Emissions*, p. 120.

70. The only exception relates to the setting of interim standards, if the administrator grants a one-year delay. These interim standards are to take costs into consideration, but the decision as to whether a delay should be granted was not supposed to include cost considerations.

71. See U.S. EPA, *The Economics of Clean Air*, Report of the Administrator (Washington, D.C., March 1970); L. B. Barrett and T. E. Waddell, *The Cost of Air Pollution Damages: A Status Report* (Research Triangle Park, N.C.: U.S. EPA, Air Pollution Control Office, February 1973); and T. E. Waddell, *The Economic Damages of Air Pollution* (Washington, D.C.: U.S. EPA, Office of Research and Development, May 1974).

72. NAS-NAE, op. cit.

73. Ibid., Vol. 4.

74. In the case of an initial fee intended to cover the lifetime of the vehicle, a correction factor for the expected years of life of the two vehicles would also be necessary.

75. J. Horowitz, "Inspection and Maintenance for Reducing Automobile Emissions," *Journal of the Air Pollution Control Association* 23 (April 1973): 273–276; and R. D. Gafford and T. A. Hule, *Effectiveness of Short Emission Inspection Tests in Reducing Emissions Through Maintenance* (Ann Arbor, Mich.: U.S. EPA, Office of Air and Water Programs, July 1973).

76. If the relationship between the initial emissions of the car and its eventual average emissions could be perfectly predicted, then the initial emissions might be used as the basis for the new-car effluent fee for each individual car.

77. These annual effluent fee figures were determined by subtracting the low-fee schedule from the high-fee schedule, dividing the difference by the lifetime mileage—110,000—and using an annual discount rate of 5 percent.

78. If the hot-start test on a "green" engine (that is, the car as it rolls off the assembly line) were to prove to be an equally unreliable guide to the emissions

properties of that car on the same hot-start test in the future, our suggestion for a uniform hot-start test would lose some of its advantages.

79. Except insofar as information from the industry helped determine the size of the effluent fee.

80. Harrison, op. cit., pp. 120–123.

81. Harrison, op. cit.; and Dorfman, op. cit.

82. At $10 per hour (wages and fringes), two assembly-line workers spending three minutes each on a test would involve a labor cost of $1 per car inspected. The National Academy of Sciences (*Report by the Committee on Motor Vehicle Emissions* (Washington, D.C., February 1973), pp. 82–83) indicates that the inspection equipment costs $2000 per lane; this lane could process 80,000 cars in one year. Gafford and Hule (op. cit., p. I-16) estimate the cost of state inspection systems at $1.16–$1.35 per car, but this figure does not include the value of the time of the worker who would have to be inside the vehicle.

83. Of course, the fuel saving cannot exceed the sum of the repair costs and the time and inconvenience costs. Otherwise, the owner would undertake the maintenance without the prod of the effluent fee.

84. Gafford and Hule, op. cit., p. I-16.

85. Though some repairs to decrease emissions may sacrifice fuel economy, other repairs, such as timing and carburetor adjustments, frequently improve fuel economy.

86. A car traveling 12,000 miles per year at 15 miles per gallon consumes 800 gallons per year. At 60¢ per gallon, the cost is $480 per year, of which 2 percent is $9.60.

87. In states where inspection is performed in private facilities, the maintenance is frequently done at the same time as the inspection, so that extra time costs are small.

88. Gafford and Hule (op. cit., p. I-16) put the cost at $1.16–$1.35 per car.

89. See NAS-NAE, op. cit., vol. 4, p. 100.

90. Dewees, op. cit., pp. 94–95; White, "American Automotive Emissions Control Policy," pp. 242–243.

91. NAS, *Report by the Committee on Motor Vehicle Emissions,* chapter 6.

92. The reported figures are for calendar years, whereas the standards applied to model years. The difference should be small.

93. NAS, *Report by the Committee on Motor Vehicle Emission,* pp. 92, 98.

94. NAS-NAE, op. cit., vol. 4, p. 98.

95. Gafford and Hule, op. cit., p. I-16.

96. NO_x emissions were increased, but the experiment was not designed to encourage repairs that decreased NO_x emissions.

97. NAS-NAE, op. cit., vol. 4, p. 417.

98. See the excerpts from his testimony in White, ''Effluent Charges as a Faster Means of Achieving Pollution Control,'' pp. 122–123.

99. Ibid.

100. Committee for Economic Development, *More Effective Programs for a Cleaner Environment* (New York, April 1974).

Comment

James A. Fay

The proponents of effluent fees have a rational case. Our present regulatory approach, by law or political necessity, must treat equally each manufacturer and consumer. But some manufacturers are better able to develop cheaper abatement equipment than are others, and some consumers live where less abatement is required than do others. Why should we not use economic incentives to provide a better-tuned response to the need for pollution abatement so that the desired goal can be reached sooner and at less overall cost than is now being achieved by a uniform, inflexible regulatory policy? Despite the evident lack of public interest in this rational approach, Mills and White bravely propose an automobile effluent fee program to cure the numerous regulatory ills to which our national pollution control program is heir.

The general deficiencies of our present policies and programs are clearly set forth by the authors. Progress in reducing emissions has been too slow and costs are too high. Incentives to develop improved control systems have been distorted by rigid emission standards. Congressional legislation has preempted all considerations of balancing abatement costs against health benefits. Most of all, the imposition of inescapable standards at well-defined deadlines has encouraged brinkmanship and confrontation. The ensuing deferral of deadlines has injured the credibility of the national program.

The authors also have more detailed criticisms of specific aspects of the present program. From studies to date, they conclude that the costs probably exceed the benefits and that very likely the minimization of total abatement plus social costs will not be achieved under present goals. One study shows that present abatement costs are regressive with respect to consumer income distribution. There are also significant technical difficulties. The original intent of the legislation to require the manufacture of control systems having lifelong reliability is not being accomplished because emissions of vehicles in use exceed the applicable standards, presumably because of inadequate maintenance by owners. Economical emission test procedures have not been developed to track each vehicle from production line to scrap yard so as to provide an enforceable means of assuring continuing emission control.

In Mills' and White's view, effluent fees would avoid most of these drawbacks. Brinkmanship tactics would no longer be needed since consumers rather than bureaucrats or judges would decide whether or not a given model automobile would be produced. Manufacturers would concentrate their efforts on cheaper and better technology instead of lobbying Congress and the administration for less stringent standards. States could apply emission fees to vehicles in use if local air quality demanded stringent control over a vehicle's lifetime. Both federal and state governments would receive increased income. For any given level of air quality, the aggregate annual costs of automobile pollution control would be less than will be the case for the regulatory approach. If fees are properly set, the level of control that will result will minimize total societal costs, even though air quality will probably be lower than what is now thought to be needed.

Few would argue with the authors' enumeration of the defects of our present national regulatory program. In exchange for the current policy, however, the authors offer an alternative with many advantages and seemingly no disadvantages. If they are correct in their views, the scant attention hitherto paid by government policy makers to such proposals has been a serious and costly mistake. On the other hand, the repeatedly poor reception given to the effluent fee trial balloons may reflect a well-calculated estimation of their drawbacks, which are only faintly recognized by their proponents. The latter is my view, which I will expand in these brief remarks.

The major obstacle to public acceptance of an effluent fee system for automobile pollution control is the increased payments by consumers. Effluent fee payments (presumably passed on to the consumer by the manufacturer) will equal or exceed the cost of pollution control equipment (and its maintenance). (In the examples of tables 8.11 and 8.12, optimum fee payments are one to two times control costs.) Aggregate annual payments will probably be $2 to $3 billion. It seems quite possible that these transfer payments from one sector of the economy to another will result in inefficiencies that will more than balance the hoped-for efficiency gains from the effluent fee system.

Over time, there has emerged a clear national consensus that the costs of pollution abatement should be paid by the consumers of goods and services whose production gives rise to pollutant emissions. This is the basis of cost allocation in most federal and local control programs. Only to a small extent are control costs sometimes born by the taxpayer through tax abatements or subsidy payments from general revenues. It will be exceedingly difficult to gain acceptance of the proposition that consumers should pay more than the cost of abatement, especially two or three times that cost.

It is not consoling to argue that effluent fees will replace other tax revenue. Based upon their experience that a new tax does not replace an existing one but adds to the total tax burden, most consumers will regard effluent fees as an excise tax payment and will resist it heartily. If present automobile pollution control costs are regressive, as indicated by the studies the authors cite, then effluent fees will make control payments even more regressive. Should federal effluent fees be instituted, there will be strong pressure to reduce more progressive revenue sources, such as personal income taxes, and compound the regressive nature of the fee system. Although economists classify an effluent charge as a fee, the public regards it as a tax and judges its desirability in comparison to other taxes or revenue-producing measures.

It is hoped that the potentially damaging confrontation between the automotive industry and Congress over the setting and timing of emission standards would not occur if effluent fees were to replace fixed emission standards. But would not a conflict still persist over the amount and timing of the effluent fee schedule? In time of economic downturn—which may become chronic in future years—the automobile manufacturers will apply intense pressures to reduce effluent fees to keep down automobile price increases and to maintain industry employment. Industry scientists will argue that the fee-setting authorities have overestimated the marginal health benefits of emission reduction and, therefore, fees should be lowered. New technology will be developed in anticipation of lower rather than higher fee schedules with the expectation that political pressure by the industry will produce the lower fees, as it now produces a stretch-out of the emission standards. The pushing and shoving between the automotive industry and the federal government is an intrinsic element of pollution control and will not be decreased greatly by the substitution of effluent fees for emission standards.

The imposition of additional effluent fees by states with especially severe urban air pollution would have little effect upon emissions unless manufacturers were able to produce and distribute to statewide markets vehicles with several grades of control systems. Given the competition among states and regions to spur economic and population growth, however, it is unlikely that individual states will agree to measures that will increase personal transportation costs vis-à-vis a competitive state or region unless the health benefits are overwhelmingly obvious. The sad history of state air-pollution-abatement efforts has provided not only the impetus for federally controlled abatement programs but also the support for nationally uniform standards. Although efficiency gains may support the argument that more local option should be given, political history suggests that this would be a retrograde step toward less abatement.

The determination of a schedule of effluent fees will be more difficult than choosing emission standards. While it is theoretically defensible to set the fee equal to the marginal health benefit expected to ensue from a marginal reduction in emissions, as the authors suggest, from a practical viewpoint it is virtually impossible to do so because of our very uncertain quantitative understanding of the link between emissions and the cost of health effects. This is especially true when atmospheric chemical reactions intervene to modify the effects of emissions of primary pollutants by formation of toxic secondary pollutants. It would be much easier to set effluent fees so as to achieve a desired reduction in emissions. In doing so, however, it would be necessary to guard against the invention of a perverse but cheap technology that might increase emissions of one pollutant while reducing that of another and thereby partially frustrate the ultimate purpose of emissions control.

The considerable attention given by Mills and White to the deterioration of emissions as automobiles age and presumably undergo abuse by their owners deserves some comment. Since the intent of the present control program is to reduce average emissions by 90 percent or more below uncontrolled levels, it is not necessary that emissions not exceed the test standards during a vehicle's lifetime, or even averaged over that lifetime, but only that deterioration of controlled vehicles should not be worse than that of their uncontrolled predecessors. (Tables 8.5–8.7 appear to show that deterioration of controlled vehicles is no worse than that of uncontrolled ones.) But the imposition of effluent fees and maintenance requirements on older vehicles would be an economically regressive policy in that the necessary expenditures would be paid by low income groups who are the predominant owners of older automobiles. Furthermore, such expenditures are probably less cost effective than would be improvements to the new-vehicle control system. In other words, it is most likely more cost effective and less regressive to reduce new-car emissions and permit greater deterioration in the older vehicles. This policy would shift some of the abatement costs to the more affluent new-car owner, who is better able to pay for them. Effluent fees for vehicles in use would be counter-productive in both respects.

It is all too easy to accuse Congress, the federal bureaucracy, environmentalists and the automotive industry of a know-nothing attitude toward the use of effluent fees. The nearly universal dislike of such proposals reflects not ignorance but a shrewd assessment of the political disadvantage to each party in abandoning the present system, which has clearly perceived political costs and benefits, for one that appears to have no political benefits for anyone. The issue is not whether effluent fees would result in increased economic efficiency—it is almost irrefutable that they would—but whether the somewhat uncertain gains in economic efficiency that might be achieved are worth the political costs of restructuring an ongoing program.

The dominant characteristic of automobile air pollution abatement is that it is a political contest. The rules by which scientific, technological, and economic factors determine the kind of automobile we buy and the quality of the air we breathe are set by the federal government in response to the pressures of the automotive and transportation industries and their labor organizations, consumer and environmental groups, and public health advocates. Although it is as desirable that abatement programs make use of efficient management techniques (such as effluent fees) as it is desirable to use efficient control technology, the fine tuning of the program to include these gains does not now appear to justify rewriting the rules of the game. Only if there is a clearly allocatable advantage of the effluent fee proposal to one or more of the contending groups (which seems not to be the case) is there a likelihood that an advocate for their adoption can prevail over the status quo.

Comment

Richard L. Garwin

I come to this writing and to these views as a physicist with more than twenty years of experience in science policy and in the management of industrial science and technology. Control of motor-vehicle emissions occupied me during four years of service on the National Research Council Committee on Motor Vehicle Emissions (CMVE) where, among other concerns, I was primarily involved with the technology of control. I endorse the view of the history of this field related by Mills and White in their paper.

The U.S. motor-vehicle manufacturers have had mixed interests. On the one hand, they could be expected to try to meet the emissions standards at the least manufacturing cost irrespective of operating cost (until EPA-mandated testing of fuel economy and the increased price of fuel). On the other hand, the manufacturers could be expected (and fulfilled these expectations) to lobby vigorously against the imposition of emission limits in order to avoid increasing the manufacturing costs of vehicles and thus (in view of the price elasticity of demand) reducing the number sold. The main problem has arisen because one of the chief arguments against the imposition of limits (or for the easing of limits) has been to show technological infeasibility of the standards. Thus, U.S. manufacturers long avoided assembling in a single car (''best car'') the control technologies that they have demonstrated individually and that could be projected with assurance (''best technology'') to work.

Foreign manufacturers seeking a share of the U.S. market had little confidence in their ability to lobby against the imposition of emission standards and have, as a result, been the source of some of the earliest-certified, least-cost (both manufacturing and operating) control technologies, some of them with performance much better than required by the standards at the time the vehicles were certified. Those engines that particularly come to mind are the Honda CVCC (a dual-carburetor stratified-charge engine) and the Bosch oxygen-sensor feedback-controlled fuel-injection system with three-way catalyst. The latter was certified in 1976 by Volvo.

Before proceeding further to detail the inadequacies of the motor-vehicle emissions control program and to recommend detailed improvements, I wish to emphasize that I believe the incentives are all wrong under the absolute limits prescribed for emissions from new cars or from cars that have been in service for some time (as mandated by the Clean Air Amendments of 1970). As I have indicated, such absolute limits simply invite manufacturers to delay until they can demonstrate that they have little chance of responsibly planning to meet the limits at the time required; therefore (the manufacturers argue) if jobs are not to be lost and the economy to suffer, the limits must be relaxed. This incentive to delay could be removed and manufacturers given a positive incentive to reduce emissions as low as economically feasible by the imposition on manufacturers of an effluent tax on motor-vehicle emissions. Thus, if the desired limits on hydrocarbons, carbon monoxide, and nitrogen oxides are, respectively, 0.4, 3.4, and 0.4 grams per mile, then a car (that is, each generic car of a tested vehicle-engine combination) that emitted an excess 10 percent of one of the regulated pollutants would have to pay $100 per vehicle. One that was 15 percent over would have to pay $150; one that was high by 20 percent in one pollutant and 30 percent in another would have to pay $500. In order to convert the tax from a money-maker for the government, and perhaps unintentionally reduce sales of motor vehicles, car types that produce less emission than the standard would receive rebates in like amount. If the standard is far under the average emissions of vehicles manufactured in a given year, a substantial tax would have been imposed on the industry (the customer). But there is a continuous differential to provide the incentive for the manufacturer to reduce the emissions from his vehicles so that he can sell cars in competition with other manufacturers. In similar manner, the requirements of the 1970 Clean Air Act that cars in *use* meet the limits could be reflected in a tax of similar nature and magnitude imposed on the manufacturer following testing by the EPA of sufficient number of cars in use (for 50,000 miles).

Some manufacturers, of course, may choose not to meet the standards in a given year, perhaps in order to introduce in the following year greatly improved technology. This would be their right—they could still sell cars but at a competitive disadvantage compared to other manufacturers. Yet they would not be able to argue that Congress and the EPA were making it impossible (or illegal) for them to manufacture and sell cars.

Some Details

Cost of NO_x Control by PEGR

Long after technical data were available to the contrary, opponents of increasing the stringency of NO_x standards maintained that a very large fuel-economy penalty was experienced through the use of EGR. This was a reflection of the very earliest crude EGR, which was simply an orifice between exhaust and intake manifolds. Such an orifice provided highest flow rate of exhaust gases when the intake of the engine was *least,* with serious impairment of fuel economy at high EGR levels.

In contrast, proportional EGR (PEGR) provides a constant fraction of recycled exhaust gas at a reasonable cost.[1] For example, it has been estimated that it is possible to achieve equivalent fuel economy without EGR at an air-fuel ratio of about 16.7 and with PEGR at an optimized air-fuel ratio of 14.7.[2] In fact, the penalty for PEGR is not fuel economy at all but some increase in hydrocarbon emissions.

EPA Delays

I am dismayed by the intensity with which the EPA promulgates control plans that are unacceptable, infeasible, and often unnecessary. Relying on a simplistic model of dispersal of sulfates produced by oxidation catalysts, the EPA seemed well on the road to barring or limiting the use of catalysts. Not academic seekers after truth and better modeling, but General Motors, which had a massive investment in catalyst technology, demonstrated (in this case by experiment) that vehicle motion reduced the high local concentration predicted by the EPA model.

Emissions Averaging

It is of course the total emissions from a fleet of cars (the number of cars times the average emissions, by definition) that is important in air quality. Insistence on holding *every* car to the emissions standard lends force to the manufacturers' argument that the standards are unachievable. If the effluent tax were introduced, there would be no standard, and the tax would be assessed, obviously, on the average for the fleet. If one retains firm standards, I believe it important even at this late date to require only that the *average* emissions meet the standard. At the same time, it is obviously important to specify a "short test" that will at least verify that the emissions control equipment has not been disconnected. Such short tests could be made mandatory for licensing and relicensing of vehicles. Such tests would be facilitated by having ports conveniently available in the engine exhaust pipe, ahead of the after-treatment facility. Both sampling at this position and injection at this point could be used to verify the operability of the emissions control components.

EPA Transportation Controls

It does no service to the cause of environmental quality to mandate a 90 percent reduction in gasoline usage in the Los Angeles area. What *would* really improve air quality would be a ban or heavy tax on old vehicles, pre-1970 vehicles that emit thirty times the limit of carbon monoxide, and even 1970 vehicles that emit ten times the proposed limits on the three controlled pollutants. The social cost of such local licensing or control over old vehicles would be far less than that of reducing vehicle-miles traveled without such discrimination.

Other Sources

It has been argued that it is ridiculous to squeeze passenger-vehicle emissions so hard without controlling truck emissions and, therefore, that passenger-vehicle standards should be eased. Alternatively, it has been argued that trucks should be more vigorously controlled. I tend toward the latter view, from considerations of both cost-effectiveness and equity.

General Comment

It seems to me that the most important change we can make in our motor-vehicle emissions control regulations is to change from the present firm standards to an effluent tax. This switch would have the major effect of shifting the manufacturers' efforts from fighting the standards to controlling emissions and to reducing the cost of doing so.

Notes

1. The Consultant Report of September 1974 to the CMVE, *Manufacture Ability and Costs,* p. 142, estimated a $20 sticker-price increase for the most advanced EGR. This situation was emphasized in the *Report of the Conference on Air Quality and Automobile Emissions,* 5 May 1975, to the Committee on Environmental Decision Making (National Research Council).

2. Marks and Niepoth, *Car Design for Economy and Emission,* SAE Automotive and Engineering Manufacturing Meeting, 13–17 October 1975, 750954, Figure 14.

Comment John R. Meyer

I find myself very much in agreement with the central theme of Mills and White that an effluent tax potentially could be a much better way of achieving cleaner air than the present program of standards, as arbitrarily established and bureaucratically enforced. I do believe, though, that they may well be oversimplifying or at least overlooking some of the difficulties inherent in effluent fees. We economists do tend to assign much too low a cost to information gathering, transaction expenses, and other costs of implementing and using a market system.

In addition, we are always open to the charge that "using prices or taxes to achieve a certain end is all very well, but it does overlook the political complications." Curiously, at this point in history, political objections may not be a great obstacle to implementation of an effluent fee program for controlling auto emissions. In the current confusion, with Congress and the motor manufacturers "eyeball to eyeball," seemingly engaged in a game of "brinkmanship," effluent fees might well be a simple way out of the impasse. Fees represent an impersonal or third-party solution: handing the problem over to the market for settlement! Avoidance of direct confrontation when difficult decisions are necessary has from time to time been found appealing in the political process.

The simplifications that trouble me in Mills and White derive, rather, from what one might term the details of establishing just what are the benefits and costs of various alternative means of attaining cleaner air. These are very difficult to establish in any simple, irrefutable form. To begin, there are the well-known and oft commented-upon lacunae in our knowledge. Do moderate amounts of NO_x in the air really have a harmful effect on health? To what extent will catalytic mufflers deteriorate after, say, 50,000 miles of use? How much would it cost to reduce NO_x emissions from stationary sources, particularly in those cities burdened with high NO_x levels, as compared with cleaning up mobile emissions?

Beyond these questions involving seemingly direct findings of fact, there are complicated issues of establishing just what should be the proper baseline for making comparisons between different programs. Thus, one might argue as Mills and White do, that one should use a "historical counterfactual" as the base from which to calculate costs of implementing these programs. For example, fuel, maintenance, and perhaps other costs for a pollution control program could be measured by comparison with a world in which at least some technological change would have continued or occurred in the absence of the program. To illustrate, Mills and White argue that some reductions in fuel consumption would have come about in the mid-1970s regardless of the emissions control program, in response, they believe, to higher fuel costs after 1973. (As a matter of fact, the historical pattern of start-up time for implementation of new automobile technologies strongly suggests that the higher fuel costs that emerged in late 1973 could not have influenced automobile production practices much earlier than 1976 or 1977.)

Use of a counterfactual to define a base for establishing program costs is indeed a complex undertaking. Not surprisingly, those of us involved in doing the NAS-NAE study argued a great deal about this approach. Perhaps I should confess immediately that in these discussions I was probably the most ardent proponent of the counterfactual approach favored by Mills and White. After a great deal of discussion, however, my colleagues changed my mind. The essential problem, as one might have expected, was establishing a reasonably objective counterfactual. By contrast, measuring what actually happened had the great advantage of being fairly directly and objectively determined. Establishing the world of what might have been, with and without the pollution program, with and without other price changes and technological occurrences, moved one into highly subjective speculations. In the NAS-NAE study, therefore, we simply reported the differences between 1970 and subsequent years; furthermore, we attempted to make the calculations self-evident and obvious so that those who, like Mills and White, wanted to construct estimates on a different basis could do so.

More generally, determining a proper baseline for comparison is really the major difficulty in assessing the Mills and White proposal. In essence, they set up what amounts to a straw man by using the 1970 Clean Air Amendments as the basis against which to measure the efficacy of their effluent fee program. Obviously, one can do better than the program embodied in the 1970 Clean Air Amendments. Indeed, recalling the political climate of 1970, one might wonder how anything remotely sensible—and the 1970 amendments do meet this test—even emerged.

Today there is obviously not much point in debating the historical virtues or disabilities of the 1970 amendments. The problem is devising a better program for the future. It has become increasingly evident that the 1970 standards almost surely push to a point where marginal costs exceed marginal benefits in effluent control. The basic question, then, for Mills and White is whether their proposal represents a reasonably close approximation to the lowest-cost way of meeting the legitimate goals of automotive effluent control.

In this connection, several questions can be raised. For example, do the marginal benefits really exceed the marginal costs when the effluent fee program is applied to cars in use rather than limited to new cars only? Or how realistic are their cost estimates for effluent testing, especially for older cars? Would the supply elasticities of procedures and personnel needed for testing really be great enough so that the actual costs would be as low as Mills and White project? Are their estimates of the costs for testing new cars, while probably subject to less error than those for older cars, really all that reliable? Have Mills and White made sufficient allowance for overhead, inventory, and other indirect costs in such undertakings? Above all else, would their program really be cheaper than some obvious alternatives?

To illustrate, ostensibly harmful NO_x levels are apparently found in only a few major urban centers in the United States. In some of these, moreover, NO_x seems more attributable to emissions from stationary sources than from mobile sources. In such circumstances, as noted in the NAS-NAE study, a simple two-car strategy has much appeal. Under this approach, cars achieving NO_x levels of 0.4 gram per mile might be mandated for, say, only California and perhaps a few other parts of the southwestern United States. In the NAS-NAE study we also pointed out that if the national NO_x standards were not set much below 1.1 or 1.2 grams per mile, diesel, stratified charge, and electronic fuel injection (which also happen to have rather desirable energy-conservation characteristics) might provide very-low-cost means of achieving most of the air quality improvement sought by the 1970 legislation.

In short, the essential problem is determining how we move from where we are now to a better course for achieving society's environmental goals. As compared with the crude but nevertheless important beginning made with the 1970 amendments, our goal now should be to identify which of many obviously better courses is both best and politically acceptable. Realism, I might add, now takes on added urgency, because with the passage of time new investments are made for achieving the mandated interim goals. These investments, in turn, become a vested interest that to some extent must be recognized in working out a sensible transition to a new and better program. Accordingly, the sooner these important issues can be settled, the less costly the eventual solution is likely to be.

9 Concluding Comments

The conclusion to the conference was a symposium on the Clean Air Act in which the participants outlined briefly how they would like to see the act changed. The comments of Richard Garwin and Frank Speizer are included elsewhere in this volume. The comments of J. Clarence Davies III, James E. Krier, Allen V. Kneese, and Roger Strelow make up this chapter.

Obstacles to Cleaner Air
J. Clarence Davies III

Introduction

In considering some obstacles to cleaner air, I shall give token recognition to disciplinary lines by identifying the three major obstacles to improved air pollution control as (1) inadequate science, (2) inadequate economics, and (3) inadequate administration. But all three in fact involve political and administrative problems, as well as economic and scientific questions.

Inadequate Science

A fundamental obstacle to improved air pollution control is the inadequacy of the scientific base for determining the degree and nature of the hazard from various air pollutants. The inadequacy of the scientific data for establishing standards is fairly common knowledge, but even more fundamental is our low level of confidence that we are controlling the right pollutants. The 1975 Annual Report of the Council on Environmental Quality (p. 326) confessed, ''It is becoming increasingly evident that the air pollutants upon which our standards and monitoring have been focusing do not represent all the important parameters of air quality. In some cases, they may not even represent the most important or informative ones.''

The number of air pollutants that have been the subject of regulatory concern is quite small. This is largely an accident of scientific and bureaucratic history. There are probably just as many air pollutants as water pollutants, but the old Air Pollution Control Administration chose to concentrate its research efforts on just a few pollutants because so little was known about the nature of the air pollution problem. The legacy of this decision is still with us. Thus, although we have only six criteria air pollutants, plus four "hazardous" air pollutants, as compared with at least thirty or forty recognized water pollutants, the real number of air pollutants that will eventually have to be recognized and regulated is probably quite large.

The evidence supporting the standards now established for the criteria pollutants is weak. The concept of thresholds for damage caused by the criteria pollutants is a politically convenient fiction. The standards for total suspended particulates should be divided between large and small particulates, but they are not. We should probably be regulating sulfates rather than sulfur dioxide. Ambient levels of nitrogen oxides were measured incorrectly for many years. The scientific basis for all the standards is inadequate.

The obvious need is to increase the amount of research on the effects of air pollutants. A small error in the level at which a standard is set has large costs for society, so the benefits from increased research are likely to be great.

The narrow focus of control efforts can also be partially corrected by amendments to the Clean Air Act. Some of the amendments considered in the last session of Congress required that criteria and standards be established for additional pollutants. But the legislative route is somewhat clumsy for this purpose because of the heavy dependence on EPA research.

One cannot talk about additional air pollution research without considering the generally unhappy history of EPA's Office of Research and Development (ORD). The office has suffered acutely from the standard administrative problem of not being sufficiently responsive to the needs of the line programs within the agency. Palliatives involving coordination and reporting systems have not proved effective. Two solutions are possible: eliminate ORD and distribute the research to the other parts of EPA or provide the other offices within EPA a degree of budgetary control over ORD. The latter may be more appropriate because of the confusion and lowered morale that would be caused by ORD's demise and because of the large element of research needs common to all the EPA programs. I would suggest that 30 to 40 percent of the research budget be given to ORD and that the remainder be given to other EPA programs for the purpose of "buying" research from ORD. Such a system is now operating in the Fish and Wildlife Service and seems to hold promise for mission-oriented agencies such as EPA.

Inadequate Economics

As several of the other papers in this volume have shown, the current approach to air pollution control is wasteful and inefficient. Society is spending far more of its resources than is necessary to achieve the air quality we want. This inefficiency poses political as well as economic problems. From a political standpoint, enforcement is made more difficult by a lack of faith that the program is just and efficient. This lack of faith is compounded by the fact that, other things being equal, the more expensive it is for a polluter to clean up the more likely he is to resist installing the necessary control measures. It is further compounded by the polluter's calculation that the law is not likely to be enforced against him.

Perhaps I have worked with economists so long that I have been brainwashed, but I subscribe to the idea that various forms of effluent charges provide a solution to many of these problems. Unlike the economists, I would put almost as much emphasis on the advantages of charges as a more effective enforcement tool as on their economic advantages. We have seen the Connecticut plan and it works. One of my problems with the paper by Spence and Weitzman is their assumption that imposition of a standard guarantees that the standard will be met. In fact, most of the incentives under a standard-setting system are conducive to noncompliance. Perhaps the greatest advantage of a charge or fee system is that it will provide the proper incentives for polluters to work for the public interest instead of against it.

The problems of a charge system should not be overlooked. How much the charge should be will be an important and difficult decision. To try to set the charge at a level where marginal costs will equal marginal benefits is a hopeless task, given great ignorance about costs and an almost total absence of quantitative information about benefits. More likely, the charge will be based on trying to achieve some desired level of ambient air or water quality. A charge system also will not avoid the problem of the geographic area to be covered. Should there be a single national charge or should the amount vary on a state, regional, or local basis? I agree with Spence and Weitzman that once a charge is established it will be very difficult to change the amount. In this respect, charges are not more flexible than standards. Under a charge system compliance monitoring will be a demanding task. The monitoring requirements would not be significantly different from what they should be under the present standards system, but more will be at stake and thus the temptation to collect or provide false or misleading data will be greater than it now is. Despite these difficulties, a charge system would still be much more efficient and effective than the currently used approaches to pollution control.

A problem related to the economics of air pollution control involves the trade-offs that must be made between cleaner air and other national goals,

such as energy conservation, economic development, and nati
Making these trade-offs is in large part a congressional respon
the choice will also be significantly influenced by the organiza
ture of the executive branch. The president is committed to reo
executive branch, but we will not get improved policy analysis or better trade-offs by creating more super-departments like HEW or DOT. What are needed are smaller, more cohesive administrative agencies that have improved policy analysis and control from above. I have described how this goal could be achieved in another paper.

Inadequate Administration

The most important question in the current administration of air pollution control is the balance or division of labor between the federal government and the states. The historic tensions of the federal system now exist against a background of two major and conflicting currents: on the one hand the ever-increasing centralization of the economy and, on the other, the recognition of our inability to successfully manage large centralized organizations that perform a variety of different functions.

In air pollution control, as in many other programs, the central problem is reconciling national rules with state and local diversity. Since the passage of the first air pollution law in 1955, the trend has been toward centralization, toward the exercise of ever greater powers by the central government. This trend was climaxed by the 1970 Clean Air Act Amendments, but now the pendulum seems to be swinging the other way.

There are significant advantages to centralization and national uniformity. First, national standards provide the basis for a national minimum level of welfare. Particularly when human health is involved, there are good grounds for arguing that the national government has an obligation to protect all citizens, regardless of where they live or where in the country they may travel. Second, national uniformity simplifies the establishment of standards. The decision has to be made only once instead of fifty times. This approach is especially important if the knowledge and the skilled people required to set standards are in short supply. The political ramifications of a national standard, as compared with a variety of state or local standards, are also significant. Third, national standards minimize interference with interstate commerce. If national manufacturers are faced with a large number of standards, at least if the standards apply to products (such as automobiles or detergents), the price of the product may greatly increase or it may even be economically impossible to manufacture it. Also, diverse state or local standards have allowed polluters to weaken both standards and their enforcement by threatening to move to another location. There are both advantages and disadvantages to allowing pollution control to be a factor in an industry's calculation of where it locates its plants, but the exercise or the threat to exercise the power to move elsewhere unless pollution control requirements were relaxed was one major reason for imposing national standards. The fourth major advantage of uniformity, the one most responsible for the current degree of centralization, is that it deters weak state programs from undermining pollution control efforts. It would be convenient if we could simply say that states with weak pollution control programs do not have stronger programs because their citizens place higher priority on other matters. But the strength of a state's program is likely to be a result of the state's wealth and its traditions regarding governmental style and competence rather than a direct reflection of its citizens' priorities.

The advantages of uniform national air pollution standards are balanced by an equal number of advantages to decentralization. First, decentralized standard setting is likely to be less costly to society. Second, it allows experimentation at the state or local level. We have only to consider California's automobile standards or Connecticut's penalty charge system to see how important such experiments have been in teaching us how to deal with environmental problems. Third, nonuniform standards probably reduce conflict between different levels of government. The strong role of EPA under the current system has generated a good deal of resentment and misunderstanding between federal officials and state and local officials. Finally, decentralized standard setting would result in a more realistic relationship between standards and their enforcement. Enforcement is now and likely always will be primarily a state and local function. If the states or the localities also established the standards, they might be less likely to want to undermine or change the standards through uneven enforcement.

On balance, I find the arguments for uniform standards more persuasive than the arguments for decentralization. But the high cost of national standards is an important problem, and we should make every effort to find ways to reduce their impact. I think mechanisms such as the auto fee proposal that Mills and White have suggested are promising steps in this direction.

The air pollution program could also be improved if the program grants section of the Clean Air Act could be changed to give some incentive to states meeting air pollution control targets specified by EPA. Under the present system, federal money tends to flow to the states doing the worst job of clean up. It is obviously useful for federal funds to be provided to states that have weak programs in an attempt to strengthen these programs, but not all the incentives should be against states with well-funded and effective programs.

Related to the whole question of centralization is the crucial dilemma of how to deal with new industrial development. How much should be allowed, under what conditions, and where should it go? I have no grand solutions to propose, although it certainly struck me that John Quarles's recent statement that EPA would allow development if the added pollution were traded off against increased pollution control elsewhere in the same area is a perfect argument for the marketable pollution rights called for by J. H. Dales and by the Council of Economic Advisors several years ago.

Conclusion

We have made considerable progress in controlling air pollution, and thus the existing approaches should not be dismissed lightly. But the cost has been high and progress is likely to be slower in the future. We need to invest much more effort in defining the nature of the air pollution problem, we need more flexibility in meeting clean air goals, and we need to reduce the excessive costs that society is incurring for some types of pollution control. I hope that this conference has contributed to these efforts.

Restructuring the Clean Air Amendments: A Suggestion
James E. Krier

I have been asked to describe how I would like to see the Clean Air Amendments of 1970[1] restructured. Given such a large task and such little space, I must quite obviously limit my remarks. I choose to focus on one very fundamental shortcoming of the 1970 legislation—its requirement that all states meet, at a minimum, uniform air quality standards established by the EPA.

Oddly enough, the provision for uniform air quality standards has received very little attention in all the criticism that has been directed to the Clean Air Amendments. This slight is odd for at least two reasons. First, the policy represented by the standards is extremely questionable in terms of both efficiency and equity. Second, the standards are at the very heart of the legislation, and they form the foundation for several of its features that have suffered particularly troubled histories since 1970. It was to achieve the uniform air quality standards in the worst areas of the nation that Congress dictated the stringent federal controls on emissions from new vehicles. Those controls, as we know, have been relaxed or extended several times, and they probably will be again. It was also to achieve the uniform air quality standards that Congress required consideration of transportation controls. Transportation controls are needed to meet the federal air quality standards in a number of areas, and they have provoked local resistance, delay, and litigation. All indications are that Congress will soon grant more time to implement them—a matter to which I shall return.

The Clean Air Amendments required the EPA to promulgate two air quality standards for each pollutant: a primary standard to protect public health and a (usually more stringent) secondary standard to safeguard public welfare. Let me focus on the primary standards. They were to be set so as to allow an adequate margin of safety; as established, they have been conservatively based in worst-case assumptions, the idea being to protect the most susceptible part of the population in the most polluted areas of the country from adverse effects, leaving considerable room for error. Most of the standards are expressed in such terms that a prescribed ambient concentration is not to be exceeded more than one day per year. The standards apply across the nation. Under present law, they are to be met everywhere by 1977 at the latest.

I lack the time to belabor the efficiency or equity of this approach to pollution policy. These are matters I have considered at some length elsewhere,[2] and I hope it will be enough here simply to summarize a few points of principle and fact. To justify uniform air quality standards in terms of efficiency one would have to assume that the costs of a given level of pollution and a given level of control are the same everywhere, and clearly that is not the case. Because the costs of pollution and the costs of control vary across the country, it is difficult to see how a uniform standard can begin to take the varying costs into account. An efficient standard for Iowa is hardly likely to be efficient for California, New York, or Colorado. To require adherence to the same stringent standard everywhere will in many areas result in the imposition of control costs that are much larger than the pollution costs avoided.[3]

As we shall see momentarily, just that has happened, and the result is hardly surprising. Congress wanted to protect the health of citizens everywhere, and it chose to ignore the costs of pursuing that goal. Perhaps it felt that considerations of fairness required such an approach. If so, it overlooked two points. First, using limited resources to pursue pristine air quality levels everywhere in the name of fairness necessarily means diverting those resources from other worthwhile enterprises that, in terms of fairness, should also be pursued. In other words, a step toward a result considered fair in the context of one problem area, if accompanied by positive economic costs, might mean unfairness in another. Second, many of the costs of programs aimed at dramatic improvements in air quality are borne by residents of the areas in question, and if the costs are unduly high, those people will be worse off despite better air. That result is rather problematic in equity terms.

These are points of principle, but there are also a few facts to consider. According to a 1973 estimate, the number of days in excess of the federal standard for oxidant in the Los Angeles area could be reduced from the then prevailing 250 to about 40 by, among other measures, an approximate 33 percent reduction in gasoline consumption.[4] To achieve the federal requirement of no more than one day per year in excess of the standard, on the other hand, would entail a reduction of at least 90 percent. In other words, the price of about 40 more days of improved air quality, after 210 days would already have been accomplished under the more modest proposal, would be an almost three-fold increase in the controls on gasoline consumption! As another measure of the costs of meeting the federal oxidant standard in Los Angeles, consider a 1973 estimate by Rand that attainment would require "not only the maximal fixed-source and retrofit controls and the biggest bus system allowed, but also . . . a $1.28 per-mile surcharge to produce an almost 95 percent VMT (vehicle miles traveled) reduction. Nearly 50 percent of the uncontrolled trips are foregone, with an associated social-cost proxy of nearly $5 billion."[5]

Los Angeles, of course, is a hard case, but it is not unique. The most recent estimates suggest that from twenty-nine to thirty-one areas require transportation controls to achieve some of the federal air quality standards, and it is likely more will be added to the list.[6] The costs of these controls in many instances would be high; moreover, a 1973 report by TRW suggests that they would be distributed in a significantly regressive fashion.[7] It is precisely for these reasons that the House and Senate recently passed bills permitting the 1977 deadline to be extended by up to ten years to meet federal ambient air quality standards in all areas where attainment by that deadline would have "serious adverse social or economic effects."[8] House and Senate alike recognized that a large number of areas could meet the federal standards by 1977 only through the use of transportation controls and that many of those controls would simply be unreasonable. In the view of the House committee, "changes and restrictions in transportation systems may impose severe hardships on municipalities if imposed too quickly. . . . Implementation of many of the (transportation control) measures is impracticable within the time frame permitted under the current Act. Some of the measures may never be practicable." The Senate committee held similar views.[9] Both committees seemed aware that the Clean Air Amendments had gone awry in ignoring costs, although they articulated this judgment in somewhat different ways. The House committee wished to "insure the protection of public health and the environment" but "at the same time take into account the energy and economic needs of the Nation." The Senate committee was more to the point and noted that many of the transportation controls promulgated by the EPA in order to satisfy the 1970 legislation had "imposed vast economic and social costs, for relatively small improvements in the quality of the environment."[10]

Differences between the House and Senate bills just mentioned were resolved in a conference committee bill that would have permitted extensions up to 1987, provided the areas in question implemented all reasonable controls in the meantime.[11] The conference committee bill, however, died when the Senate adjourned after a five-hour filibuster. Nevertheless, it is promising evidence of a new congressional sensitivity to considerations of fairness and efficiency that apparently was lacking when Congress enacted the Clean Air Amendments. With one small but very important alteration it could represent the best policy for the future. The alteration, of course, would be to abandon uniform standards entirely, rather than simply put them off. If something like the conference committee bill were to become law, doomsday would not be repealed but only delayed. Every region would once again have to achieve air quality meeting the federal standards, notwithstanding that this would not be worthwhile in some areas and would be impossible in others.

It is true, of course, that the conference committee bill would have granted time to the states, but then so too did the Clean Air Amendments—seven years, in fact. That span has quite clearly been shown to be an inadequate one in which to achieve the unreasonable, and ten years more would surely prove no better. Within those years, states would be expected to allocate scarce resources to achieving ill-founded ends; as doomsday approached, they could be expected to devote them to fighting unreasonable demands. The same sort of waste, debacles, and delays that burdened the 1970 legislation would likely result again. (There could then be, of course, yet another extension, but this is a questionable strategy by which to make policy—it encourages gamesmanship and results in unnecessary expense.) Congress should concede that the nation is not uniform and never will be and that no legislation can or should try to make matters otherwise. It should, in my judgment, abandon uniform standards in favor of something like the management standards proposal that grew out of work by Kenneth Heitner and myself at Cal Tech's Environmental Quality Laboratory several years ago.[12]

The management standards approach would recognize the need for federal air quality standards, but it would not make the mistake of assuming that "federal" necessarily implies "uniform." Rather, the approach would take into account the reality that uniform standards do not and probably never will fit some areas, and that a little tailoring is necessary. The tailoring would not be particularly difficult. Let me sketch just the barest outline of how we recommended going about it.

We suggested beginning with a requirement, much like that of present law, that the EPA promulgate uniform primary ambient standards. Unlike present law, however, the standards would be expressed only in terms of concentration levels and the practice of providing that these levels are not to be exceeded more than one day per year would be abandoned. The latter would be done simply because getting down to that one day (as opposed to five or ten or twenty days) would not be worthwhile (or possible) in many instances, and to proceed as though this is not the case asks only for trouble. The management standards approach would permit regional variations in the number of days in excess of standards and in the schedule of required improvements.

For each pollutant the EPA would be required to develop criteria to determine whether a particular region fell into one of two classes: class A regions in which it is feasible, in the judgment of the agency, to achieve no or virtually no days in excess of (for example) the present primary ambient concentration standards by 1977 in accord with present law, or class B regions in which the foregoing is not feasible in the judgment of the agency. Feasibility criteria would deal with constraints imposed by technological, economic, administrative, political, and other considerations—considerations that would bear on the issue of what schedule of compliance would just approach (but not exceed) that point where any more demanding schedule would not be worthwhile in light of its costs and consequences. These are the sorts of considerations the EPA has taken into account in reviewing state implementation plans and requests for extensions, and the sorts of considerations that the EPA would have to take into account under the terms of the measure proposed by the 1976 conference committee bill discussed earlier. Thus federal policymakers can hardly claim that the considerations are beyond human capacities or that they would result in administrative costs significantly higher than those that attend the present program.

As the next step, we recommended that the EPA be required to promulgate, within a specified period of time, a schedule of management standards for each pollutant, expressed in terms of the *minimum* percentage reductions (in the number of days per year in excess of the prescribed ambient concentration standard) required by a series of given dates for class B regions. The percentage reductions required of each region by the schedule would depend upon the average number of days annually each region's ambient air quality, based on the best data available, exceeded the ambient standard. Because the schedule would set forth the generally applicable minimum requirements, it would be based on worst, not best or average, areas. Then the expectation would be that most areas could achieve more than the minimum, and each area would have the burden of convincing the EPA otherwise. Our proposed schedule would operate such that the worst the quality of a region's air (in terms of number of days in excess of the standard annually), the more the region would generally be required to improve both in absolute and relative terms. In other words, the more serious a region's problem, the more resources the region would have to devote to it if necessary to comply with the management schedule. And, of course, the schedule would be subject to change as new information, new technology, and so forth developed.

I hope this quick sketch suggests the drift of our proposal. It is an approach that would insist, in essence, on constant air quality improvement up to the point beyond which more would not be worthwhile, rather than on pursuit of the impossible dream of uniformity in a world of varying wants and circumstances.

Notes

1. Pub. L. No. 91-604, 84 Stat. 1705 (1970), codified at 42 U.S.C. sections 1857–1858a (1970).

2. See J. Krier and E. Ursin, *Pollution and Policy* (Berkeley and Los Angeles: University of California Press, 1977), chap. 15.

3. See generally A. Teller, "Air Pollution Abatement: Economic Rationality and Reality," *Daedalus* 96 (1967): 1082.

4. See Environmental Protection Agency, Los Angeles Hearings, 6 March 1973, Figure 1 (statement of L. Lees).

5. B. Goeller et al., *Strategy Alternatives for Oxidant Control in the Los Angeles Region,* Report Number R-1368-EPA (Santa Monica, Calif.: Rand Corp., 1973) p. xiii.

6. See U.S. Senate, Committee on Public Works, Clean Air Amendments of 1976, Senate Report No. 94-717, 94th Cong., 2d Sess. 29 (1976) (hereafter cited as Senate Report); and U.S. House of Representatives, Committee on Interstate and Foreign Commerce, Clean Air Act Amendments of 1976, House Report No. 94-1175, 94th Cong., 2d Sess. 190 (1976) (hereafter cited as House Report).

7. TRW Transportation and Environmental Operations, *Socio-Economic Impacts of*

the *Proposed State Transportation Control Plans: An Overview* (prepared for the Environmental Protection Agency, 1973), pp. 202–203 especially.

8. See Senate Report, p. 28; and House Report, pp. 189–192.

9. House Report, pp. 190–191; and Senate Report, pp. 5–6, 29–31.

10. House Report, p. 22; and Senate Report, p. 6.

11. See U.S. House of Representatives Report No. 94-1742, 94th Cong., 2d Sess, 49–52, 104–106 (1976).

12. K. Heitner and J. Krier, "An Approach to Air Quality Management Standards," *Air Pollution Control Association Journal* 24 (1974): 1039.

A Commentary on Needed Changes in the 1970 Air Quality Act Amendments
Allen V. Kneese

Introduction

Before giving my specific comments I will present a sketch of how the 1970 amendments came to be. As several of the papers in this conference have pointed out, the history of the process is very important for understanding the nature of this law.

For the most part, air quality legislation prior to 1970 was modeled on water quality legislation with a lag of a few years. The major expection to this generalization arises from the fact that much of the overall air pollution problem is attributable to the products of a single giant oligopolistic industry—automobiles. Indeed, it was a combination of Los Angeles smog and an acute episode of industrial pollution in Donora, Pennsylvania, in October 1948, during which some twenty people died and about six thousand fell ill, that stimulated the process out of which finally came the Clean Air Amendments of 1970.

In 1950, researchers at the California Institute of Technology established a link between automobile emissions and photochemical smog in the Los Angeles Basin. A short time later the Los Angeles Air Pollution Control District began calling for action from the automobile companies and the state government. Despite company claims that the requisite technology was not available, a study group that was set up under the auspices of the Automobile Manufacturers Association reached a cross-licensing agreement for emission control devices. Over the same period a number of unsuccessful resolutions were introduced in the Congress calling for federally sponsored research on the air pollution problem. Senators Thomas H. Kuchel of California and Homer E. Capehart of Indiana took a leading role in this effort; in 1955 Senator Kuchel introduced legislation authorizing a federal program of research, training, and demonstrations. In the meantime, President Eisenhower had received a report from an interdepartmental study committee recommending the same steps. Congress passed the legislation and the president signed the first federal air pollution law in July 1955.[1] The level of activity authorized by the 1955 Air Pollution Control Act, however, was very low—$5 million annually for five years to support all its functions.

By this time the problem in California had worsened, and the state took the initiative in establishing automobile emission controls. A law requiring recirculation of crankcase blowby (to reduce hydrocarbons by about 20 percent) on new 1963 cars induced the industry to begin installing the simple crankcase device on some 1961 models. In 1963, over the objection of the automobile industry that such technology did not exist, California legislation required exhaust control devices on vehicles once two such devices had been approved by the state Motor Vehicle Control Board. When four devices produced by independent manufacturers were approved in 1964, the industry discovered that it could indeed introduce its own devices on cars sold in California starting with the 1966 model year. In that year the first California emission standards were set.

Meanwhile, back in Washington, nothing much happened for quite a while. The main reason was that air pollution was widely regarded as an exclusively state and local problem. This was the official position of the Department of Health, Education, and Welfare toward the 1955 Act. Accordingly, eight years elapsed between passage of this act and the first permanent air pollution legislation, although in 1959 the 1955 act was extended for four more years.

In 1962, President Kennedy asked the House to pass a bill sponsored by Senator Kuchel that had passed the Senate in 1961. It authorized the surgeon general to hold hearings on particular interstate air pollution problems. Some features relating to the research program and grants to state and local governments were added. The House again deferred action. Meanwhile, another major incident, a deadly smog that hit London in the winter of 1962, underlined the dangers of air pollution.

A recommendation by the administration in February 1963 finally produced the Clean Air Act, which President Johnson signed in December of that year. This law for the first time gave the federal government enforcement powers. These powers followed closely the pattern of the procedures earlier legislated for water pollution. At the request of a state, HEW could call a conference on air pollution problems in a particular region or air shed and then hold hearings. If there was no satisfactory result, HEW could bring court action. In interstate cases, HEW could act on its own initiative. The bill also specifically mentioned the need for additional attention to the auto exhaust problem.

Hearings held in 1964 by the Senate Public Works Subcommittee on Air and Water Pollution underlined the inadequate attention that had been given to automobile emissions in federal legislation. The administration held that voluntary cooperation should be sought from the industry, and so it opposed enforcement legislation proposed by Senator Edmund S. Muskie of Maine in 1965. This position was widely denounced in the press, however, so the administration reversed it. Thus a second title to the 1963 act was passed in 1965 as the Motor Vehicle Air Pollution Control Act, which authorized HEW to set emission standards for automobiles as soon as practicable.[2] The first standards, set for 1968 models, were roughly the same as those applied in California in 1966. Many people felt that the federal program was unimaginative and lagged behind the progressive California program.

Exacerbating the matter, the automobile industry took a series of bewildering actions that destroyed—almost as if intentionally—the favorable public image that it had so long held. The attempt by General Motors to intimidate Ralph Nader backfired spectacularly, and its president was forced to apologize before a congressional committee and a national television audience. During the same period the Los Angeles County Board of Supervisors charged that the Automobile Manufacturers Association committee, established ostensibly to exchange emission control information, was really a setting for collusion to prevent or delay controls. They cited evidence and asked the attorney general to take action. The ensuing justice department investigation ended in 1969 with a consent decree that provided for an end for possible conspiratorial activities, although it did not officially concede their existence. The year before, representatives of the industry had given testimony on alternatives to the internal combustion engine that, to put it mildly, was highly deceptive.[3] The image of the industry had hit rock bottom. These events contributed heavily to the political climate in which the 1970 act was passed.

The 1967 Air Quality Act

In the move toward control of air pollution, a dramatic incident once again proved to be a factor. A four-day inversion episode in New York in 1966 was estimated to have caused eighty deaths. A month later, a National Air Pollution Conference was held. HEW hoped to use the conference as a stimulus to legislation embodying regional control organizations and national emission standards. Senator Muskie, chairman of the Pollution Subcommittee at that time, conceded that stronger legislation was needed but opposed national standards. In 1967, President Johnson delivered a message to Congress dealing primarily with air pollution matters and, despite Muskie's opposition, proposed legislation inculding national emission standards for major industrial sources and establishment of regional air quality commissions for enforcement.

After hearings that reinforced Senator Muskie's apprehensions about national emission standards, the Senate Public Works Committee delayed a decision and reported a bill that provided for a two-year study of such standards and transformed the regional agencies from devices to enforce them into organizations involved with the states in setting them. The pattern of the 1965 Water Quality Act was followed. HEW was charged with issuing "criteria" that set forth the relationship of concentrations of specific pollutants in the atmosphere to damages to "health and welfare." Ninety days after publication of the criteria, each state had to file a letter of intent that within six months it would establish standards for ambient air quality and that within six more months implementation plans would exist for each of those pollutants in the air sheds over which it had jurisdiction. The secretary of HEW could establish such standards himself if the state failed to comply. The final version, which left these elements intact, was passed and signed by the president in November 1967 as the Air Quality Act.[4] The act also authorized a greatly expanded research effort and for the first time set national standards for automobile emissions.

Thus, at the end of 1967, the federal law pertaining to the setting and enforcement of air pollution standards for nonmobile sources had roughly the same structure as water pollution legislation had attained in 1965. It was not long before this approach came under severe fire.

Experience with the enforcement conference process was similar to that with water pollution, perhaps worse. Between 1963, when the process was initiated, and 1970, eleven conferences were initiated, all but one on interstate cases. Only one case was heard by the courts; it concerned a rendering plant in Bishop, Maryland. Most of the other enforcement conferences included the whole of major metropolitan areas and their recommendations dealt largely with desirable organizational arrangements. These conferences anticipated the requirements under the 1967 act.

Furthermore, HEW was slow to provide the criteria that were the first step in the state-regional approach dictated by the act, and the states in turn were slow to act once criteria had been issued. By 1970 not a single state had a fullscale plan of standards and implementation in effect for any of the pollutants, and a Nader study estimated that the process would not be concluded until well into the 1980s.[5] This report not only roundly condemned HEW's National Air Pollution Control Administration and all its works but also contained an attack on the Subcommittee on Public Works. Other senators and committees were trying to push into the environmental arena. And the president boarded the now fast-rolling environmental bandwagon. It was in 1970 that the first Earth Day was proclaimed. The credibility of the automobile industry was shattered. This was the dramatic political setting for the Clean Air Amendments of 1970.[6]

The 1970 Amendments on Automotive Emissions By 1970, congressional framers of legislation had come to the conclusion that motor-vehicle emissions would not be lowered to levels sufficient to protect public health unless Congress specifically established emissions standards and set schedules for meeting those standards. The congressional standards set in the amendments of 1970 were intended to assure attainment of health-related air quality levels according to calculations supplied by the National Air Pollution Control Administration in the Department of Health, Education, and Welfare. The fact that the deadlines for meeting those standards were 1975 and 1976 model years gave some recognition to the need of the industry for lead time to develop the necessary control technologies and equipment. The standard for automobiles sold during model year 1975 and thereafter called for a reduction in hydrocarbons and carbon monoxide emissions of 90 percent from levels produced by 1970 cars, which already had achieved a modest degree of control. Similarly, in model year 1976 a 90 percent reduction in nitrogen oxide emissions was required compared to the 1971 standard.

Congress also authorized the administrator to grant a one-year delay of these standards. The administrator did grant a delay; thus the 1975 requirements for HC and CO were pushed back to 1976 and the 1976 requirement for NO_x was pushed back to 1977. When the administrator granted this one-year delay, he set interim standards for 1975, as was required by the law. One set of standards was set for the forty-nine states, a more stringent standard was in effect for California in 1975, and both were more lenient than the full 90 percent reduction requirement.

In June 1974, Congress amended the Clean Air Act by adopting the Energy Supply and Environmental Coordination Act. That act further delayed new-car emission standards and carried over the interim standards prescribed by the administrator through model year 1976. In addition, Congress authorized the administrator to grant an additional one-year delay of the HC and CO standards to model year 1977. Furthermore, Congress postponed the full 90 percent NO_x reduction requirement until model year 1978. Thus the automobile industry already has received a moratorium of three years (through 1978) from the initial compliance date written into the 1970 act. This delay came as a consequence of both legislative and administrative actions. At the present time, the automobile industry is striving hard to obtain further delays in the implementation of the statutory standards. The Clean Air Act also specifies that a fine of $10,000 per vehicle be levied on any manufacturer that is not in compliance with the standards. This extreme penalty, which, if applied, would shut down the industry has been one of the reasons that there has been a reluctance to enforce deadlines. Furthermore, the control that has been achieved up to this point has come at the cost of considerable fuel penalties vis-à-vis what could be achieved presently if such controls were not implemented. If the original 1976 NO_x requirements were implemented now, it appears they could be achieved only at the cost of substantial additional fuel penalties.

Since the 1970 amendment there have been several detailed studies of government policies on automotive emissions control. They are carefully reviewed in Chapter 8 of this volume. I have read several of these studies as well as the Mills and White paper and concur in the conclusion reached by Mills and White that, although a certain amount of progress has been obtained in emissions control, it has been a slow and costly process involving tinkering with present technology rather than any substantial change in the basic technology that would lead to power plants inherently low in emissions. Several such technologies have been known since the 1920s and 1930s; the stratified-charge engine, fuel injection, the diesel engine, lean-burn engines, and other devices can be employed to regulate the combustion process more precisely. More recent developments include the use of sophisticated electronics and computer technology. The succession of gradually tighter requirements in the current legislation has encouraged small modifications of the standard internal combustion engine rather than more fundamental changes. Another well-documented aspect of the current situation that is related to what has just been said (again see Mills and White) is that automobiles that meet the test requirements in the legislation in their prototype stage do not perform nearly so well on the road. In part, the reason is that the manufactured models cannot perform as well as the prototypes, but probably even more important is poor maintenance and tampering with control devices.

Some Alternatives How to solve the twin problems of insuring maintenance and stimulating technology?

Fifteen years ago economists at the Rand Corporation proposed the outline of an answer that, administrative and distributive problems aside, appears to be the economically ideal solution.[7] The proposal is that cars in service be tested periodically and assigned a smog rating indicated by a seal or coded device attached to the car. When the driver purchased gasoline he would pay a tax over and above the basic gasoline tax, which would vary with his smog rating. Charging the tax to the final user of the car has the advantage that it is able to stimulate responses all along the chain from driver decisions to the manufacturer. An individual could reduce his smog tax bill in several ways:

1. Tuning or overhauling his engine to reduce emissions and obtain better gas mileage would be an economical alternative to paying the tax. Since poor tuning of automobiles on the road is an important contributor to emissions, this incentive could be quite important.
2. The car owner has many options that would allow him to drive fewer miles per year—living closer to his job, using mass transit, or participating in a car pool. Standards based on emission per vehicle-mile do nothing whatsoever about miles driven, but a smog tax could affect this extremely important variable as well as emissions per mile.
3. Control devices could be installed on older cars. In 1970, in a market test, General Motors offered control kits for pre-1968 models at about $20 installed, but no one bought them. Clearly it was nonsensical to expect anyone to make this investment because, without assurance that others would buy the kits, any one person's effect on the situation would be negligible. Similarly, no one would buy the kit if he was sure that everyone else would do so—his air would be equally clean whether or not he bought the kit, so why bother. This phenomenon is what economists refer to as a "free-rider" problem in the economic theory of public goods. A smog tax would introduce a new and persuasive element into this calculation. Also, because consumers would demand them, the manufacturers would have an incentive to design automobiles that had better smog ratings not only when they rolled off the assembly line but throughout their lifetimes. In the long run, this is probably the most important incentive effect of all. Because such a smog tax would not require specific deadlines, manufacturers would have a greater incentive to invest in research on basic technologies than to tinker with existing engines in an effort to meet precisely timed deadlines.

As I see it, there are two difficulties with this otherwise excellent proposal. The first is the question of whether an effective administrative structure could be established at reasonable cost, and the second is the problem of the possibly adverse redistributive effects of a smog tax because it is likely that, in general, poorer people have the older and more polluting vehicles. I will say more about this later.

A tax on the final users of the automobiles could be combined with an emissions fee on new automobiles. This system would provide a direct motivation for both the manufacturer and the automobile user to respond to economic incentives in an efficient manner.

In their chapter, Mills and White present a proposal of this nature and provide some quantification of probable results. This proposal merits serious consideration because of the detail with which it is worked out and its sensitivity to distributional issues and administrative aspects. It is the best paper I have seen on the subject of automotive emissions fees. I would just like to suggest some modifications that should be less costly to administer and/or can address the distributional issues more directly.

Possible Modifications to Take Account of Administrative Problems and Distributive Effects The Mills–White proposal merits very serious attention from policy makers, but there are some difficulties with it. First, as the authors acknowledge, there may be a substantial time and inconvenience cost in administering the program of fees on automobiles already in use. Second, because it is probably true that poorer people drive the most polluting cars, the effect of fees on the automobile users will likely be somewhat regressive with respect to income. There are several conceivable ways to come to grips with these problems in framing legislation. One alternative to the fees on automobiles in use proposed by Mills and White would be to levy a pollution surcharge on the sale of gasoline and provide rebates to those who offer their vehicles for inspection and meet emission standards specified for cars of a particular age. A surcharge of, let us say, 10 cents a gallon, would yield revenues of $100 per vehicle, on average. Since not all automobile owners would find the time and convenience cost of bringing cars in, keeping necessary documents, and so on worth the rebate, the actual amount available for rebates would be in excess of $100 per car minus whatever the administrative costs the program would be. The noncoercive aspect of this proposal is attractive.

In general, it would seem that poorer people would be more willing to incur the time and inconvenience cost, so the program would most probably be progressive in its cost incidence vis-à-vis the Mills–White proposal. The problem of the oldest cars, which cannot be brought into compliance without a very large expenditure of funds, say, for engine replacements, could be handled by initially making the emission standard zero for cars beyond some age, say eight years. But the age of cars to which standards apply should be increased progressively to encourage the introduction of emission reduction technologies that have high durability.

A More Radical Revision of the Mills–White Proposal If the modification just proposed still appears to involve undesirable administrative difficulties, a further modification suggests itself. It is this modification that I tentatively propose be given the most serious consideration by policy makers, primarily because it leaves most of the present regulations intact and, therefore, is probably more tractable politically.

Under this proposal the present standards for new automobile emissions would be retained with one exception. The 1977 NO_x standard (2 ppm) should be reduced to the 1976 NO_x standard (3.1 ppm). The reason is that near-term technologies could provide significantly better fuel economy.[8] To foster further progress in emissions control, all automobiles sold should be subjected to a fee based on the Mills–White formula, *except* that the fee should not be based on the testing of prototype automobiles prior to marketing but on a random sample of each car model tested after one year of general usage. There are two reasons for suggesting this lagged fee arrangement. The first is that it would provide a much more accurate and equitable basis for the fee because it would be based on actual performance and not on the highly artificial circumstances of a prototype test. The second and related reason is that it would provide an incentive for manufacturers to develop approaches to emissions control that have greater durability and, perhaps, to establish maintenance programs to help keep the emissions low.

After a year or more of usage, emissions probably become more a function of how the automobile has been used and maintained than of its initial characteristics, although these qualities continue to be important. My second modification would deal with this problem without having to undertake the very-large-scale testing that the Mills–White proposal and my first modification of it would require.

A fee based on a somewhat reduced scale relative to the fees on new cars should be levied at the time of sale of every used car from an organized dealership. The purpose of this proposal is to make it economically worthwhile for dealers to sell only cars that have been tuned and repaired. Moreover, such a fee program would confer higher resale value to cars containing long-lasting control technologies, such as stratified-charge engines. Initially, to avoid adverse distributive effects on poor people driving heavily worn automobiles, the fee might apply only to cars built, say, in the last eight years. Eventually the fee should apply to all cars of whatever age. This policy, in turn, would allow owners to choose to make major repairs or to scrap their cars earlier relative to the current situation, and it would permit earlier introduction of less polluting technology to the national automobile stock.

Concluding Statement on Auto Emissions I think that a well-crafted effluent fee program could vastly improve our approach to automotive emissions control. I think it could be used to stimulate the implementation of much-longer-lasting technologies, it would give the manufacturers an appropriate flexibility to plan and design automotive engines, it would provide for a better reflection of environmental costs in the price of the product, and it could be designed to deal with the very difficult and important problem of maintenance after cars have been sold. In some versions, it could also influence decisions about how much to drive and the choice of transportation modes. I agree with every major study of the matter I have seen that the approach embodied in the 1970 amendments has not been a cost-effective one and that modifications of that approach to include economic incentives are a matter of high priority.

The 1970 Amendments and Stationary-Source Emissions

The 1970 Clean Air Act Amendments sharply expanded the federal role in setting and enforcing standards for ambient air quality. The act embodies the concept of a "threshold value"—a level of ambient concentration below which it is assumed that no damage occurs to health. Materials subsequently designated to have threshold values include the main pollutants by mass, namely, sulfur dioxide, carbon monoxide, nitrogen oxides, particulates and oxidants. The notion of threshold value can be regarded as a politically convenient fiction, which permits the law to appear to require pollution damage to health to be reduced to zero—an absolutely unambiguous number.

Congress directed the Environmental Protection Agency to use scientific evidence to determine threshold values for pollutants assumed to have them and then to set those values minus an adequate safety margin as primary standards. These standards, which relate to injury to human health, are to be met first. More rigorous standards to be met later relate to public welfare and aim to protect property, crops, public transportation, and aesthetics from pollutants. The states were to prepare implementation plans assuring that the primary standards would not be violated anywhere in the state after mid-1975.

Congress did not rely solely upon the established standards for ambient air quality to control stationary source pollution. It also gave the Environmental Protection Agency power to set specific limits on emissions of certain kinds of pollutants. It recognized a category of substances called "hazardous pollutants," which are considered to have especially serious health implications (some of the heavy metals are examples). The EPA was directed to prepare a list of such substances and to issue regulations limiting their emissions by both new and existing sources to be enforced at federal level.

The act also directed the administrator to set new-source performance standards to limit the emissions of pollutants from new industrial plants to an amount no greater than that obtainable with "the best adequately demonstrated control technology."

It is now the end of 1976 and, as the paper Roberts and Farrell prepared for this volume (see chapter 4) adequately documents, primary standards are very far from being met. With respect to technology, the matter of what has been adequately demonstrated is in contention and in litigation. Roberts, Farrell, and Ingram conclude after a careful look at the political and administrative aspects of the Clean Air Act that the act is essentially unworkable. Enforcement is largely hung up in arguments about technology, and requirements have been laid on the states that they cannot, will not, and, in many cases, do not want to meet. The whole act has become so contentious that despite major efforts to do so, the last Congress could not agree on amendments. Those amendments that were proposed were filibustered to death. The main issues were auto emissions, which I have already discussed, and nondeterioration of clean air areas. If general enforcement of the act presents severe problems, protecting clean air areas appears to me to pose considerably worse ones.

One member of the Southwest Region Under Stress Project has done some atmospheric modeling of the San Juan Basin in Colorado and New Mexico that is pertinent to this topic. The model is of the high-terrain type, which lends itself to partial calibration against field measurements made in Arizona with respect to the Navajo Plant. The much discussed Four Corners Plant is in the San Juan Basin and is already a large enough source of emissions so that primary standards are violated in a number of places. However, in general, the air of the basin is still relatively pristine. There are more or less uncertain plans to build additional power plants and synthetic gas facilities in the basin that would use the inexpensive coal that is still available there. The atmospheric dispersion analysis suggests that even after retrofitting and drastic cleanup the Four Corners Plant would still restrict the construction of other plants in the basin. However, with very careful attention to siting, it appears that a considerable increase in electric production could still be carried out within the basin. As many as seven new 2000-megawatt plants might be built without significantly aggravating the existing violations or producing new ones; this development would not meet the demands of nondeterioration, although it would come fairly close to meeting the primary standards. Nevertheless, the area that received significant concentrations of SO_2 and NO_2 would be considerably expanded. Furthermore the average concentrations throughout the basin would increase; the greatest impact would be on high terrain in the region. Note that this analysis limits itself strictly to the primary standards that have been set. At this stage of development, it says nothing about visibility effects or such matters as the fallout and rainout of sulfuric acid and toxic chemicals. The point here is that even to achieve minimal protection in the face of future development, direct regulations approaches would require an effective and carefully programmed retrofitting activity and at the same time would require the exercise of positive control on siting facilities, based on rather sophisticated modeling of a complex air basin. Institutionally, scientifically, and administratively, the states of Colorado and New Mexico are a light-year away from being able to do these things.

Needed Changes Despite a political climate that Ingram sees as not very conducive to major innovation in the area of air quality legislation, I think innovation is desperately needed and every effort should be bent to achieving it. The next logical step, indeed a step it would have been logical to take long ago—and some people did try—is to levy a flat national emissions fee on the major bulk emissions of sulfur compounds, nitrogen oxide, and particulates. I realize of course that such a flat national fee is not ideal, but the recent increase in understanding of the atmospheric effects of large emissions sources suggests that effects are very widespread. Meteorologists and atmospheric chemists are now speculating, for example, that the burning of coal in the Southwest could make a significant, even substantial, contribution to the already severe problem of the low pH of rainfall in the Northeast. Needless to say, effects on high terrain in large areas of the Rocky Mountains could be substantial. Further development in the eastern United States would no doubt aggravate the large-scale problem that already exists there. These considerations make it more justifiable to take a national view of the problem, at least with respect to large emitters, rather than the regional one that I and many others once regarded as the most appropriate.

For reasons that have been given many times, such an emissions charge promises not only to be more effective than the present ambient emissions standard approach but also considerably more efficient in terms of the cost of achieving given levels of emissions reduction, and the associated monitoring problems appear solvable.[9] Drafters of legislation for such an emissions fee could build upon the proposals that Congressman Aspin and Senator Proxmire, among others, have previously introduced. Whether such an emissions fee should be introduced as an amendment to the Clean Air Act or as quite separate legislation coming from the Commerce Committee or the Ways and Means Committee is a fair question. The Air and Water Pollution Subcommittee of the Public Works Committee may be so locked into and embroiled in the present approaches that it could not possibly develop or come to an agreement on a genuine effluent fees approach.

An exception I would suggest to a flat national charge is that a large surcharge be added to it for development in designated "clean air reserves." I would substitute this provision for the apparently nonworkable, nondegradation requirements of the present legislation. I am simply convinced that at the present time we are not institutionally or scientifically equipped to introduce the timely retrofitting and the careful siting of new facilities that the San Juan example shows would be needed to protect air quality in clean air regions in the face of large development. A stiff surcharge for developments introduced into these areas would assure that if development occurred it would be because it indeed had a large economic value. It would also assure that great attention would be paid to emissions control for those developments that do occur. In practice, if not in principle, I believe this approach would result in much more effective protection of clean air regions than trying to enforce and administer nondegradation provisions by means of direct regulation. I would further propose that revenues collected by the effluent fee system be returned to the states with some preference for those where the major emissions occur. The purpose would be two-fold: to provide some compensation for the environmental damage that would inevitably accompany development even under conditions of careful management of emissions and to permit the states to build up their scientific and administrative capabilities to the point at which more refined management might become possible.

Noncriteria Pollutants I now will turn briefly to a discussion of what are usually called "noncriterion" pollutants. These are the "hazardous pollutants" I mentioned earlier, for which the Environmental Protection Agency is supposed to identify and develop control programs. For reasons that presumably have to do with the emphasis being placed on efforts to enforce restrictions on the criteria pollutants and because of the inherent difficulty of the problem, EPA has been very slow to move on this category of substances. A few—asbestos, beryllium, and mercury—have been designated as hazardous pollutants and partial control proposals have been developed with respect to them. Vinyl chloride has been proposed as another.

Hazardous substances may be of critical importance with respect to future energy development. If I may again use an example of coal development in the Southwest, more specifically in the San Juan Basin, the point can be illustrated. All coal, and Southwest coal, in particular, contains many substances other than hydrocarbons. The Southwest is one of the most highly mineralized parts of the world and the coal that formed there also contains many other chemical substances, including mercury, cadmium, selinium, uranium, and beryllium. Except for those substances that may remain in the ash, it is hardly possible to imagine a better method for distributing them around the countryside than to introduce them into a very hot fire and send them rushing through a tall stack high into the atmosphere.

From studies of the technology employed within the Southwest Regional Project it appears that coal gasification could become the source of substantial emissions of polycyclic hydrocarbons. This class of hydrocarbons is known to have powerful carcinogenic properties.

To make things more interesting, the San Juan Basin is also the site of the Navajo Irrigation Project. This project is ultimately scheduled to include 110,000 acres, a substantial share of which would be planted to alfalfa in order to supply feed to feedlot operations. This combination of energy development and agricultural development could furnish a beautifully direct route for inserting these hazardous substances into the human food chain.

Some informed persons in the region are very concerned about these matters, but the amount of resources going into their study is minute. There is practically no analysis of generation, dispersion, and behavior in the food chain for most of these substances.

I must confess that I am somewhat at a loss to recommend improved policy in this area, but I think that there are at least two implications. First, it is very important that we achieve mechanisms that will provide effective control of the criteria pollutant emissions from power plants and other energy facilities in the hope that such control will, as a side effect, at least reduce the emission of hazardous substances. Second, it seems to me that the analysis, monitoring, and modeling of these hazardous substances merit much greater effort. Overwhelming evidence has produced almost universal agreement that most cancer is environmentally induced. The same may be true of several other major chronic diseases. At the present state of our knowledge, we are playing Russian roulette with these hazardous substances.

The "Start-Up" Problem In conclusion I want to call attention to a problem for regulation or a fee system that I have not seen discussed in print. Units of a power plant cannot run continuously for indefinite periods. They must be shut down periodically for maintenance and repair. A typical unit would make perhaps twenty cold starts in a year, lasting from eight to ten hours each time (some units are "down" much more often, however). During much of this period emissions are very high for two reasons. First, combustion is poor, so the amount of residuals generated is high. Moreover they include large quantities of hydrocarbons, probably of the polycyclic variety, which are practically absent when combustion is up to standard. Secondly, the control equipment on the plant does not function well.

Mike Williams, a member of the Southwest Region Under Stress Project who called this problem to my attention, estimates, on the roughest of evidence, that for a plant otherwise exercising high-level emissions control these start-up emissions may be as large as those for the rest of the year. Apparently they have never been measured. Also, all dispersion models have been based on normal emissions rather than on the much larger start-up emissions. This appears to be a fertile problem for further investigation.

Notes

1. P.L. 84-159.

2. P.L. 89-272.

3. "Automobile Steam Engine and Other External Combustion Engines," Joint Hearings before the U.S. Senate, Committee on Commerce and Subcommittee on Public Works, 90th Cong. 2nd Sess. (1968).

4. P.L. 90-148.

5. John C. Esposito, ed., *Vanishing Air,* The Ralph Nader Study Group Report on Air Pollution (Grossman, 1970), p. 158. An informative discussion of enforcement problems in the air pollution field is found in *Assessment of Federal and State Enforcement Efforts to Control Air Pollution from Stationary Sources,* Report to the Congress by the Comptroller General of the United States (Washington, D.C.: U.S. General Accounting Office, 1973).

6. P.L. 91-604.

7. D. N. Fort et al., "Proposal for a Smog Tax" reprinted in tax recommendations of the president, hearings before the U.S. House of Representatives, Committee on Ways and Means, 91st Cong., 2nd Sess., 1970, pp. 369–370.

8. Letter from E. N. Estes, president of General Motors Corporation to Senator Domenici, 27 September 1976.

9. See Frederick R. Anderson, Allen V. Kneese, Russell Stevenson, and Serge Taylor, "Environmental Charges: Economic, Technical, Legal and Political Aspects," Chapter 4, manuscript, Resources for the Future, Inc.

An Insider's Viewpoint on Possible Clean Air Act Amendments
Roger Strelow

I will try to emphasize in my remarks the viewpoints of someone who has served inside the Environmental Protection Agency and has attempted to make the current Clean Air Act work. Between 1974 and early 1977, I was Assistant Administrator for Air and Waste Management at EPA.

First, I would like to discuss a very fundamental aspect of the Clean Air Act, namely, the air quality management philosophy embodied in the act. This approach to air pollution control relates most specific control actions to the quality of the ambient air rather than to technological or other reference points and seems almost inescapable conceptually because public health and, to a less extent, public "welfare" are the basic concern. Since the fundamental political and legal justification for the many actions required under Clean Air legislation is protection of public health, control requirements to a large degree must be related to public health needs and hazards. In particular, the public must have some assurance that the substantial expenditures they and various industries are expected to incur will provide a significant measure of public health protection.

However, scientific uncertainties, political problems with certain types of controls, including the so-called "transportation controls," and other factors make undue reliance on relating controls to ambient air quality effects a perilous proposition. Often, for example, the relationship between a particular substance that is emitted from a smokestack or tail pipe and the resulting public health hazard in the atmosphere is extremely complex because of interactions with other substances and various natural atmospheric conditions. Current knowledge often precludes drawing any precise relationship between emissions and ambient effects.

Accordingly, I would like to suggest the possibility that the basic air-quality management framework in the act be retained but that much of the presently unproductive debate over emission and ambient relationships be reduced by adding to the act some baseline, technology-based control requirements for existing sources as well as new sources. Of course, under the current law, the new-source performance standards in section 111 require the best available technology considering costs to be used in new sources, regardless of ambient air quality effects. Something comparable to this requirement, perhaps patterned after the "best practicable control technology" requirement in the Federal Water Pollution Control Act would be a good addition to the Clean Air Act. Among other things, such an addition could vastly speed up the process of revising and improving the state implementation plans that form the basis for regulatory controls.

Second, I would like to discuss briefly the question of economic incentives. I am convinced from my experience at EPA that nothing will produce cleanup and "responsible" behavior as fast or as well as an economic incentive to control pollution. The traditional notions about effluent or emission fees or charges in the classical sense have yet to be translated into anything resembling specific programs that will produce predictable results that are acceptable to the public. Something much less ambitious is probably needed as a first step in the direction of economic incentives. Indeed, in my judgment, economic incentives will rarely replace regulatory measures. Rather, they most likely will be used for both political and legal reasons to supplement basic regulatory requirements.

The "delayed compliance fee" incorporated in the 1976 Clean Air Act bill, which was patterned after the Connecticut experience, appears to be an excellent first step. Such fees do not replace regulations but instead furnish a clear economic incentive to comply with regulations.

Some form of economic incentive may have a role to play in controlling auto emissions, as well. Indeed, there is some largely overlooked authority for this approach under current law. Thus current provisions in the Clean Air Act permit the imposition of fines based on each vehicle produced that fails to meet the standards. In any event, all of the discussion about a possible shutdown of Detroit in the absence of congressional relief for the 1978 model year is probably misguided for the simple reason that the federal courts are very unlikely to order such a shutdown even if the EPA administrator were to request it.

I have outlined these points more fully elsewhere,[1] and I am pleased to see that there is increasing interest in the use of economic incentives with a more explicit statutory base in the auto emissions area. Senator Proxmire, chairman of the EPA Appropriations Subcommittee in the Senate, has expressed to me a strong interest in the use of economic incentives for auto emissions control. In addition, Senators Hart of Colorado and Domenici of New Mexico, both members of the Senate Public Works Committee, have written to President Carter urging that his administration give serious consideration to legislative proposals for economic incentives on auto emissions. Senator Hart later introduced such a bill. Also, Congressman Waxman from California, a member of the House Interstate and Foreign Commerce Committee, has expressed considerable interest in this approach. Although it now seems unlikely that Congress either will have or will take the time to thoroughly consider such an approach in the present session, I believe it will receive increasing attention in the future.

Third, let me turn to the subject of new stationary sources. It is interesting that the first shift of attention from the initial task of cleanup—which had held the attention of most people since the 1970 amendments—to the question of new sources came in connection with protection of presently clean areas. EPA has established regulations to prevent "significant deterioration" of clean air, and Congress clearly intends to provide the more explicit guidance that EPA and others have requested for this purpose.

More recently, the agency has taken the lead in drawing public attention to the serious question of new facilities in presently dirty areas. If such areas are ever to become clean, in other words, if they are to meet air quality standards, the addition of new polluting sources to such areas must be effectively controlled. This does not necessarily mean that they must be banned. EPA has issued a ruling that permits new-source growth in dirty areas if the net effect, with offsetting emission reductions that are not already required, is air quality improvement rather than degradation. The Congress needs to address this fundamental issue and provide workable guidance. It probably presents one of the most difficult dilemmas under the Clean Air Act.

Fourth, let me address very briefly the question of procedural or structural changes in the Clean Air Act. Last year, there was some discussion in the Congress of the possible need for greater procedural safeguards and more elaborate procedural requirements relating to EPA rulemaking and other decision making. It is my view that changes of this sort are not needed, partly because of the fact that the courts have provided to date what I believe is an effective and appropriate mechanism for testing and checking agency action to insure that it meets requirements of the law and due process. It might be useful, however, to specify in the act a more explicit procedure for meaningful federal, state, and local partnership in EPA decision-making processes. This detail is particularly critical because the Clean Air Act imposes enormous requirements on state and local governments. Therefore, they need to have a strong and effective voice in EPA decision making.

Fifth, I want to return to the general subject of auto emission control. The endless debates on what the new car standards should be and when they should be met tends to divert attention from a fundamentally more important concern: the actual performance of controlled cars in use. Studies show that controlled cars would meet standards in use if they were properly maintained. However, the same studies show that many cars today are not properly maintained and therefore many of these cars pollute more than they should. There are two basic answers to this problem. One is that the manufacturers must be made to build cars so that their emission controls are less sensitive to the foreseeable mismaintenance that so many cars receive. Another is that, because it is probably impossible to build a fully maintenance-free car, certainly not in the near term, programs such as mandatory inspection and maintenance must be more widely required and adopted to ensure better maintenance. Inspection and maintenance requirements are under a cloud of legal uncertainty, which relates to EPA's ability to require states to adopt such programs, until consolidated cases now pending before the U.S. Supreme Court are decided.[2] Regardless of the Supreme Court's decision, one probable requirement for achieving effective implementation of inspection and maintenance by the states is to make expenses for such programs an eligible item under the Federal Highway Trust Fund. Indeed, the law could be amended further to require that a certain portion of the trust funds must be used for inspection and maintenance purposes. Other changes may be needed in Title II of the Clean Air Act to prevent private garage tampering and to make it easier for consumers to invoke emission-control warranty protections.

Transportation controls, which are needed and required under the act in many areas to further reduce auto emissions by reducing auto travel, must be rehabilitated. Experience has clearly demonstrated that EPA can never make these controls work without strong legal and funding support from the U.S. Department of Transportation and its programs. In addition, EPA must be given the flexibility under the act to work with state and local governments to draw up mutually satisfactory transportation control programs. EPA, DOT, state officials, and others should be authorized to arrange a series of open, public forums and analyses to define with a reasonable consensus what transportation control measures are reasonable and necessary.

There are numerous other aspects of the Clean Air Act that deserve serious attention. I have not attempted to be comprehensive but rather have focused on key problems on which I believe action is most critical.

Notes

1. *Washington Post,* 2 January 1977.

2. On 2 May 1977, the U.S. Supreme Court in EPA v. Brown, 9 E.R.C. 2074, vacated the Courts of Appeals judgments and remanded the cases, without any decision on the merits, on the ground that the Government's concession of the need to modify the regulations in question would make such a decision an advisory opinion.

Participants

Donald Allen, Yankee Atomic Electric Co., Massachusetts

Richard Ayers, Natural Resources Defense Council, Inc., Washington, D.C.

Henry Beal, Director, Connecticut Air Pollution Program

Karl Braithwaite, U.S. Congressional Commission on Public Works

Edwin H. Clark II, Council on Environmental Quality

Charles Corkin II, Department of the Attorney General, Massachusetts

Roger Cortesi, Environmental Protection Agency, Washington, D.C.

Norman Dahl, MIT, Consultant to the Provost

J. Clarence Davies III, Conservation Foundation, Washington, D.C.

Donald N. Dewees, University of Toronto, Department of Political Science

Robert Dorfman, Harvard University, Department of Economics

William Drayton, Jr., Harvard University School of Government

Susan O. Farrell, Harvard School of Public Health

James A. Fay, MIT, Department of Mechanical Engineering

A. Myrick Freeman, Bowdoin College, Department of Economics

Ann F. Friedlaender, MIT, Department of Economics and Department of Civil Engineering

Albert Fry, Environmental Protection Agency, Washington, D.C.

Richard L. Garwin, International Business Machines Corporation

Alvan S. Gordon, California Air Resources Board

David Harrison, Jr., Harvard Graduate School of Design

Charles M. Heinen, Chrysler Corporation

John B. Heywood, MIT, Department of Mechanical Engineering

Gregory Ingram, Harvard University, Department of Economics

Helen Ingram, University of Arizona, Institute for Government Research

David Iverach, New South Wales Department of the Environment, Australia; Visiting Fellow, MIT Sloan School of Management

Charles O. Jones, University of Pittsburgh, Department of Political Science

Paul L. Joskow, MIT, Department of Economics

Raphael Kasper, National Research Council, Washington, D.C.

Allen V. Kneese, University of New Mexico, Department of Economics

James E. Krier, University of California, Law School

Lester B. Lave, Carnegie-Mellon University, Department of Economics

Larry Linden, MIT, Energy Laboratory

John R. Meyer, Harvard University

Charles J. Meyers, Dean, Stanford Law School

Edwin S. Mills, Princeton University, Department of Economics

Richard R. Nelson, Yale University, Department of Economics

James L. Oakes, U.S. Court of Appeals for the Second Circuit

Marc J. Roberts, Harvard School of Public Health, Executive Program in Health Policy and Management

Daniel L. Rubinfeld, Institute of Public Policy Studies, Ann Arbor, Michigan

Jack Ruina, MIT, Department of Electrical Engineering

Paul A. Samuelson, MIT Institute Professor of Economics

Adel F. Sarofim, MIT, Department of Chemical Engineering

Joseph L. Sax, University of Michigan Law School

Kenneth Small, Princeton University, Department of Economics

Robert Solow, MIT, Department of Economics

Frank E. Speizer, Harvard Medical School

A. Michael Spence, Harvard University, Department of Economics

Richard B. Stewart, Harvard Law School

Roger Strelow, Environmental Protection Agency, Washington, D.C.

Martin L. Weitzman, MIT, Department of Economics

Lawrence J. White, Graduate School of Business Administration, New York University

Harold Wolozin, Department of Economics, University of Massachusetts

Joel Yellin, MIT, School of Humanities and Lecturer, Political Science

Index